Climate Adaptation and Resilience: Challenges and Potential Solutions

Anticipatory Governance, Planning and Dialogue

Peter Lang

Bruxelles · Bern · Berlin · New York · Oxford · Wien

Pascaline Gaborit

Climate Adaptation and Resilience: Challenges and Potential Solutions

Anticipatory Governance, Planning and Dialogue

Peter Lang

Bruxelles · Bern · Berlin · New York · Oxford · Wien

This book is based among others on the results and research of the project Climate Resilient Cities www.resilient-cities.com pilar one.

This project is co- funded by the European Union in the framework of the project.

Co-funded by
the European Union

Disclaimer: The views and opinions expressed in this publication are sole responsibility of the authors and do not necessarily reflect the standpoint of the project or of the funders' organizations.

This book's final version was received on the 9th of august 2022.

This publication has been peer reviewed.

No part of this book may be reproduced in any form, by print, photocopy, microfilm or any other means, without prior written permission from the publisher. All rights reserved.

© P.I.E. PETER LANG s.a.
Éditions scientifiques internationales
Bruxelles, 2022
1 avenue Maurice, B-1050 Bruxelles, Belgium
www.peterlang.com ; brussels@peterlang.com

ISBN 978-2-87574-643-6
ePDF 978-2-87574-644-3
ePUB 978-2-87574-645-0
DOI 10.3726/b19893
D/2022/5678/36

Bibliographic information published by "Die Deutsche Nationalbibliothek"

"Die Deutsche National Bibliothek" lists this publication in the "Deutsche Nationalbibliografie"; detailed bibliographic data is available on the Internet at <http://dnb.de>.

Table of Content

Introduction Chapter on Climate Adaptation 11
PASCALINE GABORIT

Part I Governance and Policies

Chapter 1 Ambitions and Challenges of Climate Adaptation in
Ten Cities 49
PASCALINE GABORIT

Chapter 2 Sustainable Local Economic Development and
Gender Perspective in the Context of Climate Change
Adaptation 81
UNANG MULKHAN

Chapter 3 Exploring Alternatives in Dealing with Climate
Change and Land Based Conflicts 99
ARIEF WICAKSONO AND ILYA MOELIONO

Part II Successful Climate Adaptation Models and Early Warning Systems

Chapter 4 Urban Climate Resilience and Water: Successful
Adaptive Planning, and Early Warning Systems 121
PASCALINE GABORIT

Chapter 5 System for Multi-hazard Potential Impact Assessment, Alert, Emergency Response Planning and Tracking (SMART) in Tamil Nadu India 153

KORLAPATI SATYAGOPAL, ITESH DASH, JOTHIGANESH SHANMUGASUNDARAM, RAMRAJ NARASIMHAN, SUBBIAH ARJUNAPERMAL

Chapter 6 Landslide Vulnerability, Risk and Resilience Management of Cultural Heritage Sites in the Western Slope of Lawu Mountain, Indonesia 181

MUHAMAD IKO KERSAPATI, MUHAMMAD ATTORIK FALENSKY, GINA FITRI, HERI PURWANTO

Part III Nature Based Solutions and the Circular Economy

Chapter 7 Nature-Based Solutions in Climate Adaptation: A Shift from Specific, Isolated Tools to Large-scale Global Conservation 213

PASCALINE GABORIT AND ZOÉ THOUVENOT

Chapter 8 The Circular Economy from the Perspective of Local Climate Adaptation and Mitigation 245

PASCALINE GABORIT

Chapter 9 Indonesian Women in a Circular Economy: Waste Management Programs 269

BULAN PRABAWANI, WIWANDARI HANDAYANI, DESSY ARIYANTI, DIANA NUR AFIFAH

Table of Content

Part IV Funding Climate Adaptation

Chapter 10 Sustainable Finance: The Hindered Potential of ESG Investing in Funding Climate Mitigation, Climate Adaptation and Resilience 297
ZOÉ THOUVENOT

Chapter 11 South Asia Region: Expanding Economy with Resilience and Adaptation Financing Challenges ... 319
KAMLESH KUMAR PATHAK

Conclusion

Chapter 12 Conclusion-Can Local and International Dialogue on Climate Adaptation Echo and Reinforce One Another? A Way Forward 337
PASCALINE GABORIT

Biographies and Acknowledgments 357

Contributions and Acknowledgments 367

Introduction Chapter on Climate Adaptation

PASCALINE GABORIT PHD

The aim of this book is to shed light on some realities of climate adaptation, resilience and the related challenges and needs in vulnerable and exposed territories. The book proposes a multi-disciplinary and multi-scale analysis of climate adaptation focusing civil society, governance systems and on dialogue and cooperation. It also identifies and analyzes a few solutions, policies and good practices such as early warning systems, nature-based solutions, and adaptive spatial planning. The book also focuses on conflict-solving approaches, and the importance of dialogue and accountability in creating meaningful climate attenuation solutions.

'Human induced climate change including more frequent and intense weather events, has caused widespread adverse impacts and related losses and damages to nature and people beyond natural climate variability' (Intergovernmental Panel on Climate change 2022)

'Approximately 3.3 to 3.6. billion people live in contexts that are highly vulnerable to climate change (…) Vulnerability is higher in locations with poverty, governance challenges and limited access to basic services and resources, violent conflicts and high levels of climate-sensitive livelihoods (eg. small holder farmers)…' (Intergovernmental Panel on Climate change 2022)

Climate change is one of the major systemic threats of our time. 2021 and 2022 have been marked by extreme climate events: unprecedented heat waves in Europe, in the Mediterranean, in North America, and in South Asia, droughts, desertification, and forest fires, unprecedented floods in Northwest Europe, and in Mozambique; tropical storms like Hurricane Yaas in Bangladesh[1] and storm Ana in Madagascar[2], as well as, consecutive cyclones Batsirai and Emnati of 2022, make up just a handful of the many natural disasters around the

---

[1] In May of 2021
[2] In January of 2022

world. Impacts are countless. An estimated 10 million people are suffering from famine in the horn of Africa in June 2022.

Summer of 2021 was the hottest summer on record over countries like the United States and Tunisia with temperatures reaching over 50°C in some areas. Summer of 2022 reached records of temperatures in Western Europe due to a 'double Jetstream effect' (Rousi et al. 2022), including in France, in Portugal, in Spain and in the UK. These heat waves caused the 'invisible and silent' death of several thousands of people in each of these countries, and especially in cities more severely hit by the heat island effect. There was a myriad of unprecedented wildfires. The Dixie fire in California in the summer of 2021 resulted in the loss of 390,000 hectares of land and wildlife and has been recorded as the largest single fire on record in the state. Across the summers of 2021 and 2022, major wildfires occurred across many parts of the Mediterranean region with Algeria, Southern Turkey, Greece, Southern France, Spain and Portugal especially affected, accelerating a vicious circle of climate change with the destruction of important carbon sinks^3. In the Gironde region in France, around 20 000 ha (200km²) have been burnt in the fires of July 2022, which is equivalent to twice the surface of the city of Paris. Another major consequence of climate change are floods. On the 14th and 15th of July 2021, Western Germany and Eastern Belgium received 100–150 mm of rainfall over a wide area on already saturated ground, causing flooding, landslides, and more than 200 deaths and many people homeless. At the same time, in the Henan province in China, flash floods were linked to more than 302 deaths, with reported economic losses equivalent to 17.7 billion US dollars (UNFCCC 2021). No area in the world seems spared by climate disasters. An unprecedented heat wave also hit India and Pakistan in April 2022, Niger is severely affected by water scarcity, a drought related disastrous famine is present in the horn of Africa, and floods in the North East of Bangladesh in June 2022 stranded 4 million of people in need of shelter, food, and clean water^4.

Most of the human losses linked to climate change were overlooked, as they did not account for famines, malnutrition and other health-related deaths. Indeed, the latest IPCC report (2022) is raising the alarm on the impacts of climate change on the most vulnerable areas: 'Increasing weather and climate events have exposed millions of people to acute food

---

[3] https://www.pilot4dev.com/images/EN_-_Policy_brief.pdf

[4] https://bdnews24.com/bangladesh/2022/06/22/floodwater-recedes-in-bangladesh-northeastern-sylhet-region-for-now and https://en.wikipedia.org/wiki/2022_Sylhet_floods

insecurity and reduced water security with the largest impacts observed in many locations, and/or communities in Africa, Asia, Central and South America, small islands, and the Arctic (...) Jointly, sudden losses of food production and access to food compounded by decreased diet diversity have increased malnutrition in many communities' (IPCC, 2022:11). The famine in the South of Madagascar in 2021 and 2022, which was directly related to the crop failures caused by reduced rainfall, is another example of the increased death toll that is not always attributed to climate change, but where climate change is an important factor together with poverty and lack of infrastructure. The horn of Africa is also deeply hit by famine and malnutrition, caused by violent conflicts and by severe increasing droughts. Currently the life of 10 million people is threatened by famine and malnutrition in this region of the world.

Globally, approximately 3.3 to 3.6 billion people live in contexts that are highly vulnerable to climate change (IPCC, 2022: 13). No region is spared by the increase in climate-related disasters, while the link between climate change and the direct impacts is not always possible to demonstrate directly when the climate-related event is occurring. Preliminary 'rapid attribution' studies[5] have been carried out for the heatwave in northwest America in June and July and found out that the heatwave is 'still rare or very rare in today's climate but would have been virtually impossible without climate change.' Global sea level changes primarily result from ocean warming via thermal expansion of sea water and land ice melt, for example, from glaciers, which is creating concern of sea level rise, floods, soil erosion, and further climate change and climate related events as oceans are accumulating heat and are the most important carbon sinks for GHG emissions[6]. As a reminder, oceans account for 90 % of the total accumulated heat that has been gained by the planet due to climate change. Human vulnerability to climate-related disasters is higher in locations with poverty, governance challenges, and limited access to basic services and resources, violent conflicts, and high levels of climate-sensitive livelihoods such as farmers with small plots of land (IPCC 2022). Cities paid a heavy toll for this situation, as they accommodate populations with a high density including vulnerable groups, but also because cities are exposed directly to hazards: These include heat waves with impacts on vulnerable populations such as aging

---

[5] Such as the one of WWA World Weather Attribution
[6] The ocean absorbs around 23 % of the annual emissions of anthropogenic CO_2 to the atmosphere according to UNFCCC 2021.

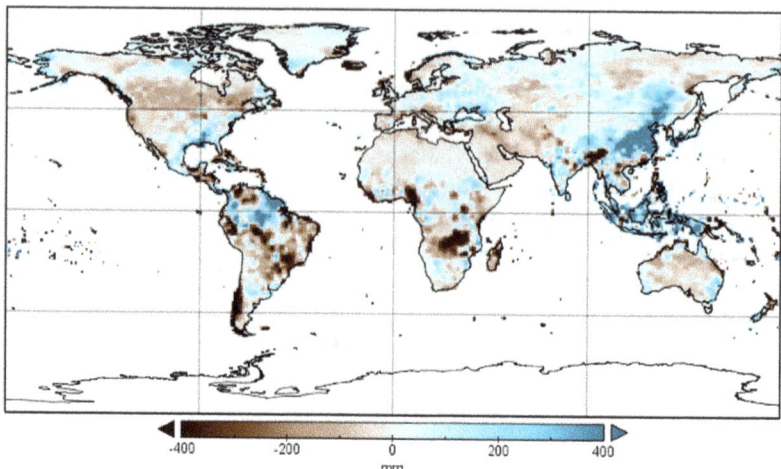

Precipitation Anomaly, 2021, Jan-Sep

Map 1. Source total precipitation anomaly in Jan-Sep 2021 w.r.t. reference period 1951–2000. Blue indicates more precipitation than the long-term means while brown indicates less than usual rainfall totals. The darkness of the color represents the amount of the deviation (Source: Global Precipitation Climatology Centre (GPCC), Deutscher Wetterdienst, Germany).

populations and children and flood exposure in the vicinity of riverbeds and coastal exposed areas. The 2022 IPCC report also explicitly mentions that vulnerability increases in informal settlements.

Southeast Asia and Africa are highly vulnerable to climate disaster risks. Africa is highly impacted by droughts, desertification, and increases in temperatures, which affect water resources, ecosystems, rain-dependent agriculture, and communities (Gaborit, 2021). South Asia (Bangladesh, Sri Lanka, Southern India) and Southeast Asia are directly exposed to floods and to sea level rise (Map 1), as well as to heat waves and increasing temperatures. 8 of the 10 countries in the world with the largest number of people living in low altitude areas are in Asia. This includes China, Bangladesh, India, Vietnam, Indonesia, Thailand, Japan, and the Philippines (Climate Central 2019). Other countries like the Maldives are equally dangerously affected. Here again, civil society organizations and governance systems such as local governments and cities are at the forefront of the climate battle. Recent studies show that a city like Manila could be entirely underwater by 2050, while cities like Jakarta are threatened by subsidence or 'sinking' below the sea level. The city lost 3–10 cm already in some areas already due to groundwater extraction among others, leading to increased floods. As we will detail later in this book, this situation of vulnerability in large, but also in medium-sized cities, can also have cascading or compound effects on water access and landslides.

In 2015, the signatory countries of the Paris Agreement agreed to fight against climate change and to limit the rise in global temperatures to 1.5 or 2°C. Seven years later, the efforts towards climate mitigation and adaptation are even more urgently needed – including scientific research on the topic. Climate change impacts are causing systemic cascading and compounding risks in different regions, where cities, civil society, and communities are at the forefront of adaptation, planning and resilience. We could assume first that there is a direct benefit for territories (and government tier levels) to invest in climate adaptation and resilience, as the investments will benefit to the protection of their own territories and communities (on the contrary to climate mitigation where the benefits are globalized, asking different countries and actors to act simultaneously). Our analysis tends to show however that in either case - of climate adaptation and climate mitigation- stakeholders and decision makers will be forced to make trade-offs to implement the necessary changes and investments.

The aim of this book is to develop different angles of approach, multi-disciplinary expertise, multi-scale analysis, and food for thought on the systemic risks, interdependencies, and possible solutions for climate adaptation and mitigation from the global level to the local level with considerations and frameworks for cities, civil society, and governments. This book is of interest for researchers, experts, decision makers, but also for any person working on climate adaptation, resilience and Dialogue. It focuses on climate adaptation challenges in Indonesia, India, Europe and other parts of the world and brings research findings, as well as a few specific examples of practical solutions and good practices. We approach interrelated systemic risks induced by climate change before considering possible solutions and success factors. Additionally, we demonstrate that governance, cooperation, nature-based solutions, green and circular economy as well as funding and financing climate adaptation can raise concerns in their implementation, even when technical solutions exist. We will focus on the Asian-Pacific Region, especially Indonesia and India, but with examples of successful developments globally.

The book is unique in that it unites different perspectives and examples from different countries and regions to develop a thorough and nuanced perspective on potential climate change solutions and the barriers to behavioral change at all levels of climate-related decision making.

The aim of the book is also to bridge a knowledge gap on climate adaptation and on the proposed solutions. It is split into 4 parts:

- Part I: Governance and Policies
- Part II: Successful Climate Adaptation Models and Early Warning Systems
- Part III: Nature-Based-Solutions and the Circular Economy,
- Part IV: Funding Climate Adaptation.

This publication is considering the challenges and solutions for the adaptation of territories to climate change. It approaches the governance actions, the cooperation of stakeholders, engineering and nature-based solutions, and finally questions of finance and funding.

The objective of the book is to create a joint publication between European and Asian researchers, on climate adaptation as a concept, and on solutions notably for cities. The authors illustrate through a cross-disciplinary analysis and complementary chapters, the coping strategies that have emerged to face climate change in particular in Indonesia and in India, but also at a larger scale.

This introduction will explore the broader picture of climate adaptation. It will first give the rationale behind the chapters of this book (part I) and then introduce the book's approach, methodology and chapters (part II). Part I. will explore the impacts of climate change and the vulnerabilities (*or why this book?*) (1), consider the systemic threats induced by climate change (2), elaborate on the concepts of climate adaptation and resilience (3), before highlighting the lack of funding (4), and the necessity to strengthen climate mitigation (5). This part also explains why the focus of the book looks towards Asia. Part II will go through the multi-scale and multi-disciplinary analysis, the field methodology, before approaching the structure of the book and the presentation of the individual chapters.

Part I The Broader Picture of Climate Adaptation

1. Why This Book? Climate Impacts and Vulnerabilities

The effects of climate change are already known and experienced in different areas of the world, with variable intensity, but it is impossible to deny that they are already occurring. Forecasts and projections are based on different scientific scenarios. The human loss calculations related to climate change appear far below the exact fatalities where climate change acted as amplifier for other causes of loss of human life when considering droughts, famines, and other related problems (e.g. in Southern Madagascar in 2021 and in the horn of Africa as well as in Niger in 2022). It is foreseeable that climate extremities and disasters such as typhoons, floods, sea level rises, and extended dry spells become more frequent. The disruptions in the ecosystems and water resources jeopardize the wellbeing of populations and can lead to displacements, unrest, and potential conflicts (Glasser 2020, IPCC 2021). In this context, islands and coastal areas like the Indonesian archipelago, Bangladesh, the Maldives and other regions such South East Asia and Africa are becoming increasingly vulnerable to climate disasters (Nicholls et al. 2007). According to the recent IPCC report, half of the world's population is now living in areas that are vulnerable to climate disasters (2022), and solutions are urgently needed to cope with this situation (and to reverse the trend when possible).

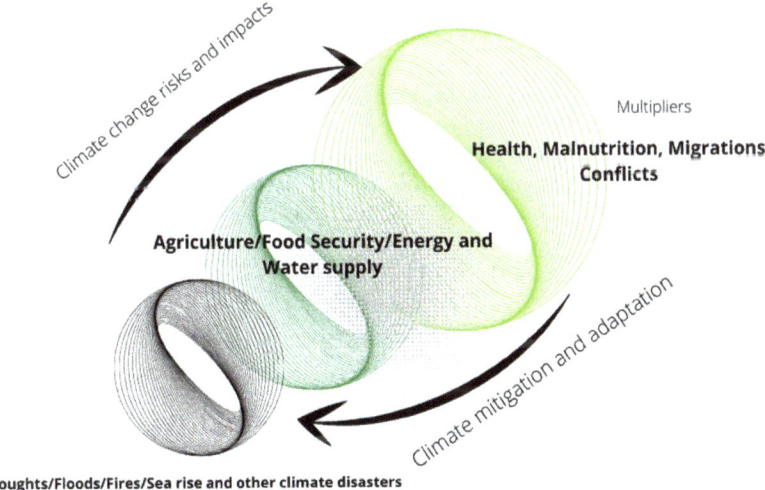

Fig. 1: Cascading effects of Climate Change

The direct impacts of Climate Change: Examples in vulnerable areas

Tab. 1: Table on the impacts of Climate Change. Source Author.

Sector	Impacts	Multipliers
Agriculture	Increase in exposure to droughts and heavy rains of different types of agriculture.	Rain-fed agriculture is highly vulnerable to droughts. Water shortages will increase
Food Security and Malnutrition	Long droughts, or climatic seasonal changes will impact crops and food production.	Malnutrition is having repercussions on children's growth, education and health. Adults' health is also exposed to food insecurity.
Disruption of the Energy Supply	Hydro-powered projects relying on riverflows are highly exposed to droughts. Other sources of energy production are vulnerable to droughts and floods.	Energy disruptions can create blackouts, creating the need for costly emergency solutions.
Cities	Cities are exposed to water shortages, floods, droughts, as well as to increased water and air pollution.	Urban expansion has impacts on land use deforestation. Ground water extraction is impacting subsidence. Transport traffic and congestion is increasing air pollution.
Access to water	Water resources and reservoirs are endangered by droughts, while the sea level rise is also causing salinization (saltwater intrusion) of drinkable water resources.	Water shortages can negatively impact agricultural production. Finally, the pollution of water is detrimental for health and leads to epidemics.
Coastal and flood exposed Areas	Coastal areas are exposed to floods and sea-level rise, as well as populations' displacements.	Populations' displacement can be accelerated by soil erosion, floods, but also by the depletion of the oceans.
Health	Climate change increases exposure to diseases such as malaria and waterborne diseases (typhoid and cholera). Heat waves increase mortality of vulnerable groups such as young children and elderly.	The increased pollution of groundwater and air is leading to premature deaths, including among younger populations and children.

Some definitions will guide us through the analysis. **Vulnerability** in terms of climate adaptation has a proper definition (Tobey 2011, Butterfield 2020). *'Vulnerability is defined as the degree to which a human or a natural system is susceptible to, or unable to cope with adverse effects of climate change'* (IPCC 2014, Tobey 2011: 2). In its terminology, the United Nations Office for Disaster Risk Reduction (hereby referred to as the UNDRR) defines 'vulnerability' as a set of conditions leading to a higher exposure to hazards: 'Conditions determined by physical, social, economic and environmental factors or processes which increase the susceptibility of an individual, a community, assets, or systems to hazards.' The theme also encompasses concepts such as the sensitivity or susceptibility to harm and lack of capacity to cope and adapt (Butterfield 2013). **Hazards** are defined in the UNDRR terminology by a 'process, phenomenon, or human activity that may cause loss of life, injury, or other health impacts, property damage, social and economic disruption or environmental degradation'. The association of the terms resilience and vulnerability have been popularized by the 2015 UN Sendai Disaster Framework for Disaster Risk Reduction which recalls the principles of climate disaster management based on prevention, preparedness, early warning, recovery, and reconstruction (as we will see in Chapter 4). Its scope includes natural disasters but also climate-related disasters such as droughts, floods, typhoons, and other extreme weather events. The advantage of addressing the questions of climate change in terms of vulnerabilities is that it encompasses different sectors such as agriculture, water, forestry, and health but also human factors and socio-economic development, including an analysis based on power.

However, the vulnerability-focused approach overlooks the potential cascading effects of climate disasters (illustrated in Fig. 1). It will also consider the impacts in terms of destabilization and systemic risks. This is why the first part of this book will approach governance, policies and conflicts. The second part of this book will be dedicated to possible solutions, adaptative planning and early warning systems, while the third part will consider nature-based solutions and the circular economy. The Theory framework is illustrated in the graph below (Fig. 2) showing that governance, socio-economic pathways and funding are needed to cope with the interrelated risks.

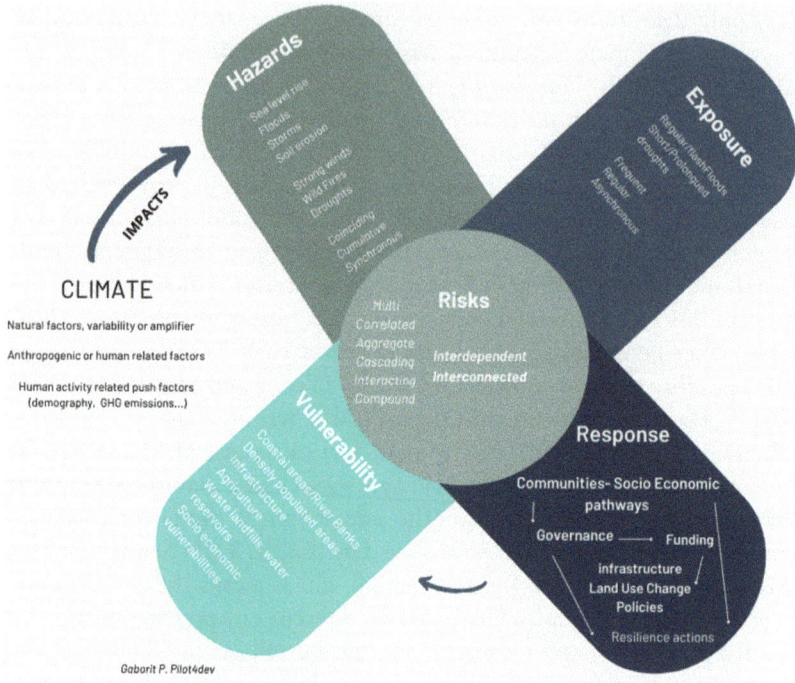

Fig. 2: Theory framework. Adapted from Simpson et al. 'A framework for complex climate change risk assessment' 2021

2. From Vulnerability to Systemic Threats: The Hidden Part of the Iceberg

Fig. 2 illustrated above gives an overview of Hazards, Exposure, Vulnerability and Response and how they articulate with the necessity to anticipate risks, and the part related to unpredictability. The questions of risks and the definition of hazards are however equally complex and interrelated systems. Behind a first level of vulnerability, a series of 'cascading' or 'compounding' risks could quickly aggravate disaster situations and lead to further disruptions and systemic threats. COVID 19 for instance symbolized to the world how a vulnerability in the form of a pandemic could develop into a systemic threat by displaying the risks in health systems, the economy, fundamental freedoms, human conflicts and geopolitics. The links between climate and conflicts (from vulnerability to systemic threats) have been documented by research and studies after analysts had found out that some territories cumulated different types of vulnerabilities in terms of security, conflicts, and climate disasters. (Ide et al. 2020, Nordqvist et al. 2018, Reveuny, 2007). The possible interrelations also have been often highlighted by actors involved in humanitarian and disaster response (Cuny, Hill 1999), and in this book Chapter 3 will attempt to provide new perspectives (Wicaksono 2022).. Over the past three decades, research has indeed produced evidence of increased relations between climate and conflicts, conflicts being fueled by climate-related conditions, and conflicts exacerbating environmental degradation. In these conditions, the situation includes coping with a range of interrelated natural and human-made hazards: conflicts and violence, displacement, natural disasters, minorities, environmental degradation, shrinking natural resources, increased health threats, poverty, inequalities, social tensions, numbers of refugees, and post-conflict fragility. It is often acknowledged that climate change is an amplifier for conflicts but is not usually recognized as the main source of conflict compared to other social vulnerabilities. An example of this is the Ethiopia Renaissance dam, when Ethiopia built a dam on the Blue Nile river with strong negative impacts in water access for the countries downstream, therefore creating a latent conflict with neighboring countries like Egypt and Sudan. In most of the studies, climate change is viewed as an amplifier for existing regional tensions. If, however, we focus on land use at a local scale, climate change, conflicts, and changes in land use appear deeply interrelated as we will see in Chapter 3 (Wicaksono 2022). Indeed, it is often hard to distinguish between chronic vulnerability caused by factors like food insecurity, poverty, and climatic variability and humanitarian crises or violent conflicts in contexts where a part of the population lives with the threat of loss of land and livelihood. The number of variables contributing to conflicts makes it difficult to demonstrate causation.

In these circumstances, it seems urgent to take action to facilitate the attenuation of climate change impacts (with climate adaptation), as well as to embed other questions such as social justice, stability, security, land use, an approach on migrations, as well as on local economic development. These points are approached with different perspectives in the first three chapters of the first part of this book on governance and policies.

3. *Climate Adaptation and Resilience: Concepts, Objectives and Limitations*

This book is about climate adaptation governance, policies and a few possible solutions, as well as about resilience. **Climate adaptation** refers to any adjustment, whether passive, reactive, or anticipatory, that can respond to anticipated or actual consequences associated with climate change (IPCC 1995). It is implicitly recognized by international organizations and researchers that future climate changes will occur and must be accommodated in policy: Adaptation to climate change is no longer a choice but a necessity, and this is why research and publications on the topic, such as this book, are so important.

The existing literature has underscored the role of local governments, cities, and civil societies or trust networks in resilience and climate disaster risk management (Grimmond 2007, Oleson et al. 2015, Rosenzweig et al. 2018). Meanwhile, the importance of giving more visibility to the local context and to resource management in development approaches has also been demonstrated (Borel et al. 2006, McCandless, Abitbol et al. 2012, 2015). The adaptation can be either reactive (after a disaster) or transformative (to prevent further disasters) see Fig. 5.

Another lens through which we have gone through this book is the **question of resilience** which will strengthen the approach on climate adaptation. Resilience is a cross-disciplinary concept referring to the ability to cope, recover, and build back from stress, shocks, and crisis. 'resilience' initially means 'bounce back'. The term has originated from the world of physics and has been since been applied to psychology (Tisseron 2018), geography (Laganier 2018), and many other areas such as disaster management. Its use in policy discourse has been soaring since the years 2000 (Brassett et al. 2013, Brinkmann et al. 2017). The increasing reference to the concept of resilience by international organizations has been concomitant with the idea that 'sustainability' was not be an easy path, and that societies, economies, cities, organizations will be affected with increasing disruptions. Resilience came therefore simultaneously as with a decrease in the faith towards continuous progress, especially in the areas of climate related disasters, lack of development, economic destabilization, political instability, and geopolitical tensions (as the current conflict in Ukraine unfortunately shows).

However, the concept has limitations: it can lead to overlook the impacts of the damage and disasters. The use of the resilience concept can lead to the denial of the grief and lack of support to the victims and resilience programs can equally be performed at the expense of some groups or stakeholders (Meerow et al. 2016, Ziervogel 2017, Féron 2021, Gaborit 2021). Climate adaptation and resilience equally lead to governance-related questions about who determines the acceptable levels of risk, under what grounds, and for the benefits of whom resilience actions or policies are operationalized. The situation of 'Only the resilient survive' in disaster management and in humanitarian aid appears strongly unethical in that case. As some authors demonstrate, in the area of security, people (as opposed to groups, states, and communities) are increasingly considered responsible for developing prevention and response strategies (Brassett et al. 2013, Chandler 2013). The complexity of the concept of resilience and its application should not disqualify its use, especially in the area of climate adaptation, and in urban resilience, as we will see in Chapter 4 of this book. Resilience as a concept can indeed bring strength to risk management practices in uncertain or unpredictable environments. Practically, however, this means that advocating resilience should be thoroughly considered, and situational complexity should be emphasized. Policy and development programs with a focus on climate adaptation and resilience should not replace prevention and assistance to vulnerable groups. For example, early warning systems should not replace disaster-resilient buildings or even in some cases, the relocation of population or activity. We could indeed not exclude what we could call a 'Titanic Syndrome,' a vicious cycle by which a focus on resilient infrastructure would overlook the necessity of any emergency response. Climate Adaptation and resilience programs should not entail less emphasis on the anticipation of risks, and especially the emphasis of systemic risks. It should not overlook systemic threats and the possibility for prevention/preparedness and transformative action to prevent the damages. For this book, we will focus especially on urban resilience (as detailed in Chapter 4). As we will see in Chapter 1 and in Chapter 4, urban resilience can be defined as the cities' capacities to respond, adapt, and recover from the pressures and crises related to climate change and other changes. These changes include subsidence, demographic growth, poor land management, insecurity or attacks, economic downturns, social unrest, unsustainable use of resources, and declining ecosystems (Meerow 2016, Ziervogel 2017, Gaborit et al. 2021).

Although an expert consensus is emerging that climate adaptation will be necessary to protect societies, communities, and ecosystems, political conversations and debates still do not pay enough attention to the implementation of climate adaptation. Instead, policymakers see climate, environment, and sustainable development as very broad and long-term concepts that are therefore not prioritized in a meaningful way in their implementation. The best evidence of this is the budget allocation for climate adaptation by multilateral donors. Despite the existence of a global Green Climate Fund[7] since 2010, the pledge of 100 billion dollars for climate action from the Paris Agreement never entirely materialized, and when the funding is available the priority is almost always entirely given to technical assistance (international experts) to the detriment of investments, support to projects, programs or even research. In addition to this, climate adaptation has been until very recently rarely prioritized in international programs.

4. The Worldwide Investments in Climate Adaptation Are Still Insufficient

How is it possible to adapt to climate change and develop resilience mechanisms without addressing the question of funding? Multilateral Development Banks (MDBs) now represent the most important sources of public climate adaptation financing. In 2019, the MDBs committed 61.5 billion US dollars in climate finance. This includes 76 % for mitigation and only 24 % for adaptation (Tall et al. 2021). For the year 2019, the largest sectors of adaptation finance flows went to energy, transport, and other built environment infrastructure (14,937 million US dollars), and to the water and wastewater systems sector (2,954 million US dollars) whereas the coastal and rivers' infrastructure accounted for only 682 million dollars (Tall et al. 2021, p. 9). NGOs and multilateral banks all point out the increasing need to invest in climate change adaptation. Other alternative sources such as the investments from the private sector are needed (Part IV). Compared with MDBs, other public sources of adaptation finance are low and cannot cover the growing investment needs. In countries, characterized by a high vulnerability index to climate disasters, the World Bank has enabled the establishment of funds to better adapt to climate related disasters. The Global Facility for Disaster Reduction and Recovery (GFDRR) and other donors, have also set up a City Resilience program which is active in 140 countries. These mechanisms aim to enhance and strengthen the resilience of cities and territories with bankable projects, but also shows that the available international budget is not always directly available for the actors who need it: local governments, communities, entrepreneurs and investors to enhance the resilience of the territory.

Part IV and especially Chapters 11 and 12 of this book will address the question of funding through different angles, the one of the private

---

[7] In the framework of the UNFCCC.

investments (for Chapter 11) and the question of the climate adaptation needs in South Asia where climate adaptation has been recognized as a necessity (for Chapter 12). But there will be no adaptation possible without mitigation and attenuation effects of climate change as reflected in Fig. 3 on the interrelation and impacts.

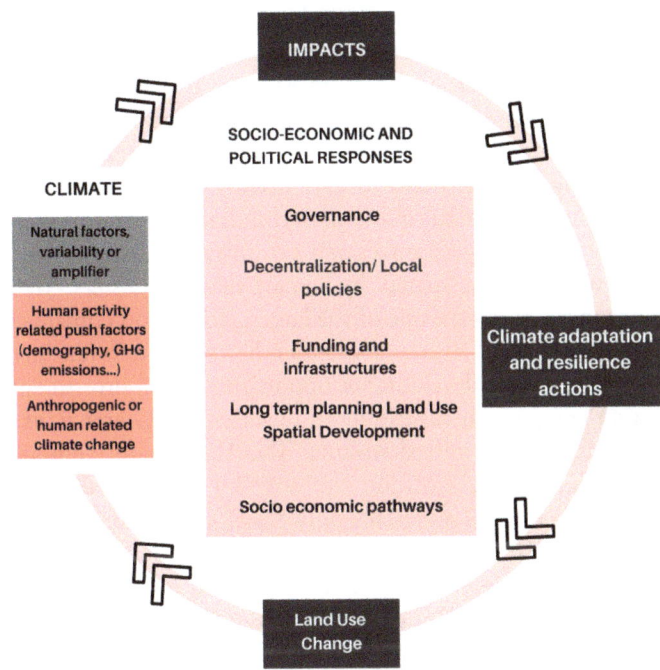

Fig. 3: Factors interrelations in climate adaptation

The latest IPCC report (2022) demonstrates that the impacts of climate change have caused widespread damages to nature and people, 'beyond natural climate variability' (IPCC, 2022, 5). This shows the urgency to limit the increase in temperature at a maximum of 1.5°C, and the need to combine mitigation of climate change and adaptation to the irreversible impacts of climate change. Adaptation and mitigation need to work hand in hand. This is also illustrated in Fig. 5.

5. A Focus on Adaptation Does Not Overshadow Mitigation and Attenuation Needs

Whereas adaptation has become a major concern, there is also a consensus that further investments are needed for climate change mitigation in order to reach the targets of the Paris Agreement to keep global warming below the level of 2° Celsius and, when possible, below 1.5° C. As we have mentioned earlier, there can be an assumption that territories' investments in climate adaptation and resilience, can be easier to justify for local government tiers and local investors, as they can benefit directly to their community. A resilient and adapted territory will be better protected against hazards and less vulnerable. On the contrary, investments in climate mitigation are made to the benefit of the common good and the planet which leads to the question of simultaneous commitments and actions. Indeed, nothing guarantees that a country 'frontrunner' in climate mitigation will be more 'spared' by climate related disasters. This question brings to the forefront one of the largest debates on climate change, including during the COP events: whether the low emitters (e.g. in Africa) should sacrifice future development opportunities and investments (e.g. in polluting energies) to contribute to lowering GHG emissions. Some comparison figures are indeed striking. The African continent accounts for 17 % of the global population, but only 3.4 % of global energy consumption, while the EU represents only 5.8 % of the global population, but accounts for 10.4 % of energy consumption[8]. Per capita, someone living in the European Union in 2019 consumed on average 9 times as much energy as someone living in Africa[9]. Asia as a continent is projected as responsible for 35 % of energy-related GHG emissions, but this figure could reach about 45 % without greater use of renewable energy (KAF 2018). The energy demand (in the ASEAN)[10] is indeed expected to triple from 2013 to 2040, and this rising energy consumption from fossil fuels including coal has caused more climate change challenges (Piseth et al. 2021). In Indonesia for instance, coal provides the majority of the electricity, and this share is increasing (Duprady et al. 2020).

Asia is going to 'make or break' climate change. There is an abundance of opportunity for Asian, like for Western countries to change

---

[8] Afrique-Europe Foundation, Online debate 20th of January 2022.
[9] Ibidem.
[10] Association of Southeast Asian Nations, is a political and economic union of 10 member states in Southeast Asia.

their energy habits to make a difference in global carbon and GHG emissions. Because of the increase in GHG emissions in Asia and because Asia holds a third of the 17 % of total annual global emissions stemming from forest and land use change, changing energy and conservation habits, anywhere but especially in Asia could meaningfully affect the severity of climate change globally (KAF 2018). Moreover, people living in low-lying Asian cities and flood zones will be some of the most affected by climate change. There is in consequence a high incentive to change global habits. Indonesia and Malaysia are endowed with major forests which experience very fast deforestation that transforms old-growth woodlands to centers of human activity, destroying what were once major carbon sinks in the area (KAF 2018:19).

The question is also a complex and risky issue for Europe. Although the European Union has committed to reduce its GHG emissions by 55 % in 2030 (compared to 1990 levels) and to become carbon neutral in 2050[11], the question of the energy transition, and the large dependence on fossil fuels, particularly natural gas, make it extremely difficult to anticipate the real evolution of GHG emissions. The recent invasion of Ukraine by Russia has highlighted the heavy dependence of some European countries on Russian natural gas and oil and helped raise awareness of fossil fuel dependency within the public opinion. Additionally, analysis shows that the extraction and transport of natural gas, like from Russia to Europe, were increasing the emissions of methane, a gas with a high impact on climate change. Recent studies and calculations report that these methane emissions, largely linked to resource extraction and transport, have been largely underestimated: Declarations of methane emissions by national governments' in NDCs have been analyzed to be 70 % lower than actual emissions (Global Methane Tracker 2022)[12].

This book demonstrates the complexity of linking economic needs, social vulnerabilities, and climate concerns, especially in the key sector of energy. To simplify, we could indeed say that the energy transition is based on a trilemma of three main factors: (1) security in the form of sovereignty, independence, and protection of infrastructure, (2) environmental factors like GHG emissions, climate impacts, and waste production, and (3) socio-economic factors like fuel poverty culture and the

---

[11] See the 'Fit for 55 strategy' which is currently debated at the European Parliament
[12] Global Methane Tracker 2022, Published by the International Energy Agency.

prices of energy. These interrelated and contradictory interests make necessary transformative change difficult (See Fig. 4 and Fig. 5).

Access to electricity is still a challenge in some rural areas including in Indonesia and in India despite a very high rate of electrification (SDG7). The access to electricity is 98.8 % of the population of Indonesia as of 2019, and 97.8 % in India according to the World Bank based on household surveys[13]. But other sources (e.g. NGO reports) mention the difficulties for the population living in rural remote areas or in the informal housing sectors to access electricity. Asia Pacific also holds the most proven reserves of coal (46.5 % of total) in the world, with China, Australia, India, and Indonesia holding the largest reserves in the region. Coal is indeed the cheapest but also the most polluting form of energy. China consumes the most coal internationally while Asian Pacific countries combined consume half of the world's coal. Other countries like India still rely 58 % on coal for their energy needs, despite a recent decline in the energy mix. Moreover, coal production is also steadily increasing in Indonesia[14]. Interestingly, Indonesia and India also account for some of the highest reserves of natural gas with 2.8 and 1.5 trillion cubic meters of natural gas reserves respectively (KAF 2018). Renewable energies are playing an increasing share accounting for 19.3 % of energy consumption in the region, with ambitions to double the production in 4 years in countries like India. In this sector, China is controlling the world's markets for the transformation of minerals such as nickel, copper, lithium, cobalt, and rare earth materials (IEA 2021), but Indonesia could become an important market player in this area[15]. Not less interesting is the fact that the renewable energy sector jobs have shifted towards Asia which accounted for 62 % of all worldwide renewable energy jobs in 2018 compared to 51 % in 2013, with photovoltaic growing the largest proportionally (Duprady et al. 2020). The palm oil biofuels production in countries like Indonesia has, however, led to conflicts in land use problems, and finally to the reduction of similar biofuels (due to international bans, but also to shortages in palm oil for other productive uses). The Philippines is the world's largest producer of geothermal energy (16 % of the electricity), while the potential for Indonesia is also tremendous (29 gigawatts, 5 % being utilized). Employment in renewable

---

[13] https://donnees.banquemondiale.org/indicateur/EG.ELC.ACCS.ZS?view=chart

[14] Indonesia: Energy Country Profile – Our World in Data https://ourworldindata.org/energy/country/indonesia#what-share-of-the-population-have-access-to-electricity

[15] Race is on for Indonesia's untapped rare earths – Asia Times https://asiatimes.com/2020/09/race-is-on-for-indonesias-untapped-rare-earths/

energy and especially photovoltaic but also geothermal is however increasing with China, the US, and India leading the job market. The figure thereafter exemplifies the trilemma of the energy transition (Fig. 4). Looking at the economic angle also shows that the whole mitigation effort cannot be transferred to cities but needs to come from a global energy transition. However, post-Covid recovery plans are not especially green in most countries (Duprady et al. 2020, UNEP 2021). The nexus between the access to energy for all (SDG7), the preservation of climate and environmental goals (SDG13), and the fight against extreme poverty (SDG1) will increase the necessity for decision makers to manage their priorities. Additionally, the climate and environmental stakes are not represented by any official parties with the exception of CSOs and regulations. Moreover, we need to integrate the fact that the most vulnerable populations are the most exposed to climate disasters.

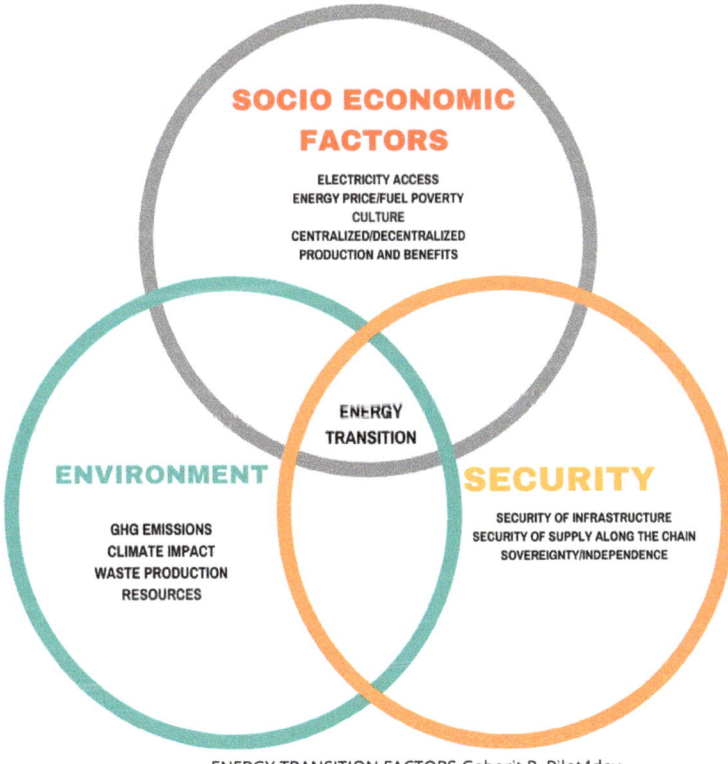

Fig. 4: Trilemma of the energy transition.

Accordingly, this situation raises some questions of accountability and the necessary prioritization for the short and long terms, but also the identification of which parties, apart from the UNFCCC, the IPCC, UNEP, activists and environmental NGOs, are defending the climate question in international negotiations considering that their messages sometimes remain unheard or unconsidered in the facts (Wicaksono et al. 2022).

This is why a global overview of solutions for both climate mitigation and also climate adaptation are important and crucial to consider. We will not systematically unlink the two aspects of climate adaptation and climate mitigation. Instead, we will focus on the climate disaster management cycle: Prevention, Preparedness, Response, and Recovery of climate adaptation, including an overview of potentially viable solutions (Parts II and III), and an insight into the different challenges such as governance and multi-stakeholders' cooperation (Part I) and funding (Part IV). To approach these important topics, our methodology is based on multi-scale analysis, focusing on international agencies, local authorities, civil society organizations, and pluri-disciplinary research. This multi-scale analysis is an important part of the book's methodology, as the methodology enables us to analyze potential leverages for transformative change.

Part II Book Methodology and Approach

1. *Overall Approach of the Book: A Multi-scale and Multi-disciplinary Analysis*

This book is an attempt to combine a multi-scale and multi-disciplinary analysis, throughout the different chapters. Multi-scale governance is a necessary step in the approach towards climate adaptation and mitigation. No system can be understood or managed by focusing on it at a single scale. All systems coexist and function at multiple scales of space, time and social organization, and the interactions across scales are fundamentally important in determining the scale's dynamics. This interacting set of hierarchically structured scales has been termed a 'panarchy' (Gunderson and Holling 2003). This panarchy shows the adaptation and interrelation of systems highlighting not only the roles of memory, conservation, and revolt but also the roles of innovation and adaptive capacity to change.

Introduction Chapter on Climate Adaptation

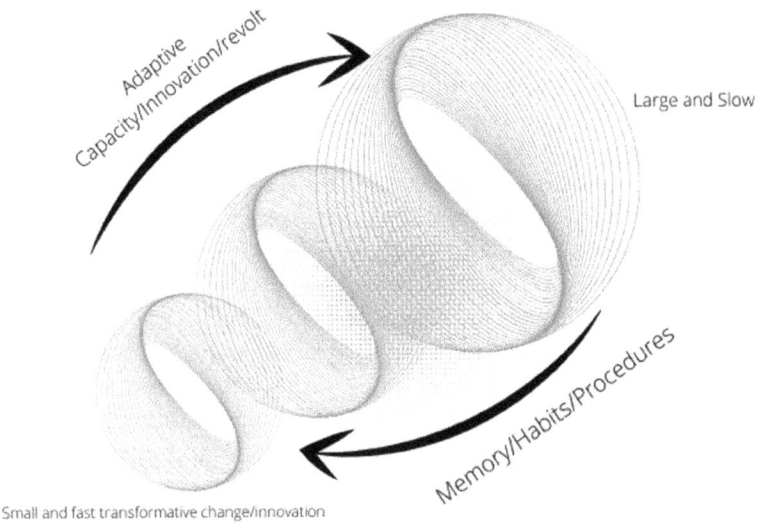

Fig. 5: Transformative change and panarchy inspired from Gunderson and Holling (2002).

Fig. 5 shows the transformative change in a multi-scale approach, with the interrelation between memory, habits, procedure reducing the opportunity for change, and the adaptive capacity and innovation leading to further change. This necessary interaction is reflected in the book's analysis of governance, policies and tools both at the local and territorial levels, as well as through national policies and international programs related to climate adaptation. The role of cities is particularly highlighted in this book throughout the different chapters. Cities are indeed considered as one of the most interesting levels to understand territories' socio-economic dynamics, to make appropriate decisions and to protect local populations and ecosystems. Urban development and concerns for the environment are undergoing permanent and constant changes and transformations: cities need to adapt to the local economy, to population needs, to legal and physical constraints and to migrations. In an ever-changing context, cities and local governments are submitted to contradictory imperatives. On one hand they need to adapt to global contexts in terms of employment, climate, and the arrival of new populations, and on the other hand they also need to provide housing, services, green spaces, and livable places, and offer good conditions of living for their local residents. In many countries, the urban population is steadily rising and creating more pressure for cities to grow. The COVID situation has slowed down but not reversed this trend. Cities will not only have to accommodate the population growth, but also to alleviate urban poverty and informal settlements, to decrease urban inequalities and to provide basic services such as safe drinking water, sanitation, and access to sustainable and modern energy for all. On top of this comes the central question of urban resilience, or the capacity of cities and civil society organizations to adapt to crisis or disasters, to develop responses, and to build back better. In a context of climate change the question of urban resilience becomes an urgent priority for cities affected by multiple hazards, including heat waves, and for the ones in coastal areas.

In this context, both at the local and at the national levels, the small and large multi scale transformation and innovation required to meet these needs will necessitate time, changes, investments but will also bring revolts and conflicts (Fig. 5)

The interrelation of scales (Fig. 5) is also reflected in very practical examples, such as the one of floods at the local level. Indeed, within cities, there are different interrelated factors, which can contribute to the floods. These factors are illustrated in (Fig. 6) and entail the loss ecosystems, the erosion of soils linked to waste and water management, or urban development. These factors are all referring to different scales and sometimes policies and practice, which come into play in the identification and in the management of floods. This is why the different chapters embrace both very practical examples, and broader analysis of climate adaptation and related policies, while proposing a multi-scale analysis.

This multi-scale and multi-disciplinary analysis is also important to understand the systemic threats, problems and disruptions which are related to climate change, and which are equally cross disciplinary, and refer to different scales, policies or to different governance mechanisms.

Introduction Chapter on Climate Adaptation

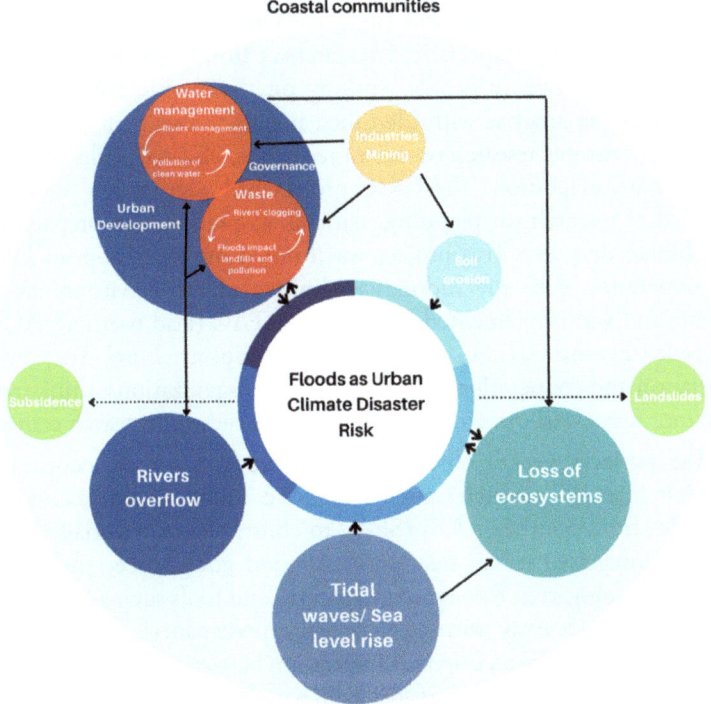

Fig. 6: The interrelations of risks in urban floods (Gaborit. P.).

2. Field Methodology and Projects

This book is linked to several projects such as the Climate Resilient and Inclusive Cities project (CRIC) and the **LIFE Adapt Island project**. The LIFE Adapt Island project promotes nature-based solutions in climate adaptation with the restoration of marine ecosystems such as mangroves and corals in highly vulnerable coastal areas in Guadeloupe, supported by the international research of a scientific committee. The LIFE Adapt Island project's experience and findings largely inspired the chapter on the circular economy and partially inspired the chapter on nature-based solutions[16].

---

[16] Other projects such as Pilot 4 Research and Dialogue or Fish CRECA have been referred to.

The research is also largely based on the collective research-action project **Climate Resilient and Inclusive Cities (CRIC)**[17]. The project involving a dozen of specialized researchers from several organizations and more than twenty people working on the coordination teams. It also crosses the analysis with identified literature and with the author's and teams' current research on urban resilience, disasters' risks reduction and climate adaptation. The CRIC project had selected ten pilot cities for further research on planning, climate adaptation and preparedness for climate disasters. This project was funded by the European Union in cooperation with the Indonesia Ministry of the Environment and Forests and was implemented by UCLG-ASPAC (lead partner), ACR+, Pilot4dev, Ecolise, AIILSG and the university Gustave Eiffel. They ensure the liaison and cooperation with cities, public organizations, think tanks, grassroot organizations, communities, universities and research centers.

The project's overall objective was to create a unique cooperation between cities and research centers in Europe, Indonesia, India and other countries from Southeast Asia (SEA), to contribute substantially to sustainable integrated urban development, good governance, and climate adaptation/mitigation through partnerships, and tools such as sustainable local action plans, early warning tools and experts panels. The project has been built to facilitate an original cooperation between cities, officials, civil society organizations and academics. It has developed trainings, online training platforms, climate action plans, and urban analysis reports in 10 medium sized cities. Moreover, the project envisioned to promote and deliver technical assistance, capacity building and research. It has been an important relay of information, and an importance source of findings as the crossroad between literature, informal dialogue and field information.

The project was based on the assumption that climate adaptation involves a robust investment from the cities, the development of ICT solutions (smart Early Warning Systems), nature-based solutions (nature buffers) as well the involvement of communities and people (in the preparedness). The CRIC cities exemplified the reality of climate adaptation and its relations to risks interrelations. For instance, Bandar Lampung is a large city and a transportation hub in Sumatra, but is also vulnerable to natural disasters, including floods, landslides, droughts, high tides and tsunamis.

---

[17] www.resilient-cities.com

3. Selection of the Contributions for the Book

A call for contributions has been circulated in November 2021 among researchers in order to receive abstracts contributing to the main research topics. We have received several abstracts, mainly from Indonesia and India. The abstracts based on the most advanced research and findings have been selected. The call proposed different areas and namely governance, early warning systems and climate and conflicts. We have been working closely with the contributors to adapt the content to the book's objective to facilitate the understanding of climate adaptation, its challenges, and to identify a few nuanced possible solutions.

How to Use This Book?

The chapters arise from different contributors and alternate between general analysis and approaches, and more specific contributions based on specific case studies (in particular in Indonesia and in India but also in Europe) with more general conclusions, and discussion points. The reader can go through one chapter only or read the entire book … start with the end or with the beginning … read one part and come back to it for further details later on. The chapters are independent from one another, and propose different angles, methodologies, solutions and definitions.

Limitations in Scope of the Book

The book will address possible solutions which have been proved by experts and researchers to have a positive impact on the climate adaptation and mitigation nexus. We will however exclude proposed geoengineering solutions such as the carbon technical sequestration, the modification of the Earth's atmosphere, or the injection of chemicals into the oceans to stimulate phytoplankton, as there is currently no evidence that these solutions would have a positive, long-term, sustainable impact without causing further harm (Wettel et al. 2018). The book also only proposes an overview of possible solutions and findings based on assumptions valid at a certain point of time (at the time of the writing). Some of them may change overtime. The different assumptions are developed in each of the chapters.

4. Introduction into the Book's Content

To guide us through the different solutions, numerous challenges, and remaining questions related to climate adaptation, this book is structured in 4 parts:

- Part I: Governance and Policies – *with an approach on Indonesia.*
- Part II: Successful Climate Adaptation Models and Early Warning Systems-*with compared approaches worldwide.*
- Part III: Nature-Based-Solutions and the Circular Economy – *as broader topics worldwide*
- Part IV: Funding Climate Adaptation – *For this, we focus on both private investments and ESG criteria, and on the funding of climate Adaptation in South Asia.*

Introduction

Climate Adaptation: Introduction Chapter, by Dr. Pascaline Gaborit

The **introduction chapter** addresses the evolution of climate adaptation, the threats, risks, and the state of the arts. It provides an overview of the different concepts and analysis, the methodology, and details different assumptions in the conversation around climate change mitigation and adaptation. Finally, it gives an overview of the expectations, and ambitions from the book, details the approach and methodologies, and introduces the different chapters. The introduction chapter does not propose fixed definitions but presents the challenges and contributions in a nutshell and sets the frame of the different research findings.

Part I: Governance and Multi-level Policies

Chapter 1: Ambitions and Challenges of Climate Adaptation in 10 Cities, by Dr. Pascaline Gaborit

This chapter will be published in a revised version in the online Journal Climate Risk Management in 2022 under the title 'Climate Adaptation to Multi-Hazard Climate Related Risks in ten Indonesian Cities: Ambitions and Challenges. Indonesian coastal cities are faced with a double imperative to urbanize quickly and to adapt to climate-related disasters. These disasters include droughts, storms, regular floods, tidal waves and water pollution. This article investigates how ten small and medium-sized coastal cities in Indonesia are developing resilience strategies to cope with the disaster risks. It approaches their level of exposure, the current impacts of climate change and the existing coping local resilience strategies or response. It also identifies key discussion points related to the implementation and the feasibility of these strategies. The research findings raise several discussion points, including the difficulties of multi-stakeholders' cooperation, the inevitable trade-offs or difficult choices, and the lack of adequate instruments in climate adaptation. Finally, this article calls for more specific, timely research on climate adaptation in cities.

Chapter 2: Sustainable Local Economic Development and Gender perspective in the Context of Climate Change Adaptation, by Unang Mulkhan

Climate change has exposed risks, structural inequalities, and exclusion in many places where the most vulnerable, including women, are often left behind. In fact, gender equality implies that the interests, needs and priorities of both women and men should be considered with the ultimate goal of ensuring women and men have equitable access to, and benefit from, society's resources and opportunities. Therefore, it is essential to encourage and facilitate women's participation in their societies, such as in local economies, in the context of climate adaptation with a robust framework, policies, and practices. However, the nature of and the extent to which women in Indonesia can and should be involved in resilience for the economy remains in question. This chapter discusses how women participate in sustainable local economic development in the context of climate adaptation. This chapter is in 3 parts: (1) The nexus between sustainable local economic development and climate adaptation from a gender perspective, (2) women as factors of resilience for the local economy in Indonesia, which includes the map of possibilities and potential challenges faced by women in their economic participation in Indonesia from structure and agency theory perspectives, and (3) core elements and mechanisms required in developing and strengthening the participation of women in sustainable local economic development in the context of climate adaptation.

Chapter 3: Exploring Alternatives in Dealing with Climate Change and Conflicts in Indonesia, by Arief Wicaksono and Ilya Moeliono

Since the start of the Industrial Revolution which began the continuous transformation of agricultural societies to become industrial, built on growth-oriented economics and the exploitation of natural wealth, climate change and conflict have become inevitable and intertwined in their consequences. As a global issue climate change is the result of a myriad of multi-dimensional and multi-stakeholder issues, many of which are or can be perceived as conflicts, such as the conflict between growth-oriented development, social justice, and environmental considerations, or even conflict between the interests of the present generation and the needs of future generations. Additionally, Indonesia's geography, historical background, and cultural diversity make it a country particularly prone to conflicts. Therefore, development initiatives in Indonesia should be based on a deep understanding of how this particular context contributes to conflicts and how those can be resolved.

Part II: Successful Examples of Climate Adaptation and Early Warning Systems

Chapter 4: Urban Climate Resilience and Water: Successful Adaptive Planning and Early-Warning Systems, by Dr. Pascaline Gaborit

We assume in this chapter that the challenges ahead of climate adaptation in cities and urban resilience are tremendous, as climate change has increasing impacts. We argue that we can learn from successful worldwide examples in adaptive urban planning, water management, grey and green solutions, technology in the form of Early Warning Systems (EWS), and people-centered approaches (contingency plans). This chapter is therefore presenting possible solutions and tools, before we find key features and lessons on how to strengthen urban climate resilience in cities exposed to hazards, including those cities exposed to multiple hazards. We also argue that the most successful examples rely on a good articulation between the local resilience methodologies developed, and national ambitious climate adaptation and contingency plans. We also claim that more attention should be given to medium-sized cities, and that there will be inevitable trade-offs and sacrificed areas and interests in the long run. Finally, the question of engaging with local populations, residents, and communities raises many questions about the possibility of a fair implementation of climate adaptation programs.

Chapter 5: System for Multi-hazard Potential Impact Assessment, Alert, Emergency Response Planning and Tracking (SMART) in Tamil Nadu India, by Dr. Korlapati Satyagopal IAS Rtd, Dr. Itesh Dash, Dr. Jothiganesh Shanmugasundaram, Ramraj Narasimhan, Subbiah Arjunapermal (RIMES)

Increasing disaster impacts across the globe, resulting in losses of life and enhanced adverse economic impacts, have meant that disaster managers are faced with an urgent need to generate and communicate potential impact advisories for saving lives, reducing impacts on moveable assets and livelihoods, and for strengthening risk reduction and mitigation strategies. Decision Support Systems are needed for multi-hazard potential impact assessment, alert, emergency response planning and tracking. The web-GIS decision support system developed in Tamil Nadu assumes special significance in this context. The performance of TN-SMART in Tamil Nadu, the southernmost state in the Indian peninsula, is one of the most 'developed' provinces in India. The State is vulnerable to weather and climate anomalies and coastal hazards and therefore the TN-SMART early warning comes up as a particular success story, amongst the top-ranked in accomplishments, when analyzed through Sustainable Development Goals.

Chapter 6: Landslide Vulnerability, Risk and Resilience Management of Cultural Heritage Sites in the Western Slope of Lawu Mountain, Indonesia, by Muhamad Iko Kersapati, Heri Purwanto.

This chapter examines the essential aspects of cultural heritage site preservations and protections: vulnerability, risk, and resilience management. Caused by climate change, landslides are the most common catastrophe in mountainous areas in Indonesia. They can occur as the impact of disturbances in the natural stability of a slope during a period of extreme rainfall and can accompany earthquakes and volcanic eruptions. Lawu mountain as one of the resting volcanoes on Java Island is structured by a massive compound stratovolcano with the eroded part in the north and parasitic crater lakes and parasitic cones on the eastern side, while the south flank is a fumarolic area. In 1995, the Temple of *Sukuh* was registered to UNESCO as a part of the world's cultural heritage and became famous for its similarity in shapes to the Mayan pyramids. Geographic Information Systems (GIS) are utilized in the environmental mapping workflow, analysis, and assessment of landslide vulnerability through the main variables such as land cover (vegetation), topographic conditions (elevation and slope), archaeological remains' distribution, and the historical records of landslide. Additionally, this chapter enhances the engagement to the multidisciplinary studies for data science, policy, archaeology, and cultural heritage preservations and protections in Indonesia.

Part III Nature Based Solutions, Waste Management and the Circular Economy

Chapter 7: Nature Based Solutions in Climate Adaptation: A Shift from Specific Tools to Large Scale Conservation Worldwide, by Dr. Pascaline Gaborit

Nature-based solutions are increasingly considered as mechanisms for climate adaptation. Territories, communities and governments are, for instance, increasingly rely on the creation of water catchment areas to mitigate the impacts of regular floods. Other nature-based solutions are rather based on innovative techniques and a strong monitoring like the restoration of coral reefs and the planting of mangroves (to protect the coast against climate events). What is the state of the art of nature-based solutions? This article will go through the literature, the current reports and focus on several case studies to assess what mechanisms could be scaled up. It will approach the question of nature-based solutions in cities, the issue of deforestation and forest degradation, and elaborate on the marine ecosystems and oceans' protection challenges. Finally, the chapter will investigate why the 'great green wall' nature-based initiative could not bring yet the expected results. It tries to demonstrate that small scale nature-based solutions cannot substitute large conservation programs (for forests, soils, land or ocean marine ecosystems).

Chapter 8: The Circular Economy from the Perspective of Local Climate Adaptation and Mitigation, by Dr. Pascaline Gaborit

Waste management is one of the cruxes of the circular economy. Plastic waste management is especially notable since plastic is a difficult and complex material to recycle and it is produced and consumed in incredibly large quantities. To deal with their plastic waste, developed countries and regions (specifically, the United States, Japan, and the European Union) tend to export it to other countries if those countries accept it. This chapter uses a qualitative analysis on how the circular economy is facing difficulties and boundaries in its application to plastics. Meanwhile, some other areas like the recycling of demolition waste could provide a promising solution for the future. This chapter will investigate how the circular economy concept can be applied in different sectors, such as in plastics and also in the construction sector. It will briefly detail how the concept of circularity is creating opportunities for climate adaptation, particularly for cities, but how it is also leading to false narratives or promises. It will approach how circular economic models could further support waste management and climate adaptation actions. This chapter proposes a global overview that considers current research and trends.

Chapter 9: Women in the Circular Economy, by Bulan Brabawani, Wiwandari Handayani, Dessy Ariyanti, Diana Nur Afifah

Women are important actors in industrialization, the economy, and business. In Indonesia in the pre-pandemic period, there were 48.75 million women working, and these numbers were trending upwards. In the education sector, women have also played a significant role in that more women have attended school than men since the 1970s. Women's participation rate in higher education is greater than men's, although women have more challenges such as gender bias, cultural and social pressure, as well as work-life balance and health issues. This book chapter will discuss the role of women in the circular economy in detail and the importance of changing business and work patterns by involving women.

Part IV: Finance and Funding

Chapter 10: Sustainable Finance: The Hindered Potential of ESG Investing in Funding Climate Mitigation, Climate Adaptation and Resilience, by Zoe Thouvenot

How to finance Climate Adaptation and how can private investment contribute to it? ESG ('Environment, Social, Governance') criteria and ESG scores are sustainable finance tools – ESG criteria are intended to ensure a clear monitoring of sustainable investments, while ESG scores are used as indicators of sustainability (environmentally, socially, and in terms of governance). This chapter is a critical review of the literature on ESG criteria in their relationship to climate mitigation, climate adaptation, and resilience. The purpose of this chapter is to identify gaps and solutions on the important topic of the mechanisms of sustainable finance.

Chapter 11: Asia Pacific Region – Expanding Economy with Resilience and Adaptation Financing Challenges, by Kamlesh Kumar Pathak

This chapter will investigate country programming and the financing needs for climate adaptation, with particular focus on South Asia. The chapter applies a blended approach to estimate the capital needs of developing countries through secondary research, and consultations from international organizations, with an objective to evaluate the climate finance needed for climate adaptation. The chapter will look into medium-term targets and financing needs by 2030 and analyze them further. The chapter will be developed with an objective to look into technical barriers and support for climate finance needed in the different sectors for meeting the goals of NDCs.

Conclusion

Chapter 12: Can Local and International Dialogue on Climate Adaptation Echo and Reinforce One Another? A Way Forward by Dr. Pascaline Gaborit

The final chapter proposes an analysis of how the dialogue at the local level could be scaled up to echo the international negotiations. The analysis highlights the intrinsic difficulties of dialogue, trust, and accountability mechanisms. It explores the reasons for the current failures, and the challenges ahead in an increasingly distrusted multilateral order. It proposes a way forward with the creation of trust networks, suggestions for a better engagement of civil society organizations, and the involvement of scientists, researchers, and policy makers.

References

Borel B., McCandless E., & Abu-Nimer M., 2006, 'Environment and Natural Resource-Related Conflicts: Moving towards Transformational Approaches', *Journal of Peacebuilding & Development*, 3(1), 1–5, https://doi.org/10.1080/15423166.2006.542264980992

Brassett J., Croft S., & Vaughan-Williams N., 2013, 'Introduction: An Agenda for Resilience Research in Politics and International Relations', *Politics*, 33(4), 221–228

Butterfield Ruth, 2020, *Vulnerability, We Adapt* https://wwww.weadapt.org/knowledge-base/vulnerability

Chandler D., 2013, 'International State Building and the Ideology of Resilience', *Politics*, 33(4), 276–286

Climate Central, 2019, ' FLOODED FUTURE: Global Vulnerability to Sea Level Rise Worse than Previously Understood', 2019CoastalDEMReport.pdf (climatecentral.org)

Cuny C., Hill R.B., 1999, *Famine, Conflict and Response: A Basic Guide*, Kumarian Press Book

Duprady J.M., Franco M.A., Hoo Poh Ying T., Len C., Singh A., Andrews-Speed P., & Yoo L., 2020, 'Resilience of Renewable Energy in Asia Pacific to the Covid-19 pandemic', National University of Singapore, and Energy Studies Institute, Konrad Adenauer Stiftung

Gaborit P., 2021, 'Vulnerabilities and Resilience to Climate Change in Tanzania' in Gaborit P., Olomi D. (Eds.), *Learning from Resilience Strategies in Tanzania, an Outlook of international Development Challenges*, Peter Lang International, https://www.peterlang.com/document/1152350, pp 155–195

Gaborit P., Olomi D., 2021, *Learning from Resilience Strategies in Tanzania, an Outlook of international Development Challenges*, Peter Lang International, https://www.peterlang.com/document/1152350

Glasser R., 2020, 'The Climate Change Imperative to Transform Disaster Risk Management'. *International Journal of Disaster Risk Science* 11, 152–154. https://DOI.org/10.1007/s13753-020-00248-z

Grimmond S., 2007, 'Urbanization and Global Environmental Change: Local Eeffects of Urban Warming'. *Geography Journal*, 173, 83–88

Gunderson L.H., Holling C.S., 2003. *Panarchy, Understanding Transformations in Human and Natural Systems* Washington, Island Press

Gunderson L.H., Allen C., & Holling C.S., 2009, 'Foundations of Ecological Resilience', in Adger N., O'Brien K. (Eds.), *Living with Climate Change: Limits to Adaptation*. Cambridge University Press. (in press)

Ide T., Brzoska M., Donges J.F., & Schleussner C.F., 2020, 'Multi-Method Evidence for When and How Climate-Related Disasters Contribute to Armed Conflict Risk', in *Global Environmental Change* n° 62

Intergovernmental Panel of Climate Change (IPCC) 2021, Climate Change 2021: The Physical Science Basis. Contribution of Working Group I to the Sixth Assessment Report of the Intergovernmental Panel on Climate Change. Summary for Policy Makers. Sixth Assessment Report.

Intergovernmental Panel of Climate Change (IPCC) 2022, Climate Change 2022: Impacts, Adaptation and Vulnerability, Summary for Policy Makers.

International Energy Agency, 2022, 'Global Methane Tracker', https://www.iea.org/reports/global-methane-tracker-2022

KAF: Konrad Adenauer Stiftung 2018, 'Asia Climate Change and Energy Security in Figures' www.recap.asia

Korhonen J., Honkasalo A., & Seppälä J., 2018, 'Circular Economy: The concept and its Limitations', *Ecological Economics*, 143(2018), 37–46

Laganier R., 2018, 'Du risque à la résilience: l'apport des sciences géographiques', in Landau B., Diab B. (Eds.), 'Résilience, Vulnérabilité des territoires urbains', Presse des Ponts

Landau B., Diab B. (Eds.) 2018. 'Résilience, Vulnérabilité des territoires urbains', Presse des Ponts

Mccandless E., 2014, 'Revitalising Our Tools to Better Engage Local Contexts and Measure and Promote Peace', *Journal of Peacebuilding & Development*, 9(1), 1–2, https://doi.org/10.1080/15423166.2014.899014

McCandless E., Abitbol E., & Donais T., 2015, 'Vertical Integration: A Dynamic Practice Promoting Transformative Peacebuilding', *Journal of Peacebuilding & Development*, 10(1), 1–9, https://doi.org/10.1080/15423 166.2015.1014268

Meerow S., Newell J.P., 2016, 'Defining Urban Resilience a Review', in *Landscape and Urban Planning*, March 2016, https://doi.org/10.1016/ j.landurbplan.2015.11.011

Nicholls R.J., Wong P.P., Burkett V.R., Codignotto J.O., Hay J.E., McLean R.F., Ragoonaden S., & Woodroffe C.D., 2007, Coastal Systems and Low-lying Areas. Climate Change 2007: Impacts, Adaptation and Vulnerability. Contribution of Working Group II to the Fourth Assessment Report of the Intergovernmental Panel on Climate Change, in Parry M.L., Canziani O.F., Palutikof J.P., van der Linden P.J., & Hanson C.E. (Eds.), Cambridge University Press, Cambridge, UK, pp. 315–356

Nordqvist P., Krampe F., 'Climate Change and Violent Conflict: Sparse Evidence from South Asia and South East Asia.' 2018, SIPRI Insights on Peace and Security, 2018/4

Oleson K.W., Monaghan A., Wilhelmi O., Barlage M., Brunsell N., Feddema J., Hu L., Steinhoff D.F., 2015, 'Interactions between Urbanization, Heat Stress, and Climate Change'. *Climatic Change*, 129, 525–541

Pisano U., 2012, 'Resilience and Sustainable Development: Theory of Resilience, Systems Thinking and Adaptive Governance', European Sustainable Development Network, https://www.esdn.eu/fileadmin/ ESDN_Reports/2012-September-Resilience_and_Sustainable_Developm ent.pdf

Piseth K., Ngoun K., Kimlong C., 2021, 'Renewable Energy and Sustainable Development in Southeast Asia: Challenges, Cooperation, and Development Models', Konrad Adenauer Stiftung

Reveuny R., 2007, 'Climate Change Induced Migrations and Violent Conflicts', *Political Geography*, 26, https://doi.org/10.1016/j.polgeo.2007. 05.001 last accessed 06/02/2021

Rosenzweig C., Solecki W., Romero-Lankao P., Mehrotra S., Dhakal S., & Ali Ibrahim S. (Eds.) 2018, *Climate Change and Cities: Second Assessment Report of the Urban Climate Change Research Network*, Cambridge University Press, New York

Tall A., Lynagh S., Bianco Vecchi C., Bardouille P., Montoya Pino F., Shabahat E., Stenek V., Stewart F., Power S., Paladines C., Neves P., & Kerr L., 'Enabling Finance in Climate Adaptation and Resilience: Current Statue, Barriers to Investment and Blueprint for Action' 2021 World Bank Group, and GFDRR Global Facility for Disaster Reduction and Recovery

Tisseron S., 2018, 'Les quatre vagues de la résilience', in Landau B., Diab B. (Eds.), 'Résilience, Vulnérabilité des territoires urbains', Presse des Ponts

Tobey J., Meena H., Lugenla M., Mahenge J., Mkama W., & Robadue D., 2011, *Village Vulnerability Assessment and Adaptation Planning, Mlingotini and Kitonga, Bagamoyo District, Tanzania*, Coastal Resource Center, University of Rhode Island, Narragansett, RI, 20 pp

UNEP UN Environment Program 2021, 'The heat is on : a world of climate promises not yet delivered', 2021 emissions gap report https://www.unep.org/fr/resources/emissions-gap-report-2021

UNFCCC State of Climate 2021, extreme events and major impacts, https://unfccc.int/news/state-of-climate-in-2021-extreme-events-and-major-impacts

Wettel K.J., Zundel T., 2018, 'The Big Bad Fix: The Case Against Climate Geoengineering' ETC Group Biofuel Watch, Heinrich Böll Stiftung

Wicaksono A., Moeliono I., 2022, *Exploring Alternatives in Dealing with Climate Change and Conflicts in Indonesia*, Chapter 3 of this book.

Zhao Q., Guo Y., Ye T., Gasparini A., et al. 2021, 'Global, Regional and National Burden of Mortality Associated with Non Optimal Ambient Temperatures from 2000 to 2019: A Three Stage Modelling Study', https://www.thelancet.com/action/showPdf?pii=S2542-5196%2821%2900081-4

Ziervogel G., Pelling M., Cartwright A.A., Chu E., Deshpande T., Harris L., … Zweig P., 2017, 'Inserting Rights and Justice into Urban Resilience: A Focus on Everyday Risk', *Environment and Urbanization*, 29(1), 123–138

Part I

GOVERNANCE AND POLICIES

Chapter 1
Ambitions and Challenges of Climate Adaptation in Ten Cities

PASCALINE GABORIT PHD

This chapter has been published in a different version under the title 'Climate Adaptation to Multi-Hazard Climate Related Risks in ten Indonesian Cities: Ambitions and Challenges', Journal Climate Risk Management, Vol 37, 2022, 100453.

Introduction

> *'The alarming projections of global warming consequences stand in stark incommensurability with the available proposed solutions'* (Russill 2008, p. 147)

There is a consensus that climate change is one of the major threats of our time. It is expected that in the next decades, climate disasters such as typhoons, floods, sea level rises, and dry spells will be more frequent, while the disruptions in the ecosystems and water resources will jeopardize the wellbeing of local populations and lead to displacements and unrest (Nicholls et al. 2007, Glasser 2020, IPCC 2021).

Due to its geographical condition and archipelagic nature, Indonesia is highly vulnerable to climate impacts. Sea level rise is a direct hazard to Indonesia, which is the 14th largest country in the world but has the 3rd longest coastline. Indonesia has 17,504 officially listed islands. This geographical situation increases its exposure to floods and storms. It is expected that climate change will amplify the intensity of rainfall which, in turn, will give rise to more floods (Vijj et al. 2017). Soil subsidence, saline infiltration, and water scarcity contribute to the country's vulnerability, which is further compounded by its rapid population growth and urbanization[1]. This situation is likely to fuel conflicts between the

---

[1] Climate Resilient and Inclusive Cities - Policy Briefs for Pilot Cities (resilient-cities.com).

authorities in charge and segments of the population, especially while the poorer population and minorities are more exposed and less likely to be informed about climate disasters.

Climate adaptation refers to any adjustment, whether passive, reactive, or anticipatory, that can respond to anticipated or actual consequences associated with climate change (IPCC 1995, IPCC 2022). The necessity of climate adaptation has been increasingly recognized both by international organizations and by research. Islands and coastal areas draw particular attention as they are becoming especially vulnerable (Nicholls et al. 2007, IPCC 2014). It has been implicitly acknowledged that future climate-related changes had to be accommodated in policy (IPCC 2021). In this framework, the role of local governments and cities in climate mitigation and adaptation has been progressively recognized as important (Grimmond 2007, Oleson et al. 2015, Raven 2011, Rosenzweig et al. 2018). Cities are considered a relevant level tier to understand the socioeconomic dynamics of the territories, to take appropriate decisions, and to protect the populations and the local ecosystems (Gaborit 2015, Wijaya et al. 2020).

In this study, we approach the case of ten medium-sized cities across Indonesia to study their climate adaptation strategies, the proposed solutions as response, but also the remaining challenges and discussion points. We argue that despite the high level of knowledge and the existing cooperation among the national and local stakeholders, the lack of available funding, adequate land use mechanisms and the insufficient international commitments are currently hampering medium-sized cities in developing enough resilient adaptive mechanisms.

I. Background and Context: Urbanization, Climate Change and Disaster Risks

The Indonesian context is characterized by both growing urbanization and an increased focus on climate adaptation by the different government tiers. Since 2012, Indonesia's urban population has grown by more than half (Salim, Hudalah 2020, Dwitama CRIC 2020). Over half of the population now lives in urban areas, and the urbanization rate is expected to keep rising. Indeed, projections show that cities in Indonesia will be accommodating at least 72.9 % of the total population by 2045[2]. In

---

[2] Indonesia Ministry of National Development Planning BAPPENAS 2018 The annual increase in the urban population is about 3 million people. Every year, the country

practice, urban migrants move to the existing cities, adding more pressure on current housing needs. The cities' growth and the increasing housing demands are therefore creating challenges, not only in terms of accommodation and housing, but also in terms of access to basic services and the enactment of inclusive policies (e.g. through the 'cities without slums program'). A city like Samarinda in East Kalimantan for instance grew 78% between 1990 and 2000 (Amri et al., CRIC, Urban Analysis Report 2020). The provision of basic services and construction needs are equally increasing simultaneously amid a period of climate-related challenges.

The country has been at the forefront of climate change issues with the National Action Plan to Reduce GHG (RAN GRK) and the Rencana Aksi National – Perubahan Iklim (RAN-API) which is Indonesia's national action plan on climate change adaptation. The latter identifies two key areas of climate change and their impacts on livelihoods. These 2 areas are: (1) increases in sea levels and (2) changes in weather, climate, and rainfall. BAPPENAS – the Ministry of National Development Planning – has drafted the RAN-API. Climate change is also integrated into the Strategy called RPJMN (*Rencana Pembangunan Jangka Menengah National/Medium-Term Development Strategy*), the national strategic development plan.

Each city is required to include and execute a detailed implementation plan integrating both climate mitigation and climate adaptation as response into their local policies and spatial development plans. A national platform for reporting data (SignSmart)[3] has also been set up together with reporting tools on the vulnerability such as SIDIK.[4] Most of the city governments and other relevant stakeholders have initiated the implementation of adaptation activities, with the support of international development agencies and NGOs. The developed programs assume that the cities (including the local governments, national agencies, planning boards, and the different stakeholders – See Fig. 3) have the capacity to develop the required mechanisms for stronger resilience to climate-related disasters. They are expected to develop adaptive capacity to increasing rainfall and sea level rise and to successfully implement the principles of the Disaster Management Cycle (UNDRR, UNISDR 2004, Glasser 2020), namely disaster prevention, preparedness, response,

---

needs to build a city the size of Surabaya, to accommodate the annual increase in the number of urban dwellers (Salim, Hudalah 2020: 176).

[3] http://signsmart.menlhk.go.id/v2.1/app/

[4] SIDIK (Vulnerability Index Data Information System/*Sistem Informasi Data Indeks Kerentanan*)

and recovery (in cooperation with the local disasters agencies *BPBD*). Moreover 160 million US dollars were allocated to the Indonesia Disaster Resilience Initiatives Project (IDRIP).[5] The funding will be used for priority investments to increase the preparedness of selected local governments to manage natural hazards and to strengthen Early Warning Systems (EWS) although it remains a very limited amount (lower than the green funds' budgets of cities like Paris), it adds up to the more important national funding schemes such as the climate budget tagging per ministry, the Indonesia Impact Fund (supported by the UNDP) and the 2018 Green Bonds green Sukuk initiative. These mechanisms aim to enhance and strengthen the urban resilience of cities which can be defined by their capacity to respond and adapt to and recover from the pressures and crises related to climate change and other changes: subsidence, demographic growth, poor land management, insecurity or attacks, economic downturns, social unrest, unsustainable use of resources, and declining ecosystems (Ziervogel 2017, Diab 2020, Gaborit et al. 2021).

Several research studies have shown that, in Indonesian cities, the level of awareness and knowledge of the different stakeholders involved in the adaptation of climate change was very high (Wijaya et al. 2020). Discussions between focus groups and the ten pilot cities in March 2020 and throughout 2021 showed the same trend.

In this article, we will use as a theoretical framework 4 bladed framework (Fig. 1), adapted from Simpson et al. 2021). The theory is reflected in the diagram below. In addition to this, Fig. 2 focuses on the response part of climate adaptation. This theory (Fig. 1 and Fig. 2) will structure the discussion points in part III, especially for the interrelation of risks and the impacts on the socioeconomic processes (governance, adaptation and resilience, socioeconomic pathways, funding, and infrastructure).

---

[5] World Bank Project: Indonesia Disaster Resilience Initiatives Project (IDRIP) – P170874.

Ambitions and Challenges of Climate Adaptation in Ten Cities 53

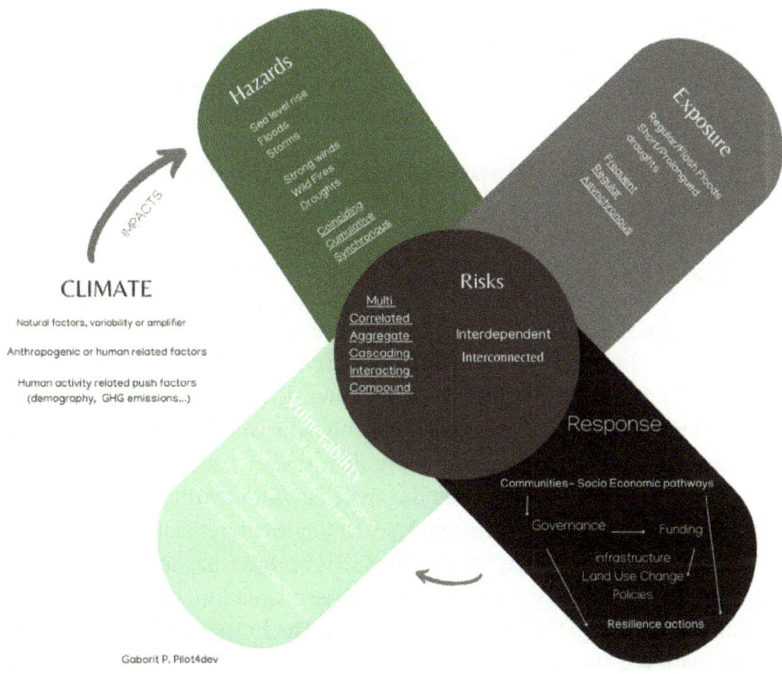

Fig. 1. Theory Framework. Adapted from Simpson et al. 'A framework for complex climate change risk assessment' 2021

The main local research focuses on the local response (policies, local coping adaptation strategies). It considers the theory of interrelated risks (Simpson et al. 2021) according to which the impacts of climate change are equally cross-sectoral as they embed questions of droughts or access to water (See Fig. 4). As an example, in several cities, such as Ternate, sea level rises in coastal areas are leading to sea salt intrusion in groundwater reservoirs and decreasing the availability of clean water (Nagu et al. 2016, CRIC Urban Analysis report 2021). Research on the policies related to climate adaptation shows that there are two categories of analysis or policies: 1) the concept of adaptation as an "adjustment" to climate change where the physical infrastructure and the communities adapt to the change in climate patterns and develop solutions (e.g. toward a more adaptive capacity of urban development), and 2) climate adaptation as a "transformational" concept in which the cities and policies transform to become more resilient (Bassett, Fogelman 2013). The reality is much more complex, as climate-related disasters transcend geographical territories, communities, and tier-level government legal jurisdictions. Part

II will present the project's methodology, the ten cities' exposure to the main climate risks, and their policies and coping mechanisms as response, before the main discussion points are addressed in part III.

II. Exposure and Adaptation of Ten Pilot Cities to Disasters Risks

1. Research and Methodology

The proposed chapter 'Climate Adaptation to Multi-Hazard Climate-Related Risks in ten Indonesian Cities: Ambitions and Challenges' is based on the collective research-action project Climate Resilient and Inclusive Cities (CRIC) spanning over 5 years and involving a dozen of specialized researchers from several organizations and more than twenty people working on the coordination teams. It also crosses the analysis with identified literature and with the author's and teams' current research on urban resilience, disaster risk reduction, and climate adaptation. The CRIC project has selected ten pilot cities for further research on planning, climate adaptation, and preparedness for climate disasters. This project is funded by the European Union in cooperation with the Indonesia Ministry of the Environment and Forests and is implemented by UCLG-ASPAC (lead partner) and ACR+ and AIILSG city networks, Pilot4dev, Ecolise and the university Gustave Eiffel. They ensure the liaison with cities, public organizations (Fig. 3), think tanks, grassroots organizations, communities, universities, and research centers.

The selection of case cities: The project has decided not to include the largest cities but to address the question of climate adaptation in small and medium-sized coastal cities which are also exposed to multi-climate related hazards. The findings are based on desk literature search, ten urban analysis reports and ten policy briefs drafted by the project's experts, local experts and five field officers and researchers. The ten studied pilot cities are spread over the country and most of them are coastal cities, meaning that they are exposed to climate change such as sea level rises, strong winds, and storms. They are medium-sized cities (entailing that their total population is below 2 million inhabitants). They represent (Kota), which is the second-level administrative subdivision in Indonesia below the province and above the district (*kecamatan*) and administrative village (*Kelurahan*). The selection of small- and medium-sized cities means that they are of secondary concern for national policies in terms of funding for climate adaptation, response building, and local investments. The selected cities benefit less from international development aid cooperation than large cities, although some of them have benefited from international aid cooperation to an extent (Kupang, Cirebon, Bandar Lampung, Pangkal Pinang).

Data Collection and Methodology: The experts based their analysis on field evidence and official documents and conducted a minimum of twelve interviews per city. The researchers also organized four focus group discussions for each city gathering at least twenty stakeholders and local officials from the cities' different departments (Kota), local disaster agencies (BPBD), civil society organizations and public decentralized administrations (BAPPEDAS), and the Environmental Agency for each focus group. The detailed organizations and their original titles are reflected in Fig. 3. Some of these focus groups and interviews took place in Bahasa Indonesia, with a transcript, while others have been organized online and benefited from simultaneous interpretation to English. This chapter is therefore based on a multi-disciplinary qualitative analysis reflecting on the actions and needs of the cities by the different cities, stakeholders triangulated with the city's available documentation, and consideration of the reports. The study is not solely based on the available documents about policies; It follows empirical research about climate adaptation (Runhaar et al. 2018) and considers the practices, policies, and realities beyond mere conceptualizations.

Tab. 1: Cities' main data.

City	Population 2021[a]	Area km^2	Density/km^2	Urban Growth in 2019
Pekanbaru	1, 227, 299	632	1 941	1,9 %
Pangkal Pinang	212,727 (2019)[b]	118	1802	2,'21 %
Bandar Lampung	1,107,542	169	6 553	1,79 %
Cirebon	322,027	37	8 703	0,9 %
Banjarmasin	740,109	98	7 552	-
Samarinda	1,030,947	718	1435	(-)[d]
Mataram	512,433	61	8 400	-
Kupang	463, 360 (2019)[c]	180	2 570	2,58 %
Gorontalo	219,399[f]	60,07	3656	([g])

[a] www.populationstat.com
[b] Mulkhan et al.. 2020 CRIC Urban Analysis Report, Pangkal Pinang.
[c] In average.
[d] 78 % between 1990 and 2000 in Amri et al. .2020 CRIC urban analysis report Samarinda.
[e] Ridwansyah et al.., 2020 CRIC Urban Analysis Report Kupang.
[f] 2019 Dillon H. et al., 2020 CRIC Urban Analysis Report Gorontalo.
[g] 67,93 % over the last 25 years in Dillon et al. 2020 opcit.

Map 1: Location of the pilot cities in Indonesia.

2. Presentation of the 10 Case Cities

The table below presents the cities and maps their geographical location on the Indonesian archipelago.

Growing cities exposed to sea level rise and floods.

The information below is based on the qualitative analysis of the urban analysis reports by the CRIC project.

1. **Bandar Lampung** is a large city and a transportation hub in Sumatra. Bandar Lampung is vulnerable to natural disasters, including floods, landslides, droughts, high tides and tsunamis. Twenty-three disasters, mainly floods, were reported between 2010 and 2019 (Priyadi et al., CRIC Urban Analysis Report 2020). Bandar Lampung experiences annual flooding due to erratic heavy rainfall which is worsened by the drainage condition and damaged dikes. As a recent example in December 2021, Sub-District/*Kecamatan* Rajabasa experienced flooding after 2 hours of heavy rainfall, which varied from 20 cm to over a meter in height.[6] Bandar Lampung also experiences seasonal tidal flooding that affects coastal communities including the fisher's village. In November 2021 only, the water inundation reached about 50 cm and affected 127 households, with the duration of inundation lasting more than 12 hours. Local communities responded to this

---

[6] Kumparan.com, 10 December 2021, Accessible at: https://kumparan.com/lampung geh/puluhan-rumah-di-rajabasa-bandar-lampung-terendam-banjir-1x5K6JAhpXv.

situation by operating pumps and constructing their own simple protection to prevent the water from inundating their house.

2. **Cirebon** is a coastal city located in West Java. It is developing into a metropolitan area with related challenges, such as spatial planning, density, building regulations, and the conversion of paddy fields into industrial and residential areas. It also faces sea level rise, which puts the clean water at risk as the seawater intrudes into the low groundwater. The illegal dumping of waste is partly responsible for the failure of past restoration programs. Cirebon is in a high disaster risk area. The city is working on drainage for floods and is planning a disaster management program. The development of Cirebon as a metropolitan area is also leading to concerns about land use as the paddy fields are converted into residential and industrial areas, threatening the city's agriculture and food safety.

3. **Gorontalo**: Located in the Maluku region, Gorontalo is an important growing area and a transportation hub, while fisheries remain an important sector. The city is prone to disasters such as earthquakes, floods, droughts, and other extreme weather events like storms. It is categorized as a medium risk area on the national disaster index. The city's urbanization adds more pressure to the agricultural land, which is further threatened by droughts and floods (Dillon et al., CRIC Urban Analysis report 2020). Upstream deforestation is a possible issue.

4. **Kupang**: Located in the South-eastern part of Indonesia (*East Nusa Tenggara*), Kupang city is vulnerable to climate disasters, including strong winds, heavy rainfall, and droughts, which cause well levels to drop. Tropical storms are also expected to increase in frequency and intensity due to climate change. The strong winds affect tourism, food supplies, shipping, property, and housing units. Droughts are just as recurrent as the city is also prone to coastal abrasion, floods, and landslides (Ridwansyah et al., CRIC, Urban Analysis Report 2020). The UNDP enacted a local disaster management program in 2015.

5. **Mataram** is a coastal city facing several disaster risks. These risks include sea level rise, extreme waves, abrasion, earthquakes, and droughts. An earthquake in 2018 triggered a tsunami wave and resulted in human casualties, injuries, and damages in several parts of the city, and displacements of the population. The city has a strong potential in tourism, but coastal areas are disaster prone.

6. **Ternate** is an island city spanning eight islands, three of which are uninhabited. Ternate island is the main island. It is a major transportation and trading hub for the province of North Maluku and East Indonesia. Ternate is part of the first six areas with the highest potential for disaster in Indonesia, including volcanic activity, earthquakes, landslides, storms, and floods. The city is undergoing rapid urbanization and requires strategies to accommodate urban growth, disaster management, and infrastructure.
   Cities located on river deltas, prone to floods from rivers with exposed riverbank areas.

7. **Banjarmasin**: Located 16 cm below sea level, Banjarmasin is the capital of the province of South Kalimantan and is called the 'City of a Thousand Rivers'. The tidal waves affect the stilt houses of the riverbank areas. The wooden stilt houses are prone to both floods and residential fires. The river's drainage is often clogged by open landfill dumping and upstream mining activities causing overflows (Amri et al., CRIC Urban Analysis Report 2020, Gaborit et al., Policy Brief 2020, Prayitno 2018). The regeneration of the city and land clearance are at stake, with open questions for the residents, and the place's identity.

8. **Pangkal Pinang** is the largest city on the island of Bangka. The city is prone to floods. It recorded 49 events in the year 2019 alone, and around 1497 houses were reported damaged (Mulkhan et al., CRIC Urban Analysis Report, 2020, Tirtariandi et al. 2020, Gaborit et al., policy brief, CRIC, 2020). Not only the pumping of water leads to the sinking of the city due to subsidence, but aslo the dry season aggravates seasonal fires.

9. **Samarinda City** is the capital city and most populated city in East Kalimantan. It is located near the possible future capital city of Indonesia. Its population grew 78 % between 1990 and 2000. The city is located 10–200 m above the sea level but is still prone to floods, droughts, fires, and landslides. The city currently faces challenges linked with rapid urbanization as its surroundings have abundant natural resources such as gas, oil, palm oil, and forests.

Inland City

10. **Pekanbaru**: With a population of over a million, Pekanbaru is a large economic center, located in Sumatra. The city faces disaster risks from floods, forest fires and haze (Mulyana et al., CRIC Urban Analysis Report 2020, Gaborit et al. CRIC Policy Brief 2020). Sustainable urbanization and disaster prevention are considered a priority.

The table below identifies some of the current priorities, local actions, needs' analysis and local policies for climate adaptation in the ten cities. The intensity in rainfall is expected to increase in the upcoming years:

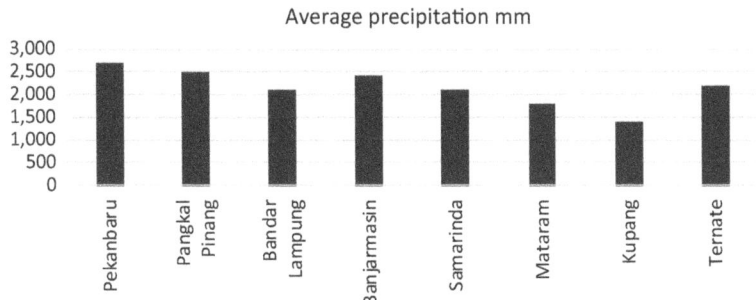

Graph 1: Average annual rainfall in the cities (Agashe CRIC 2020).

The elevation above or below sea level is also an explanatory factor for hazard occurrence.

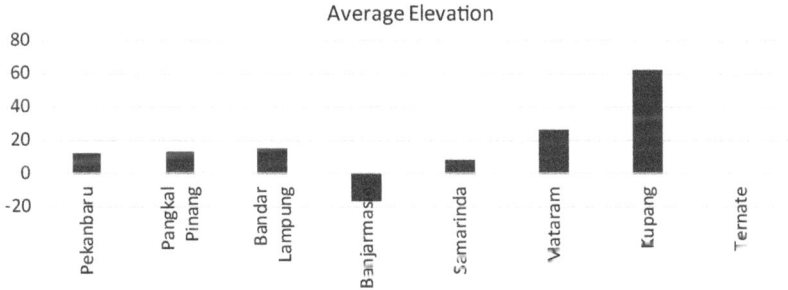

Graph 2: Elevation above or below the sea level of the 10 cities (Agashe CRIC 2020)

The hazards can be correlated, cascading and interacting as detailed in Fig. 1 and in Fig. 4.

Faced with these hazards, local authorities have adopted local and resilience policies which are detailed in the table below. They have also identified the needs and constraints to develop better and more efficient policies for climate and disaster risks.

III. Main Findings and Discussion Points

Our discussion points focus on the response part of climate adaptation including policies, planning, and possible resilience local actions and coping strategies of both local governments and local stakeholders. The Diagram (figure 2) will structure the discussion. Indeed climate adaptation strategies are challenged by several parameters which are detailed further.

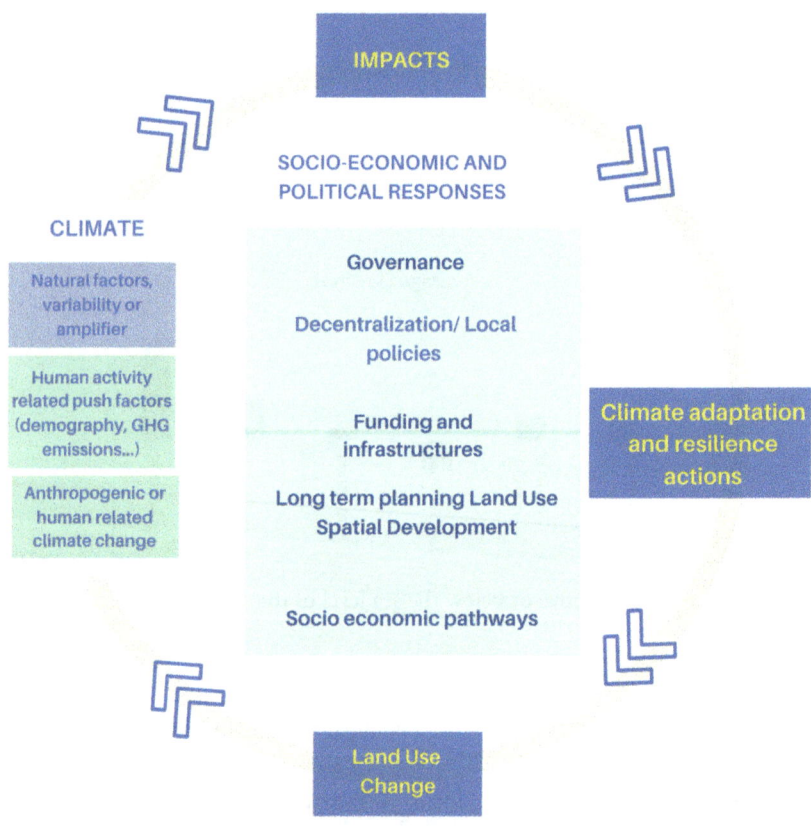

Fig. 2: The factors' interrelation in climate adaptation and resilience.

The different discussion points confirm that the priorities are severalfold, and that public authorities, especially cities, will be faced with increasing trade-offs. A trade-off is a situation that necessitates choosing or balancing between one or more desirable but conflicting plans, policies, or measures (Grafakos et al. 2019). In our case, the trade-offs can be: whether to relocate part of the population, whether to develop contingency plans and evacuation routes, selecting the level of alerts, and choosing which areas will benefit from resilience and adaptation actions and which will not.

The discussion points below are structured around the theory framework (Fig. 1) and (Fig. 2) and address the central questions of governance, the involvement of different stakeholders (1) and the decentralization process (2), the socioeconomic pathways including the conflicts created by land use and spatial development (3), the unequal impacts on vulnerable groups (4), the funding and infrastructure difficulties in setting up priorities with limited funding streams (5), and the call for more research categories for further acknowledgement of the climate change impacts (6).

Fig. 2 shows the central role of governance within the socioeconomic political processes. In this article, the challenges of governance are addressed through the cooperation of multi-stakeholders as well as the process of decentralization.

1. Governance: The Challenges for Multi-stakeholders' Cooperation

A lot has been written about how governance would be key in climate adaptation, and how political ecology would be important in understanding future policy making (Basset, Fogelman 2013, Funtowicz 2020, Koch et al. 2021). There is a consensus that the fight against climate change necessitates a genuine cooperation among all the different stakeholders (Rosenzweig et al. 2018, Wijaya et al. 2020, Gaborit 2021). International agencies have been active in promoting urban resilience, and the involvement of both local governments and community groups in the climate adaptation process (Wijaya et al. 2020). The experience of the CRIC project (and the study of the ten pilot cities) shows that there is already a consultation process among the stakeholders and a dialogue mechanism in place at the local level (*Musrenbang*). The Ministry of the Environment and Forests (MoEF) is also part of the dialogue. The knowledge level of the different stakeholders on climate issues is particularly important. It is overall recognized as a topic of important priority by the Ministry of the Environment and Forests, local and district governments as well as local planning boards or Bappedas *(CRIC-Urban Analysis Reports)*. The CRIC research[7] shows, however, that the local decision-makers and stakeholders currently lack the capacity in terms of infrastructure, funding, and information in real time to tackle climate disasters and further adaptation. The first reason for this is the astounding number of simultaneous challenges to be answered by the city governments: to ensure development planning, increase urban resilience, maintain the continuity of urban ecosystems, reduce the vulnerability of the coastal and exposed areas, decrease potential human and economic losses, and develop contingency and disaster response plans. The second issue hampering stakeholder cooperation is that the different organizations are pursuing different interests in the absence of a strong national cohesive program for each of the concerned cities. The focus group discussions show that the dialogue among stakeholders, including with official representatives, is necessary but presently insufficient when it comes to achieving climate response, adaptation, and resilience action. The case of Pangkal Pinang equally highlights the issue of a lack of trust in the cities' stakeholders and local population towards the capacity of the different organizations to implement climate disaster actions (Mulkhan et al., Urban Analysis Report, CRIC, 2020).

---

[7] Through the focus groups, interviews, discussions and reports.

Ambitions and Challenges of Climate Adaptation in Ten Cities 63

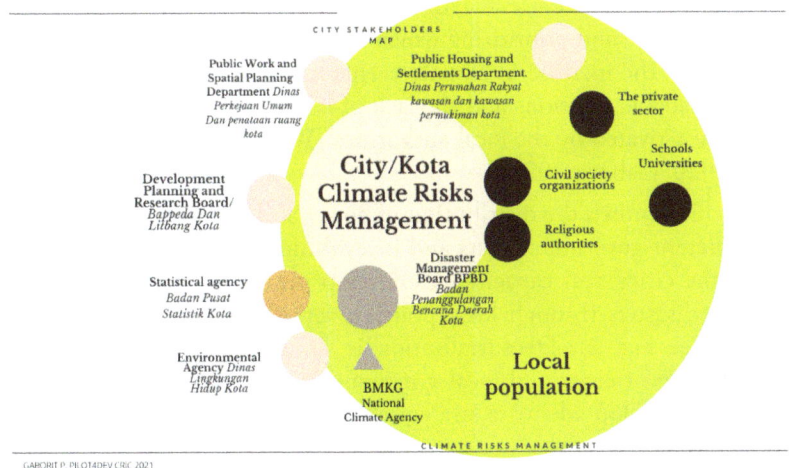

Fig. 3: Mapping the stakeholders involved in cities' climate risk management.

Additionally, the question of climate and weather data transfer in real time and stakeholder anticipation of the risk level has emerged[8] as an equal difficulty in case of climate events like floods. Real-time information transfer is needed between the meteorological agency BMKG and the local disaster agencies BPBD, and also between the local disaster agency and the local population (See Fig. 3). A level of cooperation is also needed to agree on the levels of alerts and the best alert system to set up (broadband systems, SMS, sirens, mosque loudspeakers, etc.). Therefore, advocating for a 'multi-stakeholder' cooperation, as a response to climate adaptation, is not sufficient and should be accompanied by incentives, coordination with community representatives, information sharing, and transparency to prove efficient. The consultative process and the population's inclusion, including the vulnerable groups, in the decision-making process are also both recognized as necessary, requiring a more efficient disaster risk reduction and management (Ziervogel 2017, Djalante et al. 2020, Gaborit 2021). The work with community representatives, civil society organizations, and religious authorities amongst other community leaders seems equally important. Transparency and a local engagement process may also confront communities with the lack of available choices. Complications will continue to arise due to necessary trade-offs by the decision-makers' need to choose between different priorities, different

---

[8] Within the focus group discussions.

levels of alerts, different evacuation routes and contingency plans, possible relocations, and prioritizing which infrastructure and neighborhoods to protect. The increase in climate events highlights the lack of current global available solutions and the lack of land use mechanisms that can be easily activated by the local authorities (Wirawan et al. 2020, Afrizal et al. 2020, Gaborit 2021).

In this context, the distribution of power and competencies between the different government tiers and its evolution is important to understand the cities' real capacities to develop prevention and response to climate disasters through local policies, actions, and land use mechanisms (See Fig. 2). Decentralization is simultaneously both enhancing the role of cities and local governments, without giving them the means to develop adequate strategies with more funding and land use mechanisms.

2. Understanding and Muddling through the Decentralization Process

Literature about the decentralization in Indonesia highlights that this is a complex phenomenon (Ardiansyah 2015, Rahayu 2016, Salim 2020). The governance of the different jurisdictions is riddled with high levels of complexity, uncertainty, and conflicts (Rihayu 2016, von Korff et al. 2019). Since 1998 and the early 2000s though, local authorities and districts were granted not only more autonomy, but also more competences over their budget and resources. Although research shows that cities still depend on national funding (Rahayu 2016), *cities* have been recognized as semi-autonomous actors in many different areas including the provision of local service, water management, waste management and disaster preparedness. However, land use and ownership are still mainly recognized as state-owned or nationalized in unbuilt areas and cannot be used by cities as a leverage to adapt the planning to climate-related events such as floods. The power, local budgets, and responsibilities have partly shifted toward the direction of cities, together with the responsibility to protect the populations and constructions, as well as natural areas. However, the cities share the competences in terms of planning and have few leverages in terms of land use. Some authors consider that decentralization in disaster risk reduction (DRR) is an achievement. According to Djalante et al. (2020), 'in line with the decentralization in the development and planning approach, the responsibility for Disaster

Risks Reduction, and Disasters Risks Management is shared across different levels of government, from heavy reliance on national governments to greater responsibility of local governments' (Lassa 2013, Djalante et al. 2020). The research on the ten cities confirms this trend of responsibility transfer but also highlights the difficulties experienced by the local government officials to find the necessary investments for climate adaptation without financial support from the ministries. The focus group discussions especially show that the decentralization process is hampered by a complete dependency on national funding for climate adaptation and inclusiveness programs (such as *'cities without slums', 'Kampung Iklim'* environmental, and climate-related programs, etc.). The forms of local governments are also complex: cities are referred to as including the local government, the BAPPEDA local planning board, and other agencies such as environmental or local disaster management agencies. Without increased funding mechanisms and investors and capacities to manage the land, cities may face a lack of preparation as well as capacity shortages in implementing possible smart solutions, early warning systems, and adaptive urban planning. Access to real-time and accurate data in terms of climate-related events is also a challenge. Indeed, planning in times of uncertainty is most definitely not an easy matter.

The next discussion point will address the question of adaptation and resilience actions (Fig 2) and reflect on the difficulties for cities' to find a response to all the interrelated priorities, especially in a situation of funding gap.

3. Adaptation and Resilience: Priorities Are Several Fold and the Funding Needs are Tremendous

There are various interrelated problems linked to climate change and adaptation such as water or waste management (for instance, to avoid the clogging of the rivers' drainage systems). Some of the problems are escalating, such as droughts and strong winds, while others, such as flash floods, are repetitive while the question of predictability is complicated. Priorities are severalfold to develop appropriate coping mechanisms as there are many interrelated factors (see Fig 4). Flash floods and puddles are extremely present in most of the pilot cities, but especially in Banjarmasin, where riverbank areas are regularly flooded, and Bandar Lampung and Pangkal Pinang, due to their urban morphology. Cities are faced with contradictory imperatives: while the ten cities are faced with floods, housing needs are simultaneously rising. Consequentially, the number of buildings is increasing, thus sacrificing many water catchment areas which were also necessary to adapt the floods. Waste management and rivers' drainage, as well as the protection of water reservoirs emerge among other important priorities together with housing, economic needs, social issues and the prevention of further floods. But the capacities and the local funding remain very limited. The Detail Engineering Design Pangkal Pinang, which includes a retention lake in the city center, was approved in 2020 but could initially not be implemented due to budget constraints. The lack of funding was also reflected in the findings from the focus group discussions of the CRIC project for the ten different pilot cities. This question of funding was indeed very central to the debates, in the hope that national and international funding could complement the city's budgets for the programs. It was, however, clear that the different ten pilot cities were confronted with numerous priorities and challenges: waste management (Mataram, Cirebon, Samarinda), early warning systems, and floods (Bandar Lampung, Pangkal Pinang and Ternate), or simply access to water and water management (Banjarmasin, Gorontalo, Kupang). Urbanization, transportation, waste, river clean-up, water, and also economic development and social inclusion were at the forefront of the cities' concerns, together with disaster risk prevention.

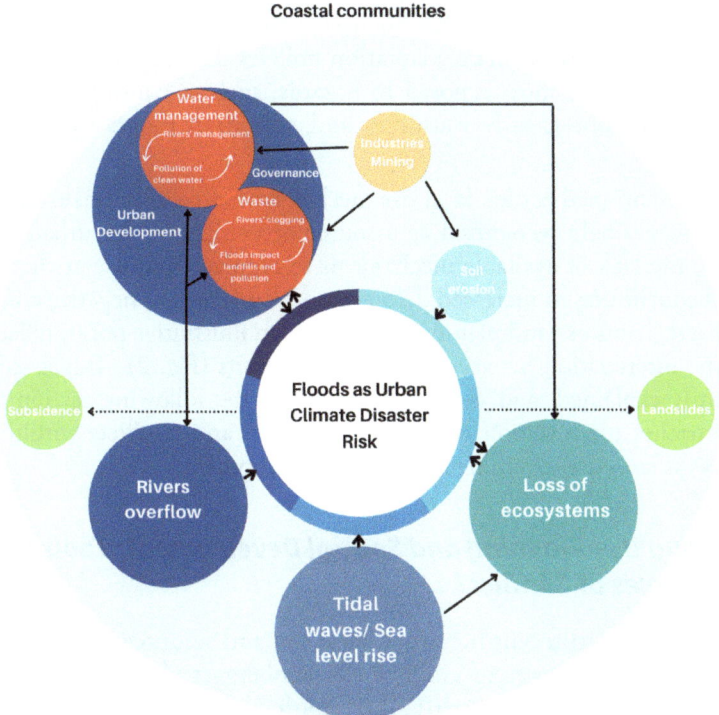

Fig. 4: Mapping the factors' interrelations of floods (climate disasters in cities).

In this context, the available city budgets, including the funding of national government programs, would quickly prove insufficient in tackling climate needs. Furthermore, different studies show that mitigation actions, when compared to adaptation actions, receive the main portion of global climate finance flows from international multilateral development aid organizations and development banks (Grafakos et al. 2019). This portion was superior to 96% between 2010 and 2011 and was still 24% in 2019 (Buchner et al 2012, Chan 2019, Tall et al. 2021). In 2019, the MDBs indeed committed 61.5 billion US dollars in climate finance. This included 76% for mitigation and only 24% for adaptation while private investments in this sector are just beginning (Tall et al. 2021). This shows that medium-term investments for the development of local climate adaptation strategies still rely not only on local political will but also on a limited local funding.

This lack of available funding leads to risky trade-offs as leaders focus on what cities can easily tackle, such as domestic waste, to the detriment of more long-term climate adaptation policies that aim to protect populations that are most exposed to hazards, such as contingency plans, evacuation routes, early warnings, and relocation programs, amongst others.

As mentioned earlier, land use mechanisms are also at stake. Indeed, it is increasingly recognized as a source of conflict at the urban level, while the lack of available mechanisms to regulate land use at the local level contributes to increased deforestation, and leads to negative climate impacts. Land use and planning constitute an illustrative point, reflected in the interrelation between factors and impacts (Fig. 2). This is subsequently explained, and detailed more under the following section that approaches urbanization needs, deforestation, and conflicts within the scope of land use, planning and spatial development.

4. Land Use Planning and Spatial Development: Tools or Sources of Conflict?

According to recent literature, land use and resources in Indonesia often become sources of conflict (Wollenberg et al. 2009, Wirawan et al. 2020, Afrizal et al. 2020). These conflicts are closely tied to unsustainable planning and the rapid increase in demand for land by a variety of interests, particularly large-scale industrial expansion, palm oil plantations, and growing urbanization. According to other analysts on urbanization and disasters, 'largely uncontrolled urbanization has led to high disaster and climate vulnerability in Indonesia' (Djalante et al. 2020: 2).

The intensity of the demand for land is in stark contrast to continued uncertainty over the legal framework for land use and ownership, as well as ongoing spatial planning efforts (Wirawan et al. in 2020). In this context, land use policies, which would be necessary for the cities to find the best balance or trade-off between urbanization, climate adaptation, and climate mitigation have limited actual authority. National spatial planning policies are based on broad environmental concerns (such as the protection of forests, ecosystems, and natural buffers to climate disasters). These considerations are reflected in the regulations related to planning, land use, and spatial management (Ardiansyah 2015, Tombourou 2013, Resosudarmo 2012). However, in their current implementation, land use policies are subjected to a range of pragmatic adaptation when confronted with economic realities, urbanization needs, and industrial interests (Wirawan et al. 2020, Afrizal et al. 2020). As a result, local authorities cannot rely on clear land use coping mechanisms for climate adaptation, such as redesigning areas for water catchment, building dikes, designing natural buffer areas, or regenerating riverbanks. Nevertheless, these authorities are often called as mediators in possible land use conflicts, such as between the industry and private owners. This makes it difficult in these circumstances to develop a consistent planning and adaptive approach to cope with the climate adaptation and growing urbanization.

In Banjarmasin, for example, the cleaning of rivers would be necessary to reduce the rivers' overflows. However, this cleaning investment would require anticipatory designation of rivers as public utilities and assets to the city government. As a result of unclear jurisdiction, private land ownership along the riverbanks, and land designation, which entails designation over the rivers, the necessary public works to clean the rivers and reduce the water overflow cannot be implemented. This is an example of barriers related to land use management, which could be leveraged or changed for more successful climate adaptation.

Again, considering Pangkal Pinang, the question of trust is critical for the city to act on land use change and to implement adapted local adaptation policies. The interviews performed among the cities' stakeholders and the local population show that a lack of trust will increase the challenges for the public institutions when initiating long term planning and land use change (Mulyana et al., CRIC, Urban Analysis Report 2020).

As we see from the analysis above, a clearer role of cities in the land use and spatial development could be the necessary tool to find the correct balance and trade-offs between the needs of growing urbanization, like the development of housing, climate adaptation like land clearance in threatened areas, protection of natural buffers such as mangroves, and climate mitigation such as the protection of forests and seas as carbon sinks. Unfortunately, the current regulations are not specific enough to provide enough competences to the local authorities on this important matter as well as enough control and enforcement authority to the national authorities.

Among the different socio-economic pathways, the impacts of climate change on vulnerable groups is also particularly important.

5. Socio Economic Pathways: Disaster Impacts on Vulnerable Groups

The unequal impacts of climate change on vulnerable communities and the inequity of intervention have been widely documented in the existing research (Marino et al. 2012, Baztan et al 2020). Within the ten CRIC cities, the populations living in coastal areas are directly exposed to the impacts of climate change. The country's population densities are indeed highest in the coastal regions, which increases the risks of hazards for most of the people living below an elevation of ten meters (Perwaiz et al. 2020). Exposure is clearly related to socioeconomic vulnerability. The growing urbanization needs and the lack of affordable, safe land increases the less affluent communities' exposure to climate hazards. People living in informal settlements are also increasingly at risk, especially with regard to flooding.

The local research in our project demonstrate that the people living on the riverbank areas (Banjarmasin) or the coastal fishers' communities are directly exposed to regular floods and damages (Cirebon, Bandar Lampung, Ternate). In cities like Bandar Lampung, 12.4 % of the population lives in informal settlements, while in Banjarmasin, the old city neighborhood depends on wood stilt housing which is more vulnerable to flash floods. In fact, 24 % of the population lives below the poverty line . The socioeconomic effects of climate-related impacts are not to be underestimated. Poverty indeed remains a determinant of vulnerability as people have less access to information, knowledge, and recovery facilities. Additionally, droughts, regular floods, and storms may thrust more people into poverty due to lost livelihoods, housing, business, or work facilities and damaged infrastructure. Furthermore, regular floods and droughts have various repercussions on public health (Perwaiz et al. 2020). Our project's interviews and research show that local governments are aware of the social vulnerabilities in their territories and are investing in slum improvement programs and relocations. The majority of the ten cities implement the national 'cities without slums' program. Nevertheless, the urban analysis reports and focus group discussions highlight that the cities' local governments focus on short-term preparedness and immediate priorities and cannot integrate all socio-economic, cross-boundary, and public health aspects as parameters for long-term policies. The short-term and long-term impacts of climate change on vulnerable groups and public health are additional complexities which could be taken into account both at a research level and when developing future policies.

6. More Categories Are Needed to Assess and Support Climate Change Adaptation

There is currently an abundant and wide range of international articles about disaster prevention and disaster management in Indonesian cities (Leitmann 2007, Djalante et al. 2017, Perwaiz et al. 2020) including other articles on the adaptation to climate change. However, the research covering both of these aspects within their specific contexts and developments are still lacking. As illustrated by Baztan et al (2020), 'climate science may need to be linked more directly to local communities, to their capacities, and their contexts of vulnerability' (IPCC 2014, Vaughan et al 2016, Baztan et al 2020). Some authors rightly suggest developing assessment frameworks to integrate the adaptation of different risks into policies, such as an Adaptation-Risk Policy Alignment (ARPA) (Sainz de Murieta et al. 2021). This framework would be used to assess whether and how climate change adaptation policies integrate risk knowledge and information. But this is still a very new approach, and it is currently used mainly to assess the climate adaptation in large, cosmopolitan cities and not yet in coastal, small-, and medium-sized ones. The ACCRN network[9] and the different resilient cities network are developing matrices on urban resilience which could then be used by local governments and practitioners. Their methodology and approach also focus on large cities and equally highlight the necessity of funding and the need for real time information.

The question of real time information is, however, extremely complex for resilience and response policies, such as in the implementation of early warning systems. The project's regular focus group discussions and interviews confirm that current information developed by the climate agencies can be slightly too technocratic to grasp for local communities and stakeholders as it is also illustrated by literature (Klein 2017, Funtowicz 2020), and may not provide the necessary and timely information needed to reduce or to respond to the risks (Baztan et al. 2020). As we have seen previously, the impacts of climate change affect several sectors simultaneously (Fig. 4). Seawater intrusion is polluting the groundwater in many coastal areas, causing possible issues in terms of water scarcity. Landslides are also becoming more frequent as a result of urbanization, as well as floods and climate change. Several cities also

---

[9] Asian Cities Climate Change Resilience Network (ACCCRN).

experience seasonal flooding. Yet, as we have seen in the example of the ten pilot cities, other factors also come into play such as the city's morphology, the river pollution from upstream mining activities, or the clogging of the drainage systems due to illegal waste dumping (Fig. 4). Although there is no 'one size fits all solution,' more research is urgently needed to study the impacts and possible solutions for the adaptation of coastal cities to climate change and to identify the relationships between the different risks and the policies.

Conclusion

Coastal cities in Indonesia are faced with a double imperative: to solve urban growth and to protect the coastal areas and riverbanks from floods and other climate-related hazards. They will need to become resilient by adapting to climate change and other changes, such as subsidence, demographic urban growth, poor land management, unsustainable use of resources, possible social unrest, conflicts, poverty, insecurity, and declining ecosystems. Both growing urbanization and adaptation to the risks related to climate change require time, land planning, financial investment, good governance schemes, and capacities. The repercussions of climate change are very much intertwined with cities' water management, access to clean water, and sustainable waste management. Cities are faced with uncertainty and cascading risks and hazards. To tackle these challenges, it is important to consider the interrelation between the factors and the impacts (Fig. 4). This interrelation is even more important in terms of governance, resilience actions, socioeconomic pathways, and funding (Fig. 2). A multi-stakeholder approach and a sound organization of priorities are both essential. As we have demonstrated, however, the cooperation between stakeholders is complex, and cannot be taken for granted (Fig. 3). Strong incentives for each segment or key stakeholders are needed to achieve impacts beyond technocratic programs. Local governments play a key role in terms of climate adaptation and cities, yet have limited financial capacities. Land use can be an appropriate leverage to find a balance (or trade-off) between the urbanization needs, necessities of climate adaptation, such as natural buffers and land clearance, and efforts towards mitigation, like the protection of carbon sinks such as forests or marine ecosystems. The land use situation remains unclear, however, and generates social conflicts regarding land ownership. Cities are increasingly carrying the burden of finding the right and most appropriate strategies within the decentralization framework. Nevertheless, priorities within the different cities are severalfold. Failures in more generalized and efficient climate adaptation will lead to more disaster impacts, damages, and victims. The populations living in informal settlements as well as coastal areas are more directly exposed. Finally, as illustrated in Figs. 1, 2, and 4, more field research is needed to systematically assess climate disaster risk adaptation and resilience, and to properly address the question of uncertainties in light of multiple risks and hazards across disciplines.

References

Afrizal, Berenschot W., et al., 2020, 'Resolving Land Conflicts in Indonesia' Review essay, Bijdragen tot de taal, land en volkenkunde 176 (2020), 561–574

Agashe Y., 2020, CRIC Comparative Analysis, Internal Report

Amri M., Jamalianuri, Risanti D., 2020, 'Urban Analysis Report Samarinda', https://www.resilient-cities.com/en/?preview=1&option=com_dropfiles&format=&task=frontfile.download&catid=41&id=34&Itemid=1000000000000, last accessed 31.03.2021

Amri M., 2020, Jamalianuri, Risanti D 'Urban Analysis Report Banjarmasin', https://www.resilient-cities.com/en/?preview=1&option=com_dropfiles&format=&task=frontfile.download&catid=41&id=33&Itemid=1000000000000, last accessed 31.03.2021

Ardiansyah F., Akbar Marthen A., & Amalian N., 2015, 'Forest and Land Use Governance in a Decentralized Indonesia: A Legal and Policy Review', Occasional paper *CIFOR, PELANGI*

Basset T., Fogelman C., 2013, 'Déjà vu or Something New? The Adaptation Concept in the Climate Change Literature., *Geoforum* 48, 42–53

Baztan J., Vanderlinden J.P., Jaffrès L., Jorgensen B., & Zhu Z., 2020, 'Facing Climate Injustices: Community Trust-Building for Climate Services through Arts and Sciences Narrative Co-production', *Climate Risk Management*, 30(2020), 100253, https://www.sciencedirect.com/science/article/pii/S2212096320300437, last accessed 08.08.2021

Buchner B., Falconer A., Hervé Mignucci A., Trabacchi M., 2012, 'The Landscape of Climate Finance 2012, Climate Policy Initiatives', https://climatepolicyinitiative.org/publication/global-landscape-of-climate-finance-2012, last accessed 12.01.2021

Chan S., Amling A., 2019, 'Does Orchestration Is the Global Climate Action Agenda Effectively Prioritize and Mobilize Translational Climate Adaptation Action?' *International Environmental Agreement* 19, 429–446

Diab Y., 2020, Resilience and Early Warning Systems, Paper project CRIC Climate Resilient and Inclusive Cities, www.resilient-cities.com

Dillon H., Alnur Angelica A., Firas Khudi A., 2020, 'Urban Analysis Report Gorontalo', https://www.resilient-cities.com/en/?preview=1&option=com_dropfiles&format=&task=frontfile.download&catid=41&id=35&Itemid=1000000000000, last accessed 31.03.2021

Dillon H., Alnur Angelica A., Firas Khudi A., 2020, 'Urban Analysis Report Ternate', https://www.resilient-cities.com/en/?preview=1&option=com_dropfiles&format=&task=frontfile.download&catid=41&id=38&Itemid=1000000000000, last accessed 31.03.2021

Djalante R., Garschagen M., Thomalla F., & Shaw R. (Eds.) 2017, *Disaster Risk Reduction in Indonesia,* Springer International Publishing

Dwitama P., 2021, 'Policy Brief', https://www.resilient-cities.com/id/?preview=1&option=com_dropfiles&format=&task=frontfile.download&catid=45&id=70&Itemid=1000000000000, last accessed 31.03.2021

Funtowicz S., 2020, 'From Risk Calculations to Narratives of Danger', *Climate Risk Management,* 27, 100212 https://reader.elsevier.com/reader/sd/pii/S2212096320300024?token=B6CBEA02054999EE756B04F666D4701C3905CDF460510AD1F52DF26A57DFF692EC212672A58C4D7A49A536076F984FAB&originRegion=eu-west-1&originCreation=20210808171814, last accessed 08.08.2021

Gaborit P. (Ed.), 2015, *European and Asian Sustainable Towns*, Peter Lang International

Gaborit P., 2021, 'Vulnerabilities and Resilience to Climate Change in Tanzania', in Gaborit P., et Olomi D. (Eds.), *Learning from Resilience Strategies in Tanzania: An Outlook of International Development Challenges,* Brussels, Peter Lang International, https://www.peterlang.com/document/1152350

Gaborit P., Aleksic A., Marengo P., Diab Y., Pathak K., 2020, Policy Briefs Ten Pilot Cities, https://www.resilient-cities.com/en/knowledge/175-policy-briefs-for-pilot-cities-2, last accessed 31.03.2021

Glasser R., 2020, 'The Climate Change Imperative to Transform Disaster Risk Management'. *International Journal of Disaster Risk Science* 11, 152–154. https://doi.org/10.1007/s13753-020-00248-z

Grafakos S., Trigg K., Landaeur M., Chelleri L., & Dhakal S., 2019, 'Analytical Framework to Evaluate the Level of Integration of Climate Adaptation and Mitigation in Cities', *Climatic Change* 154, 87–106

Grimmond S., 2007, 'Urbanization and Global Environmental Change: Local Effects of Urban Warming', *Geography Journal* 173, 83–88.

Intergovernmental Panel of Climate Change (IPCC) 1996, Climate Change 1995: Impacts Assessment of Climate Change. Working Group II. Contribution to the IPCC fourth assessment Report. Summary for Policy Makers. World Meteorological Organization Geneva

Intergovernmental Panel of Climate Change (IPCC) 2021, Climate Change 2021: The Physical Science Basis. Contribution of Working Group I to the Sixth Assessment Report of the Intergovernmental Panel on Climate Change. Summary for Policy Makers Sixth Assessment Report (ipcc.ch)

IPCC (Intergovernmental Panel on Climate Change) 1995, 'IPCC Second Assessment Climate Change: A Report of the Intergovernmental Panel on Climate Change' 1995, WMO, UNEP, https://www.ipcc.ch/site/assets/uploads/2018/05/2nd-assessment-en-1.pdf

IPCC, 2014, Climate Change 2014: Synthesis Report. Contribution of Working Groups I, II and III to the Fifth Assessment Report of the Intergovernmental Panel on Climate Change, in Core Writing Team, Pachauri R.K. & Meyer L.A. Eds., IPCC, Geneva, Switzerland, 151 pp

IPCC (Intergovernmental Panel on Climate Change) 2014, 'Climate Change 2014: Impacts, Adaptation, and Vulnerability' IPCC Working Group II Contribution to AR5, IPCC Cambridge UK and New York USA

Intergovernmental Panel of Climate Change (IPCC) 2022, Climate Change 2022: Impacts, Adaptation and Vulnerability, Summary for Policy Makers.

Klein R.J.T., Adams K.M., Dzebo A., Davis M., Kehler S., 2017, 'Advancing Climate Adaptation Practices and Solutions: Emerging Research Priorities', Stockholm Environment Institute

Koch L., Gorris P., Pahl Wostl C., 2021, 'Narratives, Narration and Social Structure in environmental Governance', *Global Environmental Change*, 69, July 2021, 102317, https://doi.org/10.1016/j.gloenvcha.2021.102317, last accessed 23.07.2021

Lassa J.A., 2013, 'Disaster Policy Change in Indonesia 1920–2010: From Government to Governance?' International Journal of Mass Emergency Disaster 31(2), 130–159

Leitman J., 2007, 'Cities and Calamities: Learning from Post-Disaster Response in Indonesia', *Journal of Urban Health: Bulletin of the New York Academy of Medicine*, 84(1)

Marino E., Ribot J., 2012, 'Adding Insult to Injury: Climate Change and the Inequities of Climate Intervention', *Global Environmental Change*, 22 (Issue 2), May 2012, 323–328, https://doi.org/10.1016/j.gloenv cha.2012.03.001, last accessed 23.07.2021

Mulkhan U., Mayaguezz H., Tisnanta H.S., Kurniawan N., 2020, 'Urban Analysis Report Pangkal Pinang' https://www.resilient-cities.com/en/?preview=1&option=com_dropfiles&format=&task=frontfile.download&catid=41&id=40&Itemid=1000000000000, last accessed 31.03.2021

Mulyana W., Ardhyarini N., Pratiwi H., 2020, 'Urban Analysis Report Mataram' https://www.resilient-cities.com/en/?preview=1&option=com_dropfiles&format=&task=frontfile.download&catid=41&id=39&Itemid=1000000000000, last accessed 31.03.2021

Nagu N., Lessy M.R., Achmad R., 2016, 'Adaptation Strategy of Climate Change Impact on Water Resources in Small Islands Coastal Areas: Case Study on Ternate Island North Maluku' Conference Paper, The First International Conference on South Asia Studies, Kne Social Sciences, pp 424–441

Nicholls R.J., Wong P.P., Burkett V.R., Codignotto J.O., Hay J.E., McLean R.F., Ragoonaden S., & Woodroffe C.D., 2007, Coastal Systems and Low-Lying Areas. Climate Change 2007: Impacts, Adaptation and Vulnerability. Contribution of Working Group II to the Fourth Assessment Report of the Intergovernmental Panel on Climate Change, in Parry M.L., Canziani O.F., Palutikof J.P., van der Linden P.J., & Hanson C.E. (Eds.), Cambridge University Press, Cambridge, UK, pp 315–356

Oleson K.W., Monaghan A., Wilhelmi O., Barlage M., Brunsell N., Feddema J., Hu L., & Steinhoff D. F. 2015, 'Interactions between Urbanization, Heat Stress, and Climate Change'. *Climatic Change*, 129, 525–541

Perwaiz A., Parviainen J., Somboon P., & Macdonald A. 2020, 'Disaster Risk Reduction in Indonesia' Status Report UN Office for Disaster Risk Reduction, Asian Disaster Preparedness Centre

Prayitno B., 2017, 'Sustainable Customized Consolidation Design of Kuin Riverside Kampong Regeneration in Banjarmasin Indonesia', SHS Web of Conferences 41

Priyadi H., Setyorogo H.D., Anastasya C., & Gunawan I., 2020, 'Urban Analysis Report Bandar Lampung' https://www.resilient-cities.com/en/?preview=1&option=com_dropfiles&format=&task=frontfile.download&catid=41&id=37&Itemid=1000000000000

Priyadi H., Setyorogo H.D., Anastasya C., & Gunawan I., 2020. 'Urban Analysis Report Cirebon', https://www.resilient-cities.com/en/?preview=1&option=com_dropfiles&format=&task=frontfile.download&catid=41&id=31&Itemid=1000000000000, last accessed 31.03.2021

Rahayu P., 2016, 'The Governance of Small Cities in Decentralizing Indonesia: The Case of Cirebon City and Its Surrounding Regions', Rijksuniversiteit Groningen

Raven J., 2011, 'Cooling the Public Realm: Climate-Resilient Urban Design Resilient Cities', in Otto-Zimmermann K. (Ed.), *Cities and Adaptation to Climate Change: Local Sustainability*, Springer, Vol. 1, pp 451–463

Raven J., Stone B., Mills G., Towers J., Katzschner L., Leone M., Gaborit P., Georgescu M., & Hariri M., 2018, in Rosenzweig C., Solecki W., Romero-Lankao P., Mehrotra S., Dhakal S., & Ali Ibrahim S. (Eds.), *Climate Change and Cities: Second Assessment Report of the Urban Climate Change Research Network*, Cambridge University Press, New York, pp 139–172

Resosudarmo B., 2012, 'Forest Land Use Dynamics in Indonesia', Technical Report, Australian National University

Ridwansyah M., Telupere F., Rhogib Asfahani D., Nur Qalbi U., & Farras Kanzil A. 'Urban Analysis Report' Kupang https://www.resilient-cities.com/en/?preview=1&option=com_dropfiles&format=&task=frontfile.download&catid=41&id=32&Itemid=1000000000000, last accessed 31.03.2021

Rosenzweig C., Solecki W., Romero-Lankao P., Mehrotra S., Dhakal S., & Ali Ibrahim S. (Eds.), *Climate Change and Cities: Second Assessment Report of the Urban Climate Change Research Network*, Cambridge University Press, New York

Rousi, E., Kornhuber, K., Beobide-Arsuaga, G. et al., 2022, 'Accelerated Western European Heatwave Trends Linked to More-Persistent Double Jets over Eurasia', Nat Commun 13, 3851, https://DOI.org/10.1038/s41467-022-31432-y

Runhaar H., Wilk B., Persson A., Uittenbroek C., & Wamsler C., 2018, 'Mainstreaming Climate Adaptation: Taking Stock about "what works" from Empirical Research Worldwide', *Regional Environmental Change* 2018, 1201–1210

Russill C., 2008, 'Tipping Point Forewarnings in Climate Change Communication: Some Implications of an Emerging Trend', *Environmental Communication* 2(2), 133–153

Sainz de Murieta E., Galarraga I., Olazabal M., 2021, 'How Well Do Climate Adaptation Policies Align with Risk-based Approaches: An Assessment Framework for Cities', *Cities* 109 (2021), 103018

Salim W., Hudalah D., 2020, 'Urban Governance Challenges and Reforms in Indonesia: Towards a New Urban Agenda', in Dahiya B., Das (Eds.), *New Urban Agenda in Asia-Pacific*, Springer Nature Singapore Pte Ltd.

Simpson N.P., Mach K.J., Constable A., Hess J., Hogarth R., Howden M., Lawrence J., Lempert R.J., Muccione V., Mackey B., New M.G., O'Neill B., Otto F., Pörtner H.O., Reisinger A., Roberts D., Schmidt D.N., Seneviratne S., Strongin S., Van Aalst M., Totin E., & Trisos C.H., 2021, 'A Framework for Complex Climate Change Risk Assessment', in *One Earth* 4, April 23, 2021, https://doi.org/10.1016/j.oneear.2021.03.005

Tirtariandi el Anshori, Enceng and Jasrial, 2020, 'Flood Disaster Mitigation Plan Analysis (A case study in Pangkal Pinang City)' 3rd International Conference on Social Transformation, Community and Sustainable Development (ISCTCSD-2019), *Advances in Social Science, Education and Humanities Research*, volume 389

Toumbourou T., Mudar M., Werner T., & Bebbington A., 2020, 'Political Ecologies of the Post-mining Landscape: Activism, Resistance and Legal Struggles over Kalimantan's Coal Mines' in Energy Research and Social Science, Vol. 65, July 2020, 101476, https://doi.org/10.1016/J.erss.2020.101476 last accessed 31/03/2022

Vaughan C., Buja L., Kruczkiewicz A., & Goddard L., 2016, 'Identifying Research Priorities to Advance Climate Services', in *Climate Services*, 4 (2016), 65–74, https://reader.elsevier.com/reader/sd/pii/S2405880716300358?token=FD8F0E6C3171D0B321CA6E9362EBB30A8BA736B6D440B92148EF7FDB16E34D1AFE495E4660BB82255C795AA3D8225B0F&originRegion=eu-west-1&originCreation=20210808172435, last accessed 08.08.2021

Vij S., Moors E., Ahmad B., Arfanuzzaman M., Bhadwal S., Biesbroek R., Giili G., Groot A.M., Mallick D., Regmi B., Saeed B.A., Ishaq S., Thapa B., & Werners S.E., 2017, 'Climate Adaptation Approaches and Key

Policy Characteristics: Cases from South Asia', *Environmental Science and Policy* 78(2017), 58–65

Von Korff Y., Daniell K.A., Moellenkamp S., Bots P., & Bijlsma R., 2019, 'Implementing Participatory Water Management: Recent Advances in Theory, Practice and Evaluation', Ecology *and Society*

Wijaya N., Nitivattanon V., Prasad Shrestha R., & Minsun Kim S., 2020, 'Drivers and Benefits of Integrating Climate Adaptation Measures into Urban Development: Experience from Coastal Cities of Indonesia', *Sustainability* 2020, 12, 750, Mdpi.com

Wirawan B., Utama S., Rachmawati Suratno I., Rizal Tambunan J., Muthadir A., Mulyana W., Setiono I., Wicaksono A., & Afiff S.A., 2019, 'Spatial Planning, Land Tenure and Claims and Increasing Conflicts over Land Claims in Sumatra and Kalimantan, Economic Development, Policy Dynamics and the Pace of Investment', Conflict Resolution Unit, July 2019. Translated by Larry Fisher

Wollenberg E., Campbell B., Dounias E., Gunarso P., Moeliono M., & Shell D., 2009, 'Interactive Land-Use Planning in Indonesian Rain-Forest Landscapes: Reconnecting Plans to Practice', *Ecology and Society*, 14(1), 35

Ziervogel G., Pelling M., Cartwright A.A., Chu E., Deshpande T., Harris L., ..., Zweig P., 2017, 'Inserting Rights and Justice into Urban Resilience: A Focus on Everyday Risk', *Environment and Urbanization*, 29(1), 123–138

Chapter 2
Sustainable Local Economic Development and Gender Perspective in the Context of Climate Change Adaptation

UNANG MULKHAN PHD

Introduction

The harsh reality of climate change is causing more intense and unpredictable impacts through several disaster events such as droughts, floods, hurricanes, changing weather patterns, etc. Thus, as a real phenomenon, climate change should be approached holistically (Klein et al. 1999; Gallopin 2006; Adger et al. 2009; Bowen et al. 2012; Basset & Fogelman 2013), in order to allow human and natural systems to continuously function well through mitigation and adaptation. Many researchers have simply defined mitigation as an effort to reduce climate change itself, such as decreasing the concentration of greenhouse gases in the atmosphere through reducing emissions (e.g Adger et al. 2009; Morecroft & Cowan 2010). Adaptation is a form of adjustments in ecological, social, and economic aspects of the consequences, impacts, and risks of climate change (Morecroft & Cowan 2010). Adaptation, therefore, involves many aspects of measurement including technical, institutional, sectoral, environmental, legal, educational, and behavioral (Fussel 2007).

Many researchers underscore that climate change through mitigation and adaptation policies should incorporate gender mainstreaming strategies (see Enarson 1998; Denton 2002; De Haan & Zoomers 2005; Walby 2005; Behrman et al. 2012). The threats posed by climate change such as environmental degradation, land-use change, and water scarcity can affect and create further disparities in gender-related inequalities (Denton 2002). The aim to promulgate gender mainstreaming in the conversation around climate change is to prevent or lessen the extent to which the consequences of climate change lead women and marginalized people into further deprivation (Denton 2002). For example, Dankelman (2002) argues that during mitigation and adaptation it is important to consider not only technological solutions, but also the need to involve political and socio-economic aspects when creating policies and solutions. This need to consider multiple contextual issues raises the question of how to have appropriate policies and practices that are more context-specific since there is no one size fits all policy for adapting to climate change (Adger et al. 2009; Alston 2014). However, there are relatively few conceptual and empirical research approaches intended to address climate-related conflicts in a context-sensitive manner (Loh & Shear 2005; Hay & Mimura 2006; Brooks et al. 2009; Gabrielsson et al. 2013).

Furthermore, there is a growing interest in discussing and formulating the idea of climate change adaptation (CCA) from a gender perspective (e.g. Denton 2002; Dessai et al. 2004; Fussel 2007), and in relation to local economic development (e.g. Barberia & Biderman 2010). Furthermore, community-based adaptation is suggested by researchers to be a promising strategy in the issue of climate change adaptation when considering the gender aspect (Jerneck 2018). 'Gender' does not merely entail the social attributes and opportunities associated with being a female or a male: It is also about the relationships between women and men. Therefore, 'gender' is not the equivalent of 'sex'. Sex refers to the biological differences between females and males. Thus, theoretically, gender is socially constructed and is subject to ongoing socio-cultural negotiations such as aspects like intra-household bargaining power (Zalewski 2010; Jerneck 2018) and gender itself is a primary way of signifying power relationships (De Haan & Zoomers 2005; Demetriades & Esplen 2009). A society or culture can construct and prescribe roles, behaviors, and identities for women and men, in the sense of what is expected of, allowed for, and valued for, in given contexts.

Gender has been discussed by researchers from several aspects such as from equality and fairness in access to socio-economic opportunities, exclusion or integration, exploitation or empowerment, and marginalization or participation. Often, gender roles and characteristics affect power relations between women and men in societies, resulting in inequality in opportunities for them. Since there are major differences in societies and cultures, gender roles can be considered highly context specific. A society and culture set gender roles using some situations such as household structure, access to resources, and ecological conditions (UNICEF 2017). Thus, it is interesting and important to discuss how to understand and formulate climate change adaptation through local economic development with gender issues and sustainability as essential aspects in diverse values and across contexts and cultures, such as in Indonesia.

The objective of this chapter is, therefore, to discuss the potentials of climate change adaptation through local economic development with a gender perspective. This chapter accordingly draws on theoretical considerations from the fields of environment, gender, and development studies. Consequently, although this chapter employs a developing country, Indonesia, as its case study, the concept may have wider applications. The chapter is divided into five sections. The next section discusses the nexus between sustainable local economic development (SLED) and climate change adaptation (CCA) from a gender perspective. In the third section of the chapter, there is an exploration of current practices and case studies of women as factors of resilience for the local economy in Indonesia. This section includes the map of potentials and challenges faced by women in the economic participation in Indonesia using structure and agency theory. The fourth section is the explanation of core elements and mechanisms required in developing and strengthening the participation of women in SLED in the context of climate adaptation. In the final section, the chapter concludes that the approach of climate change adaptation through SLED needs to consider gender perspectives in each specific context.

I. The Nexus between Sustainable Local Economic Development and Climate Change Adaptation from Gender Perspective

The Local Economic Development (LED) concept has evolved from the focus on the attraction of outside investment and hard infrastructure investment to the retention and growth of existing local businesses to inward-looking shifts with a growing emphasis on the enhancement of local capacities and competitiveness (Jacob & Stott 1992; Nel & Humphrys 1999; Nel 2001; Helmsing 2003; Hopwood et al. 2005; Brodhag & Taliere 2006; Adeel & Safriel 2008; Baumgartner & Korhonen 2010; Rogerson & Rogerson 2010; Betzet et al. 2012). Thus, as an integrated strategy, LED is expected to address complex challenges by promoting economic development in local areas with their own uniqueness. In practice, LED aims to enable local people in building upon local resources, and to make use of their advantages and specific knowledge to their territory (Dale & Newman 2009). Very broadly speaking, current practices of poverty alleviation programs have also been shifted from 'charity' to building self-reliance locally. This kind of model and strategy of development can help avoid a 'beggar mentality' during massive aid intervention. In short, LED is defined as a cooperation between public, private and non-governmental sector partners where they work collectively to create economic growth and employment through local resources.

There are some characteristics of local economic development. LEDs must firstly be locally owned. Secondly, the process of participatory development must be undertaken within a given territory or local administrative area. Thirdly, LED entails a process of partnership with both public and private stakeholders. Therefore, LED utilizes local resources and competitive advantages to create decent employment and sustainable economic growth. However, the ultimate goal of LED is to pursue social goals of poverty reduction and social inclusion. Therefore, LED includes activities under social economy which can involve a social movement in more sustainable communities (Newby 1999; Walo 2017). In this sense, the design and implementation structures of LED should create space for dialogue among local stakeholders to enable them to actively participate in the decision-making process. In this way, LED is more about the 'inside-out' approach because people and their communities have their own local context (Walo 2017).

Furthermore, the natural environment becomes a major consideration to deal with the causes and consequences of climate change in the context of local economic development (Morecroft & Cowan 2010). The main goal of adaptation is to considerably reduce or moderate many of adverse impacts of climate change and exploiting opportunities (McCarthy et al. 2001; Morecroft & Cowan 2010). Adaptation action can create unintended impacts on other natural and social systems (McCarthy et al. 2001; Adger et al. 2005). These include processes, practices, and structures to tackle the impacts of climate change (Morecroft & Cowan 2010). However, adaptation is a complex process embedded in layers of social relations and decision-making power (De Haan & Zoomers 2005). For this chapter, however, adaptations are more focused on actions undertaken by human beings including social, economic, and environmental aspects.

Researchers also argue that adaptation on climate change can be approached through sustainable development with the balance of social, economic, and environmental considerations so that a holistic approach to sustainability can be achieved (see Morecroft & Cowan 2010). Tackling climate change and its impacts, therefore, should be placed under the framework of sustainable development strategies. Sustainable development mainly entails meeting present-day societal needs without compromising the ability of future generations to meet their needs. Sustainable development is obviously crucial in both adaptation and mitigation.

The concept of Sustainable Local Economic Development (SLED) has been interpreted as not only about economic growth but also about equitable, inclusive, and resilient strategies and practices (Swinburn 2006; Rangarjan et al. 2012; Bek et al. 2013; Fineberg 2013; Wilson & Polter 2020). In the context of climate change adaptation, SLED can be linked to self-reliance strategies involving collaboration with local stakeholders (Walo 2017; Nel & Binns 2002; Nel & Rogerson 2005). SLED is also a consistent concept to embrace Sen's (2000) ideas on development which is people centered. For Sen, development should entail the expansion of human freedoms, capabilities, well-being, and opportunities for all individuals (Sen 2000). This is relevant to SLED in emphasizing endogenous resources and environmental sustainability criteria and putting forth local social and cultural realities (Morecroft & Cowan 2010). SLED is, therefore, the result of a process in which the integration of three pillars of sustainable development, namely, social, economic, and environmental in order to pursue a sustainability that is embedded in existing local social structure. Also, this approach is viewed in relation to power institutions, authority, and society.

Gender perspective on SLED policy can be mainstreamed by using Walby's (1997) ideas of having three stages of gender equality policy: (1) by providing equal treatment for women through, for example, legal statutes that deliver rights to women; (2) by having positive actions for women such as leadership training; (3) by mainstreaming gender issues as a commitment to a comprehensive assessment of organizational structures, policies and practices. Based on ECOSOC (1997), the gender mainstreaming is defined as: 'the process of assessing the implications for women and men of any planned action, including legislation, policies or programmes, in all areas and at all levels. It is a strategy for making women's as well as men's concerns and experiences an integral dimension of the design, implementation, monitoring and evaluation of policies and programs in all political, economic and societal spheres so that women and men benefit equally, and inequality is not perpetuated. The ultimate goal is to achieve gender equality.'

Above ideas can be coined with the term of SLED through five principles of SLED which are: quality of life, fairness and equity, participation and partnership, environmental preservation, and thought for the future (Newby 1999). SLED, therefore, provides practitioners a new approach that takes into account the holistic picture of economic development, for the betterment of the community. Although Newby (1999) has identified those five principles of sustainable local economic development, those principles need to be further investigated on how gender perspective can be embedded. Moreover, UN WomenWatch (2011) notes that 'in many contexts, women are more vulnerable to the effects of climate change than men [and] are especially vulnerable when they are highly dependent on local natural resources for their livelihood. Those charged with the responsibility to secure water, food and fuel for cooking and heating face the greatest challenges [and] when coupled with unequal access to resources and to decision-making processes, limited mobility places women in rural areas in a position where they are disproportionately affected by climate change. It is thus important to identify gender-sensitive strategies to respond to the environmental and humanitarian crises caused by climate change.' Thus, it is necessary to enhance the capability of women to have more access to resources and to have skills for disaster preparedness, mitigation, and rehabilitation (Alston 2014; Jerneck 2018), meaning that gender-sensitivity in climate strategies is a crucial aspect during and after climate disasters.

SLED is relevant to the agenda of Sustainable Development Goals (SDGs) 2030 agenda. The SDGs 2030 was adopted in September 2015 by the United Nations General Assembly, defining 17 goals and 169 targets to be achieved in 2030. SLED can be linked to the issues of poverty eradication, increasing employment and decent work, and strengthening partnerships. In other words, this concept and practices can be instrumental for sustainable development with efforts to improve the living standard, well-being, and local economic prosperity of communities. Decent work is a critical goal to be achieved in the SDGs agenda. Promoting and investing in women's economic empowerment towards equality can help women achieve these goals. Access to decent work for women is a fundamental pillar for the creation of territorial systems that combine competitiveness and economic efficiency, promoting inclusion and social cohesion.

Conclusively, it is believed that SLED represents a promising strategy for the CCA (e.g. Agrawala & Carraro 2010; Adhikari 2013). Whereas gender perspectives have been discussed in the literature as an exogenous part of local economic development for the CCA (see Alston 2006, 2014; Arora-Jasson 2011; Gabrielsson & Ramasar 2013; Cornwall & Rivas 2015), integrating gender perspectives into SLED leads to a different understanding of climate change adaptation initiatives. It is necessary to elaborate the concept of SLED more fully in the context of climate change adaptation from a gender perspective. Therefore, climate change and gender should be incorporated into the process of incorporating a gender perspective to any action, policy, or legislation in order to ensure that the concerns and needs of all people are addressed.

II. Women as Factors of Resilience for the Local Economy in Indonesia

Indonesia has a strong civil society tradition and democracy as a result of the nation's reform era, which began in 1998. Geographically, most areas in Indonesia are prone to natural disasters and many of the areas are agricultural, thus the majority of people work as farmers. However, there is still a complex situation in the sense of tackling poverty and economic inequality, especially in post-disaster situations. The people of the country are characterized by a strong collective culture that promotes values of harmony and togetherness and the culture of mutual help. Therefore, there is a potential to develop cohesive, integrated societies. The embeddedness of individuals, groups, and nature can build strong relationships that lead to trust and cooperation in the long term. Through these nested factors, the ecosystem of the local economy can be maintained while the above practice encourages employment opportunities in rural areas.

However, having a holistic strategy on climate change adaptation throughout local economic development from a gender perspective is still pivotal in Indonesia. Drawing upon the work of Denton (2002), gender-related inequalities are pervasive in the developing world and this can be investigated from different fields of studies. In this context, the relational aspects of gender can be understood in societies and cultures in aspects such as income and workload. Often, women receive lower incomes and have higher workloads when compared to men (Jerneck 2018). The lack of representation of women in decision-making is still a crucial problem in closing the gender parity gap in society.

Despite significant progress made in promoting gender equality in local economic development, a high percentage of women still remain in sex-stereotype occupations that are often more precarious, more vulnerable, and entail lower remuneration than men's work. In many areas in Indonesia, women are still disproportionately more affected by decent work deficits and are henceforth more likely to live in poverty. Gender gaps are a challenge in most areas in Indonesia, affecting productivity and competitiveness and leading to a practice of discrimination between males and females in terms of resources, power, and status. Although climate change is not the only factor causing gender inequalities there is an inherently inequitable balance of power relations and resource allocations during the response to climate disasters.

Decentralization also affects several consequences such as authority and responsibilities in fostering local economic development and building resilience. Decentralization of authority and resources creates another challenge. Local economic development and the creation of jobs are not only the responsibility of the central government but also the local government. Enhancing rural-urban linkages is important. Several opportunities and benefits can be explained such as the fact of decentralization, participatory development, and the competitiveness of local economies. However, urban-rural dynamic in CCA and SLED is often overlooked. Indonesia, especially in rural areas, is mostly dominated by agricultural activities. Local governments can play a large role in promoting sustainable economic development. Therefore, narrowing gaps between rural and urban communities is necessary. This can help to reduce inequality between areas and territories within a multi-level governance perspective, to introduce an integrated territorial planning framework for inclusive and sustainable growth, and to conceptualize rural-urban relations through a comprehensive framework in promoting their integration and interdependence.

In several cases, women's lack of ownership of land and resources affects their ability to protect their livelihoods in post-disaster situations (Arora-Jonsson 2011; Alston 2014; Jerneck 2018). Some researchers consider several factors such as less control over production and income, less education and training, increased workloads including taking care of their children, less access to institutional support and information, less freedom of association, and fewer positions on decision-making bodies as causes of vulnerability of women during and post-disaster situations (see Walby 2005; Dankelman 2010; Adhikari 2013; Alston 2014; Jerneck 2018). Additionally, Dankelman (2010: 59) argues that men are more likely to migrate to look for work, resulting in women having more responsibilities in their households.

In short, some challenges that need to be tackled are: how to improve understanding of gender and the goals of gender mainstreaming, how to limit impacts of urbanization, how to increase political will at state and local levels, how to enhance public participation, and how to improve and provide access to both women and men on skills and training.

III. Core Elements and Mechanisms

There is a prerequisite to execute the concept of SLED in relation to governance structures presuppose active women's participation and effective social inclusion. In the Indonesian context, looking at SLED from gender perspective, women should be supported to have access to resources without receiving any disruption from relations of power within the household. Women should have both an access to resources and the ability to exercise power. Jerneck (2018) explains strategies for the general concept of SLED such as how it is necessary to improve the enabling environment for business development, to develop capacities of the key economic institutions and actors, to determine and to process the local resources, both human and natural, to create the economic chain, and to utilize local perspectives and interests. It is necessary to allow projects to be conceived and managed by members of the community to enhance ownership on the part of the target beneficiaries, resulting collective action (Bhattacharyya 1995, 2004; Blair 2004).

It is also crucial to understand the importance of the rural–urban linkages for the better strategies amid rapid urbanization and the dynamics of local governance structures and practices in Indonesia since there is no separation between rural-urban interactions. Participation can include stakeholders from civil society organizations and the private sector. This means that building partnerships for change is vital in order to have collaboration in identifying common needs and resources.

Women are particularly vulnerable in the current climate change phenomenon, especially when they are too close to and dependent upon nature (Arora-Jonsson 2011; Alston 2014). Generally, women experience results from climate change that can usually be seen from the aspect of traditional gender relations within their households, financial matters, child rearing, reproduction, and freedom of movement outside the home (Alston 2014). Therefore, many interventions have been conducted by donors to increase their confidence to confront those issues, aiming to have other support and economic options outside the home. Towards this aim, there should be an effort to enhance the adaptive capacity of women to cope with changes and uncertainties in climate, including variability and extremes in order to avoid vulnerabilities and to promote sustainable development. For instance, it is how to have women-focused policies designed to ensure that women are supported and empowered to take action on their own behalf (Daly 2005; Alston 2014; Jerneck 2018).

Several researchers have discussed the aspect of power in the context of local economic development practices (see Cronwall 2004; Cahill 2005, 2008; Avelina & Rotmans 2009). For instance, power relations have been recognized as essential to the sustainability of development interventions (Avelino & Rotmans 2009). However, the way in which power relations affect climate change adaptation strategies in the context of gender and the local economy has remained relatively under-theorized. In fact, power relations can be seen from the processes of empowerment, redistribution of resources, and participation (Cronwall 2004; Cahill 2005, 2008). Often, mechanisms such as micro-finance and sustainable livelihood initiatives have been discussed in the area of power relations. The phenomenon of gender inequality is interesting and important when viewed and discussed from climate change adaptation policies and processes.

Recognizing the power of localities is essential to understanding how interventions can be made through exploring ability, knowledge, and locally available resources to start an enterprise or otherwise develop one's economic standing. In other words, understanding the power of the local community can help improve initiatives to provide sufficient investment capital, knowledge, and skills to initiate a business project run by individuals in the local communities. Local people can have a perspective of power in terms of resources, networks, and economic opportunities. Therefore, SLED could provide a sense of local power and ownership over the local business project(s). In this sense, power should be defined as not only something about resources but also something generated by control and reproduction from network and social interaction.

Addressing power within the practice of local economic development from gender perspectives can help better intervention when dealing with the consequences of climate change. However, local power dynamics can create a challenge to achieve better intervention (Cronwall 2004; Cahill 2005, 2008). For instance, it is a challenge to increase women's participation in participatory budgeting, which is still male dominated in public affairs. This shows that power remains central to the concerns of local economic development. Given that SLED from the perspective of gender entails structural and inter-sectional power, gender-informed approaches are essential (Alston 2014).

Support of local government to women in accessing credit and capital is also important. Several mechanisms can be considered such as supporting the establishment of microfinance and providing low-cost credit and capital in order to help community enterprises and business ventures to grow. The local government can also provide training to transform local communities as the principal defenders of resilience, the local environment and natural resources, and their cultural rights. Creating bottom up or grass root approaches are vital to mobilizing local assets and resources in order to create local products, to benefit people in the community, especially in targeting women and other disadvantaged people. This is a different form of approach compared to the traditional way of viewing local economic development through attracting businesses from outside, or foreign direct investments (FDIs). By identifying and mobilizing local resources and assets, the inclination of people to focus on what the community lacked or needed for development are considered and gathered.

Local governments should design mechanism(s) in order to strengthen asset security such as land for women, to have good governance and coordination among government agencies, and to facilitate digital inclusion. Several facilities can also be provided through specific mechanisms such as providing equal access to economic resources including finance, productive assets, markets and technical knowledge, and skills and trainings. A participatory approach to SLED requires not only a balanced representation of women and men participating in the process, but the creation of conditions in which both women and men have the same rights to raise their opinions. Furthermore, the local government can put forward its commitment on regulation to equally access local resources to help impoverished both men and women become self-reliant (van der Waldt 2016, 2018).

Conclusively, several elements and mechanisms can be considered in order to increase women's participation in SLED through gender-equitable local economic development. It is important to consider gender perspectives on how to increase decent livelihoods, jobs, and sustainable economic activities for men and women with the use of the local resources. All these need mechanisms in order to increase access to and control over resources and other actions to lessen and ideally remove gender discrimination and inequality. It is important to enhance active participation in building resilience and in promoting social inclusion. This reflects fairness and equality in local economic development projects, with the primary concern that all individuals in a community have an equal chance to grow, have access to opportunities, and to achieve higher qualities of life.

Conclusion

This chapter has discussed the nexus between sustainable local economic development and climate adaptation with a gender perspective. It is becoming clearer that there are opportunities in connecting climate change adaptation with local economic development from a gender perspective. Participation of women as economic actors is essential to build healthy and sustainable economies. In this effort, economic development can be intertwined with the management of ecosystems at the local level, where women can play an important role and improve their livelihoods as well as their ability to be resilient before and after disaster events. To do this in a manner more connected to the sustainable development agenda, it is essential to ensure that three aspects, social, economic, and environmental, are considered and practiced. The concept of SLED through a gender perspective can be a good model for climate change adaptation although it has some challenges in the context of Indonesia. Case studies of women as factors of resilience for the local economy in Indonesia can lead to successful examples. But the reticulation of climate adaptation and local economic development authorities and local actors should understand and provide solutions to problems and challenges due to power inequality, as only then can gender equality be mainstreamed in the climate adaptation agenda.

References

Adeel, Z & Safriel, U 2008, 'Achieving sustainability by introducing alternative livelihoods', Sustainability Science, vol. 3, no. 1, pp. 125–133. https://doi.org/10.1007/s11625-007-0039-4

Adger, WN, Arnell, NW, & Tompkins, EL 2005, 'Successful adaptation to climate change across scales', Global Environmental Change, vol. 15, no. 2, pp. 77–86. https://doi.org/10.1016/j.gloenvcha.2004.12.005

Adger, WN, Dessai, S, Goulden, M, Hulme, M, Lorenzoni, I, Nelson, DR, Naess, LO, Wolf, J, & Wreford, A 2008, 'Are there social limits to adaptation to climate change?', Climatic Change, vol. 93, no. 3–4, pp. 335–354. https://doi.org/10.1007/s10584-008-9520-z

Adhikari, B 2011, 'Poverty reduction through promoting alternative livelihoods: Implication for marginal dryland', Journal of International Development, vol. 25, no. 7, pp. 947–967. https://doi.org/10.1002/jid.1820

Agrawala, S & Carraro, M 2010, 'Assessing the role of microfinance in fostering adaptation to climate change', SSRN Electronic Journal. https://doi.org/10.2139/ssrn.1646883

Alston, M 2006, 'Gender mainstreaming in practice: A view from rural Australia', NWSA Journal, vol. 18, no. 2, pp. 123–147. https://doi.org/10.2979/nws.2006.18.2.123

Alston, M 2014, 'Gender mainstreaming and climate change', Women's Studies International Forum, vol. 47, pp. 287–294. https://doi.org/10.1016/j.wsif.2013.01.016

Arora-Jonsson, S 2011, 'Virtue and vulnerability: Discourses on women, gender and climate change', Global Environmental Change, vol. 21, no. 2, pp. 744–751. https://doi.org/10.1016/j.gloenvcha.2011.01.005

Avelino, F & Rotmans, J 2009, 'Power in transition: An interdisciplinary framework to study power in relation to structural change', European Journal of Social Theory, vol. 12, no. 4, pp. 543–569. https://doi.org/10.1177/1368431009349830

Barberia, LG & Biderman, C 2010, 'Local economic development: Theory, evidence, and implications for policy in Brazil', Geoforum, vol. 41, no. 6, pp. 951–962. https://doi.org/10.1016/j.geoforum.2010.07.002

Bassett, TJ & Fogelman, C 2013, 'Déjà vu or something new? The adaptation concept in the climate change literature', Geoforum, vol. 48, pp. 42–53. https://doi.org/10.1016/j.geoforum.2013.04.010

Baumgartner, RJ & Korhonen, J 2010, 'Strategic thinking for sustainable development', Sustainable Development, vol. 18, no. 2, pp. 71–75. https://doi.org/10.1002/sd.452

Behrman, J, Meinzen-Dick, R & Quisumbing, A 2012, 'The gender implications of large-scale land deals', Journal of Peasant Studies, vol. 39, no. 1, pp. 49–79. https://doi.org/10.1080/03066150.2011.652621

Bek, D, Binns, T & Nel, E 2010, 'Wild flower harvesting on South Africa's agulhas plain: A mechanism for achieving sustainable local economic development?', Sustainable Development, vol. 21, no. 5, pp. 281–293. https://doi.org/10.1002/sd.499

Bert Helmsing, AHJ 2003, 'Local economic development: New generations of actors, policies and instruments for Africa', Public Administration and Development, vol. 23, no. 1, pp. 67–76. https://doi.org/10.1002/pad.260

Betz, MR, Partridge, MD, Kraybill, DS & Lobao, L 2012, 'Why do localities provide economic development incentives? Geographic competition, political constituencies, and government capacity', Growth and Change, vol. 43, no. 3, pp. 361–391. https://doi.org/10.1111/j.1468-2257.2012.00590.x

Bhattacharyya, J 1995, 'Solidarity and agency: Rethinking community development', Human Organization, vol. 54, no. 1, pp. 60–69. https://doi.org/10.17730/humo.54.1.m459ln688536005w

Bhattacharyya, J 2004, 'Theorizing community development', Community Development Society Journal, vol. 34, no. 2, pp. 5–34. https://doi.org/10.1080/15575330409490110

Blair, R 2004, 'Public participation and community development: The role of strategic planning', Public Administration Quarterly, vol. 28, no. ½, pp. 102–147.

Bowen, A, Cochrane, S & Fankhauser, S 2011, 'Climate change, adaptation and economic growth', Climatic Change, vol. 113, no. 2, pp. 95–106. https://doi.org/10.1007/s10584-011-0346-8

Brodhag, C & Taliere, C 2006, 'Sustainable development strategies: Tools for policy coherence', Natural Resources Forum, vol. 30, no. 2, pp. 136–145. https://doi.org/10.1111/j.1477-8947.2006.00166.x

Brooks, N, Grist, N & Brown, K 2009, 'Development futures in the context of climate change: Challenging the present and learning from the past', Development Policy Review, vol. 27, no. 6, pp. 741–765. https://doi.org/10.1111/j.1467-7679.2009.00468.x

Cahill, A 2005, 'Engaging communities for local economic development: Lessons from the Philippines', Paper presented at the International Conference on Engaging Communities, 14–17 August, Brisbane, Australia, Retrieved 20 November 2007. http://www.engagingcommunities2005.org/abstracts/CahillAmanda-fifinal.pdf

Cahill, A 2008, 'Power over, power to, power with: Shifting perceptions of power for local economic development in the Philippines', Asia Pacific

Viewpoint, vol. 49, no. 3, pp. 294–304. https://doi.org/10.1111/j.1467-8373.2008.00378.x

Cornwall, A & Rivas, AM 2015, 'From 'gender equality' and 'women's empowerment' to global justice: Reclaiming a transformative agenda for gender and development', Third World Quarterly, vol. 36, no. 2, pp. 396–415. https://doi.org/10.1080/01436597.2015.1013341

Dale, A & Newman, LL 2009, 'Sustainable development for some: Green urban development and affordability', Local Environment, vol. 14, no. 7, pp. 669–681. https://doi.org/10.1080/13549830903089283

Daly, M 2005, 'Gender mainstreaming in theory and practice', Social Politics: International Studies in Gender, State & Society, vol. 12, no. 3, pp. 433–450. https://doi.org/10.1093/sp/jxi023

Dankelman, I 2002, 'Climate change: Learning from gender analysis and women's experiences of organising for sustainable development', Gender & Development, vol. 10, no. 2, pp. 21–29. https://doi.org/10.1080/13552070215899

De Haan, L & Zoomers, A 2005, 'Exploring the frontier of livelihoods research', Development and Change, vol. 36, no. 1, pp. 27–47. https://doi.org/10.1111/j.0012-155x.2005.00401.x

Demetriades, J & Esplen, E 2009, 'The gender dimensions of poverty and climate change adaptation', IDS Bulletin, vol. 39, no. 4, pp. 24–31. https://doi.org/10.1111/j.1759-5436.2008.tb00473.x

Denton, F 2000, 'Gender impact of climate change: A human security dimension', Energia News, vol 3, no. 3, pp. 13–14.

Denton, F 2002, 'Climate change vulnerability, impacts, and adaptation: Why does gender matter?', Gender & Development, vol. 10, no. 2, pp. 10–20. https://doi.org/10.1080/13552070215903

Dessai, S, Adger, WN, Hulme, M, Turnpenny, J, Köhler, J & Warren, R 2004, 'Defining and experiencing dangerous climate change', Climatic Change, vol. 64, no. 1/2, pp. 11–25. https://doi.org/10.1023/b:clim.0000024781.48904.45

Eaton, K 2001, 'Political obstacles to decentralization: Evidence from Argentina and the Philippines', Development and Change, vol. 32, no. 1, pp. 101–127. https://doi.org/10.1111/1467-7660.00198

Enarson, E 1998, 'Through women's eyes: A gendered research agenda for disaster social science', Disasters, vol. 22, no. 2, pp. 157–173. https://doi.org/10.1111/1467-7717.00083

Farmer, E 2005, 'Community development as improvisational performance: A new framework for understanding and reshaping practice', Community Development, vol. 36, no. 2, pp. 1–14. https://doi.org/10.1080/15575330509490172

Fineberg, A 2013, 'Promoting sustainable local economic development for all areas: Looking forward or looking back?', Local Economy: The Journal of the Local Economy Policy Unit, vol 28, no. 7–8, pp. 914–920. https://doi.org/10.1177/0269094213500751

Füssel, HM 2007, 'Adaptation planning for climate change: Concepts, assessment approaches, and key lessons', Sustainability Science, vol. 2, no. 2, pp. 265–275. https://doi.org/10.1007/s11625-007-0032-y

Gabrielsson, S & Ramasar, V 2013, 'Widows: Agents of change in a climate of water uncertainty', Journal of Cleaner Production, vol. 60, pp. 34–42. https://doi.org/10.1016/j.jclepro.2012.01.034

Gallopín, GC 2006, 'Linkages between vulnerability, resilience, and adaptive capacity', Global Environmental Change, vol. 16, no. 3, pp. 293–303. https://doi.org/10.1016/j.gloenvcha.2006.02.004

Hay, J & Mimura, N 2006, 'Supporting climate change vulnerability and adaptation assessments in the Asia-Pacific region: An example of sustainability science', Sustainability Science, vol. 1, no. 1, pp. 23–35. https://doi.org/10.1007/s11625-006-0011-8

Hickey, S & Mohan, G 2004, 'Participation: From Tyranny to Transformation: Exploring New Approaches to Participation in Development', Zed Books.

Hopwood, B, Mellor, M & O'Brien, G 2005, 'Sustainable development: Mapping different approaches', Sustainable Development, vol. 13, no. 1, pp. 38–52. https://doi.org/10.1002/sd.244

Hustedde, RJ & Ganowicz, J 2002, 'The Basics: What's essential about theory for community development practice?', Journal of the Community Development Society, vol. 33, no. 1, pp. 1–19. https://doi.org/10.1080/15575330209490139

Jacobs, M & Stott, M 1992, 'Sustainable development and the local economy', Local Economy: The Journal of the Local Economy Policy Unit, vol. 7, no. 3, pp. 261–272. https://doi.org/10.1080/02690949208726152

Jerneck, A 2018, 'Taking gender seriously in climate change adaptation and sustainability science research: Views from feminist debates and sub-Saharan small-scale agriculture', Sustainability Science, vol. 13, no. 2, pp. 403–416. https://doi.org/10.1007/s11625-017-0464-y

Kirkpatrick, LO 2007, 'The two "Logics" of community development: Neighborhoods, markets, and community development corporations', Politics & Society, vol. 35, no. 2, pp. 329–359. https://doi.org/10.1177/0032329207300395

Klein, RJ, Nicholls, RJ & Mimura, N 1999, 'Coastal adaptation to climate change: Can the IPCC technical guidelines be applied?', Mitigation and Adaptation Strategies for Global Change, vol. 4, no. 3/4, pp. 239–252. https://doi.org/10.1023/a:1009681207419

Loh, P & Shear, B 2015, 'Solidarity economy and community development: Emerging cases in three Massachusetts cities', Community Development, vol. 46, no. 3, pp. 244–260. https://doi.org/10.1080/15575330.2015.1021362

McCarthy, JJ 2001, Climate change, intergovernmental panel on climate change. Working group II, Cambridge University Press.

Morecroft, MD & Cowan, CE 2010, 'Responding to climate change: An essential component of sustainable development in the 21st century', Local Economy: The Journal of the Local Economy Policy Unit, vol. 25, no. 3, pp. 170–175. https://doi.org/10.1080/02690942.2010.486124

Nel, E 2001, 'Local economic development: A review and assessment of its current status in South Africa', Urban Studies, vol. 38, no. 7, pp. 1003–1024. https://doi.org/10.1080/00420980120051611

Nel, EL & Humphrys, G 1999, 'Local economic development: Policy and practice in South Africa', Development Southern Africa, vol. 16, no. 2, pp. 277–289. https://doi.org/10.1080/03768359908440077

Newby, L 1999, 'Sustainable local economic development: A new agenda for action?', Local Environment, vol. 4, no. 1, pp. 67–72. https://doi.org/10.1080/13549839908725582

Rangarajan, K, Long, S, Ziemer, N & Lewis, N 2012, 'An evaluative economic development typology for sustainable rural economic development', Community Development, vol. 43, no. 3, pp. 320–332. https://doi.org/10.1080/15575330.2011.651728

Rogerson, CM & Rogerson, JM 2010, 'Local economic development in Africa: Global context and research directions', Development Southern Africa, vol. 27, no. 4, pp. 465–480. https://doi.org/10.1080/0376835x.2010.508577

Roseland, M 2000, 'Sustainable community development: Integrating environmental, economic, and social objectives', Progress in Planning, vol. 54, no. 2, pp. 73–132. https://doi.org/10.1016/s0305-9006(00)00003-9

UN WomenWatch 2011, 'Women, gender equality and climate change', Viewed November 29, 2011, http://www.un.org/womenwatch/feature/climate_change/downloads/ Women_and_Climate_Change_Factsheet.pdf

UNICEF Regional Office for South Asia, 2017, Gender Equality: Glossary of Terms and Concepts.

Van der Waldt, G 2016, 'The role of government in sustainable development: Towards a conceptual and analytical framework for scientific inquiry', Administratio Publica, vol. 24, no. 2, pp. 49–72.

Van der Waldt, G 2018, 'Local economic development for urban resilience: The South African experiment', Local Economy: The Journal of the Local Economy Policy Unit, vol. 33, no. 7, pp. 694–709. https://doi.org/10.1177/0269094218809316

Walby, S 2005, 'Gender mainstreaming: Productive tensions in theory and practice', Social Politics: International Studies in Gender, State & Society, vol. 12, no. 3, pp. 321–343. https://doi.org/10.1093/sp/jxi018

Walo, MT 2017, 'Unpacking local economic development: A case study from Nekemte Town and its hinterlands, Oromia, Ethiopia', International Journal of Public Administration, vol. 40, no. 12, pp. 1000–1012. https://doi.org/10.1080/01900692.2016.1177832

Weaver, R & Knight, J 2017, 'Analysis of a multipronged community development initiative in two distressed neighbourhoods', Community Development Journal, vol. 53, no.2, pp. 301–320. https://doi.org/10.1093/cdj/bsx009

Wilson, D & Polter, R 2020, 'Sustainable local economic development indicator framework: A tool for property building redevelopment projects', Community Development, vol. 51, no. 5, pp. 609–627. https://doi.org/10.1080/15575330.2020.1825503

Zalewski, M 2010, 'I don't even know what gender is': A discussion of the connections between gender, gender mainstreaming and feminist theory', Review of International Studies, vol. 36, no. 1, pp. 3–27. https://doi.org/10.1017/s0260210509990489

Chapter 3
Exploring Alternatives in Dealing with Climate Change and Land Based Conflicts

ARIEF WICAKSONO AND ILYA MOELIONO

Introduction: Understanding the Context of Conflict and Climate Change

Climate change has become a concern at the highest level of governments and international organizations. People are increasingly becoming aware that climate change is an inescapably global issue. Still, presidents, prime ministers, and dictators did not give sustained attention to the technicalities of climate data until the emergence of the Kyoto Protocol in 1997 and, later, the UN Framework for Climate Change Conference (UNFCCC) in which more scientists were involved. Until then, negotiations were generally left to mid-level diplomats. They in turn had to rely on their national experts for advice on viable actions to combat and mitigate climate change. To a degree not often seen in international affairs, scientists wrote the agenda for action. A similar picture has been developing in conflict handling at the local level where the negotiation process between the conflicting parties requires the support of experts related to the subject and object of the conflicts, and at the same time requires commitment from legitimate decision makers.

At the global level, the impact of climate change is getting worse, and from a perspective of conflict, this is partly due to many unresolved climate-related conflicts of interest between and within countries. These unresolved conflicts span from the debates on nomenclature and the issue of inequality between developed and developing countries in terms of climate justice to the priority issues of adaptation and mitigation efforts. Therefore, the holding of the COP becomes an arena for conflict resolution through multilateral negotiations in order to resolve differences towards mutual agreement.

The Conflict Resolution Unit: A Brief Background

Partly in response to the global discourse on climate change, in 2015 the Conflict Resolution Unit (CRU) was initiated by the Indonesian Chamber of Trade and Industry (KADIN; *Kamar Dagang dan Industri*) as a contribution to climate change mitigation and the creation of a favorable investment climate. The chamber has been imbedded in the Indonesia Business Council for Sustainable Development (IBCSD) as its legal umbrella and its incubator with the stated intension of becoming an independent conflict resolution institution in the future. It has been established as a platform to resolve conflicts between land-based enterprises such as palm oil plantations or industrial forest estates and local or indigenous communities who live in and around the concession areas.

There are many different reasons these conflicts arise. In many cases, conflict emerges due to uncoordinated licensing processes by the government, which issues forestry and plantation permits in areas where people are already living, dislocating them. As most local communities are farming communities that need access to agricultural land, disputes over those lands are unavoidable. In other cases, the conversion of forests to plantations jeopardizes the livelihoods of indigenous communities who are depending on the forests. While most conflicts are about land and livelihoods, some companies' environmental impacts – such as the pollution and substantial use of water – were also detrimental to the surrounding communities as it reduced the availability of potable water.

During the period 2015–2020, with the support of the UK Climate Change Unit (UKCCU), The Department for International Development (DFID) CRU, in collaboration with 11 professional mediators and assisted by 36 interns has handled a total of 56 conflict cases in the forestry and plantation sectors. CRU's general conflict resolution approach has been mediation which included conducting preparatory research to develop an adequate understanding of the conflicts and build the necessary information base for the parties to constructively engage each other in negotiation, joint problem solving, and planning for the implementation of agreements reached. Through these mediation processes the majority of these cases were resolved with only a few exceptions.

However, working at the site level dealing with one conflict at a time seems a mere drop in the bucket when considering the staggering number of existing agrarian conflicts. For instance, a staff member of the Indonesian Presidential Staff Office related that there are currently more than 15,000 agrarian conflicts reported to their office while only about 1 % of those cases could be resolved as of early 2022 (Usep Setiawan, 2022, personal communication). It is clear that conflict resolution efforts need to be scaled-up through various means and methods not only to accommodate the interests of the parties involved but also for broader societal goals, one of which is the adaptation and mitigation of climate change.

This paper will make the argument for this need to scale-up and intensify conflict resolution efforts from the perspective of climate change and suggest some ideas on the methodology to be considered for this purpose. The paper may also entail wider applications for the management of conflict resolution in land and resource use.

Climate Change as the Result of Conflict

Historically, the Industrial Revolution led to the continuous transformation of agricultural societies into industrial societies characterized by growth-based economics and the exploitation of natural wealth. Since then, climate change and conflict have become intertwined phenomena. Additionally, an unintended side-effect of the industrial revolution has been the division of the world into the industrial West and the global South – the so-called 'Third World' – and the amplification of the division between the world's population of 'haves' and 'have-nots'. This rift is also a major contributor to the continued tensions surrounding the issue of climate change.

At the conceptual level those conflicts inherent to the natural resource-based economic development led to increased carbon emissions. Ultimately, climate change can be understood as the continued tension between unbridled capitalism, social justice, and environmentalism or even conflict between the interests of the present generation and the needs of future generations. At the political level the friction is between growth-based economic investments and environmental and social issues. At the policy level the discord is between science-based long-term decision making, and short-term reactive policies. While at the ground level there are numerous conflicts due to the increased competition over natural resources among many diverse stakeholders.

Thus, as a global issue climate change can be perceived as the result of a myriad of multi-dimensional and multi-stakeholder conflicts at nested levels, which accordingly become the cause of many diverse conflicts at the local level. The question then becomes, with all the complexity and urgency that exists, **how can conflict resolution approaches and methodologies be utilized to contribute to more optimal climate change control initiatives?** How can actors turn existing controversies and debates – that is those conflicts – into catalysts for constructive change for collective efforts to tackle climate change issues? The following section will be an exploration of those questions as an initial attempt to answer them from a practitioner's perspective. Some conflict resolution approaches will be proposed.

Climate Change and Conflict from a Field Level Perspective

In recent decades, Indonesia has achieved impressive economic growth, with poverty levels reduced to a single digit for the first time in the country's history. Today, Indonesia is the world's fourth most populous nation, the world's tenth largest economy in terms of purchasing power parity and is a member of the G-20. Public and private corporations have been investing in the exploitation and development of millions of hectares of land in Indonesia to produce timber, pulp and paper, food, and biofuels.

However, economic success has come at significant environmental and social costs. Expansion of agriculture and timber and oil palm plantations (both legal and illegal) has been a major source of forest degradation and habitat loss. Land-based activities, including mining, agriculture, and forestry have led to problems such as deforestation, pollution, and conflicts over land rights. In the period 2005–2015, Indonesia lost 7 % of its forest cover, a total of 1.4 million hectares (Austin K.G. et al., 2019). Since then, the level of deforestation has decreased, but remains high compared to that of other countries. Rapid land-use change and heavy reliance on fossil fuel energy make Indonesia one of the world's largest greenhouse gasses (GHG) emitters.

This global trend offers Indonesia opportunities to attract foreign and domestic investment, increase agricultural productivity and create jobs, but it also brings potential threats to small-scale producers and indigenous communities. For instance, despite the fact that palm oil production is able to absorb labor in large numbers, it does not follow automatically that the livelihoods of small land holders, plantation workers, and indigenous people around the plantation will be improved. Rapid development and expansion of palm oil plantations have resulted in well-documented, undesirable ecological and social consequences, including river pollution, loss of biodiversity, soil erosion, and a growing carbon footprint. In many places, traditional livelihoods are facing increasing pressure, as the lands previously available for hunting and gathering and traditional agriculture are shrinking in size due to the expansion of the plantations and declining water quality caused by the pollution of agrochemicals makes it difficult to find clean water for drinking and bathing.

To illustrate the situation as it relates to natural resource conflicts: The Consortium for Agrarian Reform (*Konsorsium Pembaharuan Agraria*, or KPA), reported 396 cases of agrarian conflicts affecting more than 1.2 million hectares during 2013, up from 198 cases affecting 300,000 hectares during the previous year. A World Bank study suggests that nearly 25 million hectares of all designated forest lands (or *kawasan hutan*) – more than 20 % of the total forest area, encompassing nearly 20,000 villages – is in conflict due to competing legal claims. In addition, 2015 alone saw 776 conflicts between palm oil companies and local indigenous communities (Putra K. I. and Dunphy M., 2018). In 2020, the NGO 'Sawit Watch' database lists 1053 conflicts in Indonesian oil palm plantations.

Most of those conflicts are corporate – community conflicts which were part and parcel of nearly all land-based development initiatives. The parties to the conflicts have been muddling through their problems; some have been able to settle their disputes amicably, but in a quite a few cases their confrontations have led to violence and even human rights abuses. Still many more cases remain unresolved, festering beneath the surface. Some of those conflicts also caused wider damage. In the years 1997–1998 there were many forest fires in Indonesia scorching an estimated 9.7 million hectares of secondary forests lands, and of course contributing to air pollution and GHG emissions. Many of those fires were set deliberately as a way to claim or reclaim access rights to lands which were disputed between communities and corporations (Applegate, G., et al., 2001).

According to an UNDP report, in the period of one year (2013–2014) forestry conflicts have caused losses valued at around USD 35 million in 12 provinces (UNDP, 2015). Land conflicts also have a negative impact on the environment, because they can hamper efforts to overcome deforestation and to prevent forest fires. Furthermore, disputed lands are also more susceptible to bush fires as the parties do not feel responsible for fire-prevention measures as they have not a clear and secured long-term, tenurial ownership of those lands.

While some government agencies – among others the Ministry of the Environment and Forestry and the Ministry of Agrarian and Spatial Planning/National Land Agency – have formed conflict resolution units, the efforts to deal with those numerous tenurial and natural resource conflict is advancing quite slowly and is even often regressing. Two recent government initiatives serve as examples of political reform processes which many activists and observers view as potentially harmful to the livelihoods of local communities and to the environment. The first was the so-called 'Omnibus Law', which was passed by the Indonesian parliament on October 5, 2020. The Omnibus Law on Job Creation, a reflection of President Joko Widodo's commitment to increasing investment and industrialization in Indonesia, includes several disparate measures, coupled with provisions to simplify business licensing procedures, changes to the existing labor law, more centralized decision making on environmental issues, and a framework for construction of Indonesia's new capital city, amongst other changes (Mulyana A., et al., 2021).

Additionally, Indonesia's geography, historical background, and cultural diversity makes it a country particularly prone to conflicts. It is for this reason that development initiatives in Indonesia should be based on a deep understanding of its particular context.

Ideas for Action: Conflict Resolution, Policy Mediation, and A Conflict-Sensitive Approach to Climate Change Adaptation and Mitigation

The Conflict Resolution Unit (CRU) was initially formed as a private sector initiative to contribute to efforts to control climate change through reducing the incidence of conflicts in the land-based business sector. The CRU's mission entails a very large mandate. The question that often arises is, how could conflict resolution efforts contribute to efforts to reduce GHG emissions? Or, how could conflict resolution and conflict-sensitive approaches help the world to limit global temperature rise to no more than 1.5° C?

Finally, when that contribution needs to be calculated in terms of carbon emission reductions, can such a calculation be made, and if so, would it show that the contribution of conflict resolution to the reduction of the emission of GHG is significant? Perhaps if conflict resolution methods could be harnessed to resolve the international disputes within the global climate change agenda resulting in the desired concerted efforts, then this question of significance would be answered irrespective of the possibility of carbon calculations. However, what about those local site-based efforts?

Given that one of CRU's stated purposes was to contribute to climate change mitigation and was supported by the UKCCU, the assumption was made that the resolution of conflicts over land would lead to more secure land rights. Another assumption was that with secure tenure the parties would be willing to make the necessary long-term investments to improve the management of those lands. Better land management would in turn contribute to the reduction of GHG emissions through the reduction of forest fires, land-clearing using fire, and increased continuous plant cover. However, on several occasions CRU was challenged by its donors and other external stakeholders to justify its conflict resolution activities in terms of climate change, which is to demonstrate how the resolution of land conflicts between corporations and communities contributes to the reduction of GHG emissions.

In response to those challenges CRU made some rough estimates based on the area of disputed lands which were then converted into tons of carbon emission considering possible land use and plant cover in the future. Admittedly those estimations might not really reflect reality as there were too many variables and unfounded assumptions. For instance, in many land conflicts involving plantation concession the total area of the concession was included in the calculations while the actual area disputed were only limited areas along the concession's boundary. These boundary areas amount to only a small fraction of the concession.

Regardless of the impracticality of making carbon calculations of disputed areas, intuitively we are confident that with secure tenure, the likelihood that the parties will manage their lands more responsibly and thus contribute to carbon sequestration will be greater. Hence, in the next section we will propose some conflict resolution ideas which might contribute to the reduction of CO^2 emissions and climate change mitigation and adaptation.

Site Level Conflict Resolution through Mediation

First of all, let's take a look at the field level. On the one hand, there are conflicts that contribute to climate change, such as land and forest conflicts related to deforestation, forest degradation, and forest fires. On the other hand, there are also conflicts caused by climate change, for example when natural disasters cause disasters such as droughts or floods which lead to the displacement of large groups of people. This situation triggers in turn conflicts due to an increased competition for natural resources between the refugees and the local population in the areas to which they flee. Additionally, there are efforts to control climate change which actually triggers conflicts. For example the formation of conservation areas such as natural parks and protected forests can deny communities and corporations access to the lands and its natural resources and cause conflicts.

It is precisely in these cases that conflict resolution initiatives can contribute to the efforts to address and to solve these conflicts. At the field level where the conflicts have already emerged, more conventional conflict resolution approaches and methodologies such as bringing the disputants together in mediated negotiations could be utilized. It is at this level that CRU and other mediation agencies are most often working and in which they have the most experience.

In short, theoretically and intuitively, we feel that through conflict resolution efforts at the site level, land and forest governance would be better managed and would contribute to efforts to reduce emissions from the land sector. This is in the realm of climate change mitigation activities. Here, we can see that conflict resolution works at the site level together with various other factors (confounding factors) and contributes indirectly to the efforts to control climate change.

Area Wide Conflict Resolution

Given the great number of unresolved conflicts as mentioned before, it is easy to presume that CRU and other mediation agencies in Indonesia need to scale-up their efforts by increasing their staff capacity and developing more efficient methodologies. While there is some truth in this, assuming that the necessary resources could be made available, the complexities of many of the conflicts mean that it would be next to impossible to speed up the conflict resolution process. This is because, while general characteristics of tenurial, natural resource, and environmental conflicts can be identified and some methodological generalizations can be made, all conflicts are unique due to their location, the geographic and cultural context, and the nature of the disputants. As such, a cookie-cutter approach to scale up conflict resolution efforts would be inappropriate and impossible.

Perhaps a more efficient approach would be scaling up conflict resolution activities to the landscape level, such as a watershed, a protected area, or a whole concession area. Such a landscape approach to the resolution of environmental conflicts is not new and has been described in the literature, for instance in Nancy J. Marning et al.'s 1990 publication. One attempt to use a similar approach was carried out in the Riung area in the district of Bajawa, in East Nusa Tenggara (Moeliono I., et al, 2001). In this approach an area-wide participatory action research study is conducted by an integrated team consisting of representatives of the various stakeholders. This participatory action research exercise is used to bring the disputants together, to build trust among them, to identify all pertinent issues of contention, to develop a sufficient understanding and information base of those issues and their context, and to initiate thematic negotiations.

However, while such a landscape approach might be more efficient, as it deals with a myriad of conflicts among multiple parties and seeks to formulate integrative solutions across all those issues, it also requires an intricate and thoroughly considered process which includes orientation and training of the representatives of all the parties and their continued involvement during a lengthy process of research, negotiation, and collaborative planning. Its preparation requires networking and engagement with all of the stakeholders' organizations within the landscape.

Policy Mediation

Conflict resolution concepts and methodology have been developed and utilized for a variety different purposes and contexts, including conflict resolution in the field of policy setting by government agencies where it is often referred to as public policy mediation. Public policy mediation creates a forum for deliberative negotiations among government, institutional and private stakeholders, and, when appropriate, the public (Howard S.B., and Podziba S.L., 2012).

CRU – IBCSD initially concerned itself with mediation activities to deal with agrarian and tenurial conflicts at the field-level, mostly conflicts between land-based corporations and local and/or indigenous communities. However, it did not take long to realize that those field-level conflicts are merely the tip of an iceberg of wider historical, political, and global issues as reflected in various national and local government policies and regulations. Forest and land conflicts, for instance, are the result of a complex set of governance problems: contradictory or overlapping jurisdictions, poor spatial planning processes, a historic emphasis on resource extraction skewed to large-scale corporate investments, and uncoordinated permitting and licensing procedures. Also, the lack of clarity over land allocation and law enforcement creates an ambiguous, uncertain working environment at the local level. The increasing awareness of the social, economic, and environmental impacts of these disputes has been well documented.

Thus, as many if not most of tenurial conflicts to be dealt with at the local level are caused by inconsistencies and contradictions among policies and regulations, would it not be more effective and efficient to move 'upstream' to address those policy conflicts first and deal with the field level conflicts later? Wouldn't the resolution of policy conflicts pave the way for the resolution of many field level conflicts? And wouldn't the resolution of field level conflict be easier when the underlying policy issues are managed appropriately initially?

Some conflicts regarding implementation of policies and regulations are mostly horizontal – among district-level agencies that are caused by uncoordinated implementation of policies. These might be resolvable through improved coordination. Of course, resolution of these horizontal conflicts would not be adequate as the field level conflicts still need to be managed.

Then there are the actual policy conflicts, which tend to be more vertical. These are where there are real and substantial contradictions between sectoral policies that need to be mediated. For policy mediation of these conflicts, the mediator or mediation agency and the parties must enter the realm of practical politics. When pertaining to district- or national-level regulations the parties and mediators might additionally need to involve local parliaments. Therefore, efforts to mediate policy conflicts are even more complicated when those conflicts are vertical, or among national, provincial, and district government policies and regulations.

Policy mediation is of course a whole new ball game for both policy makers and mediation agencies in Indonesia. Not only is this idea new for most of the stakeholders, but the methods being used are also still unfamiliar. Therefore, the involvement of independent mediators in policy affairs of the government is often promptly rejected. However, as Indonesia's political climate becomes more open the idea and use of policy mediation might gain more traction.

Conflict Sensitive Approaches to Planning

While conflict resolution efforts can indeed be helpful, what is more important is preventing conflicts from occurring and for that we need to adopt a conflict-sensitive approach in planning the development of activities and business enterprises which might otherwise cause conflicts. If thus far conflict resolution has been carried out reactively like firefighters who rush to the scene when the fire is already burning, it could easily be understood that preventing conflict at the planning stage would be far more useful. As the saying goes, 'an ounce of prevention is worth more than a pound of cure.'

The idea of integrating conflict resolution in the planning process is not new (Forrester, J., 1989), and obtaining free prior informed consent (FPIC) from all stakeholders who potentially might be affected or impacted by a development or business initiative should be an accepted planning principle. For this purpose, several activities which can be carried out as conflict prevention efforts are: opinion polls and social surveys, public hearings, including potential social conflicts as a risk factor in business planning, participatory research, and planning, amongst others. All are well-known methodologies which currently seem to have 'gone out of fashion,' not because they are not effective but because of fickle organizational policies.

As the methods are known and feasible and the benefits clear, conflict prevention efforts could be integrated in all planning activities and policy formulating processes, and conflict-sensitive approaches need to be included in climate change adaptation and mitigation planning. The assumption that we need to agree on is the fact that there is no intervention or policy that is neutral and does not have a negative impact on climate change. Therefore, a conflict-sensitive approach can be a key part of the safeguards of the climate change mitigation and adaptation agenda.

Constraints and Challenges

While the ideas proposed above are either in the process of being tested in dealing with actual cases or are still being considered, many challenges have already become apparent and need to be anticipated, among others:

The Nature of Climate Change

While global warming and climate change as mentioned above have become a well-established idea in development for many stakeholders, at the local level it remains an abstract and intangible issue. Some results of climate change might be observed, its causes however are not readily observable. Although people could experience crop failures due to droughts or have to evacuate their homes due to floods, the reasons for these occurrences might seem to be elusive to many local actors.

Furthermore, even when there is an awareness about the human causes of climate change, as it is a cumulative effect of countless actors dispersed all over the globe, individual behaviors of the parties in specific conflict cases doesn't seem to matter as much. Perhaps, Garret Hardin's classic tragedy of the commons is being demonstrated over again (Hardin G., 1968). In conflict resolution, this abstract nature of climate change has several consequences, some of which are listed in the following paragraphs.

Climate Change Interests vs. Interests of the Parties

One of the principles of conflict resolution is the use of interest-based negotiations rather than allowing for positional bargaining (Fisher, R., et. al. 2011). Of course, in the negotiation process disputants will engage each other in attempts to gain the fulfillment of their interests. However, reducing carbon emissions to control climate change might not be one of the parties' most pertinent issues. While one might argue that controlling climate change is in the interest of everybody, including of the parties to the conflicts over land and natural resources, it is a distant, intangible, and abstract need that might not come up in mediation agendas if left solely to the parties.

The challenge then becomes how to bring climate change into the negotiations. One idea is to involve civil society agencies in the process as is described by Crowfoot (1990), however currently the majority of environmental NGOs in Indonesia are in advocacy and have no history or experience in conflict resolution. Besides, would their involvement be acceptable to the parties? Efforts in education and persuasion will be needed not only for NGOs but also for the parties in the dispute.

Representation of the Public/Climate Change Interests

Conflict resolution fora are ideally the place and process where the parties to a conflict bring their interests and attempt to integrate these interests into an optimal mutually acceptable solution. However, if none of the parties is concerned with climate change interests and willing to represent those, then those interests might not be part of the solutions promulgated at the fora. In other words, if a stable climate is in the interest of the next generation, who is representing future generations at the conflict resolution fora? This might seem far fetched, however, if conflicts are pertaining to large areas of forest land and might have potentially significant adverse environmental impacts, would it not be necessary that those interests are represented? While the law allows for environmental agencies to claim the legal standing to represent the public's environmental interests in the courts of law, the practice cases in which this has happened are still few and far between. Moreover, as mentioned before, the participation of environmental agencies in mediation processes is not yet readily accepted in conflict resolution in Indonesia.

Time Perspective of the Parties

Climate change mitigation is a long-term proposition and might be beyond the short-term interests of the parties. To ensure the sustainability of any agreement between conflicting parties, there should always be the effort to negotiate based on the parties' long-term interests. While this is attempted, it will be highly dependent on the parties' time horizons. Corporate business plans, community growth plans, and even government development plans might only address the forthcoming 5 years, leaving long-term future planning to future policymakers. People might have long-term visions beyond 5 years, but when faced with concrete conflicts and having to prioritize their interests, short-term interests often overshadow long-term effects and visions for the future.

Science as a Reference in Conflict Resolution

The resolution of agrarian and natural resource conflicts is to large extent dependent on how well the parties understand their conflicts, and for this objective science is the reference which might be accepted by all parties. Scientific developments are often useful and necessary when there are disputes over the facts of the case and the merits of proposed alternative solutions. This is especially apparent when the conflicts are concerning adverse environmental impacts: For instance, the health effects of air and water pollution are often not readily observable and can only be shown by scientific field studies.

While climate change is currently well-founded and well-researched and, for most climate scientists, the facts are clear, this is unfortunately not the case for many stakeholders in agrarian conflicts. The causes and effects of climate change and how it is relevant to the case at hand might be puzzling to many stakeholders and bringing climate change issues into the conflict resolution process can be challenging.

Government Agencies as Parties to the Conflicts

As could be inferred from the above, many government agencies are party to agrarian, natural resource and environmental conflicts by the nature of their mandate and jurisdiction. However, in Indonesia many government officials perceive themselves to be above conflicts. The tacit principle to which many officials adhere seems to be that the government should be above all interests and parties and therefore cannot participate as a party in conflict resolution processes. In fact, this perception has encouraged some government officials to act as mediators in the role of an authoritative mediator as described by Moore (1996).

While ideally the government should encompass the interests of all parties and the public, including climate change interests of reducing carbon emissions for climate change mitigation, sectoral agencies have specific missions and responsibilities in their area of jurisdiction which may or may not be in conflict with the interests of other parties. Indonesia is part of the United Nations Framework Convention on Climate Change (UNFCCC) and declared, as part of the Nationally Determined Contribution (NDC), to reduce its emissions by 29 % independently and 41 % when internationally supported by 2030. The Indonesian government further stated its intention to mainstream it's NDC into its development planning. While all government agencies are bound by this policy, in practice at the district level their specific sectoral goals are more prominent and take precedence when their priorities conflict. Even when those sectoral agencies are defending the climate agenda, considering themselves as parties to the conflicts would help create more meaningful solutions.

Additionally, while in some cases the role of authoritative mediator by a government official might be helpful, in others the willingness of government agencies to negotiate as one of the parties at a level playing field might be conducive to the resolution of the conflicts at hand.

The Costs of Conflict

Conducting preparatory research, engaging with and preparing representative negotiation teams of all stakeholders, and bringing those parties together for negotiation and joint planning will obviously incur substantial costs. The questions which inevitably emerge are as follows: Are those costs justified? And who will pay those costs?

If a lengthy and costly process leads only to intangible results such as agreements among the parties and implementation plans of those agreements then the costs seem to be exorbitant, prohibitive, and unjustified. Profit-oriented businesses would not be willing to waste their funds on such unquantifiable results. However, if the costs of conflict resolution processes are compared to the actual long-term costs stemming from the continuance of the conflict, the perspectives might shift.

To begin to explore this question of cost-effectiveness, CRU has initiated some research on the costs of conflict. A study which focused on palm oil plantations found that conflict produces tangible costs which are direct and factual for the companies involved, and also intangible and indirect costs. Those real costs are estimated to be around 70,000–2,500,000 US dollars arising from loss of potential income and loss of opportunity. The risks may include loss of profits, staffing and legal fees, compensation, or rising production costs. These costs may represent 51 %–88 % of plantation operating costs and 102 %–177 % of annual investment per hectare. This percentage comes from the total allocation costs of dealing with conflict. Additionally, social conflicts also carry intangible or 'hidden' costs that range from the equivalent of 600,000 to 9,000,000 US dollars. These costs are indirect losses caused by the risk of escalation of conflict, reputation risk, and the risk of violence against property and people (Barreiro V. et al., 2016).

This study did not include the cost of negative environmental impacts, such as deforestation, soil erosion, water pollution and the reduction of biodiversity which might be caused by the plantations. The companies often consider these costs 'externalities' that either the government or the society at large may accommodate.

While such costs calculations might justify the investment in conflict resolution processes, the next question is 'who should pay?' The immediate answer would be 'the parties to the conflict,' as it is they who would benefit from conflict resolution. Some corporations whose operations are troubled by conflicts might be willing and able to pay, but for most communities in Indonesia this is not the case simply because they do not have the necessary funds. Some communities might consider fund raising activities from their members; however, this too might not be feasible given the economics of the villages. Even when village communities are able to accumulate funding, the amount might be insignificant to that of the corporation, which entails bias: If the process – including the mediators – is paid for mostly by the company, would this not influence the fairness of the process? Will the process be skewed towards the interests of the party whose contribution is the most significant? And if there are any suspicions of possible unfairness, would the parties be willing to participate in the process?

One alternative to consider is the formation of a blind trust fund. The question then becomes who will be willing to pay into this fund as those payments will be non-attributable and what would the governance structure be like? Another alternative is the use of public funds, which require government decision-making and allocation which possibly could be justified in terms of reducing environmental risks and climate change mitigation. However, thus far, none of these alternatives have been pursued.

Conclusions

While the idea of climate change as a problem of the global commons has been promoted by scientists and activists, in reality this idea still seems to be elusive. Regardless, as mentioned earlier, climate change can be constructed as a multi-party, multi-issue conflict among the members of those commons. One of the basic tenets for governing these commons sustainably, as identified by Elinor Ostrom (1990), is the availability of conflict resolution mechanisms and methods that are appropriate and easily accessible to members of those commons, it follows quite logically that climate change issues and conflicts should be dealt with through conflict resolution systems and approaches built into these commons.

While the existing emission accounting methodology and the contribution of conflict resolution efforts at the field level are difficult to quantify in the context of mitigation activities, they have the potential to be elaborated upon in climate change adaptation initiatives through conflict resolution at various levels and preventive or precautionary activities. These activities should utilize conflict sensitive approaches in the planning of various development efforts and business enterprises.

Thus, to keep up with the rate and scale of emerging conflicts, we must explore how conflict resolution methodologies have been emerging and could be adapted and developed to better contribute to climate change mitigation initiatives in the context of Indonesia's development. Many of the conflict resolution principles are clearly applicable to climate change mitigation and adaptation initiatives at wider geographical scales and higher-level policy formulation.

The proposed ideas could contribute to conflict resolution methodologies and conflict-sensitive approaches to efforts to control climate change. This requires commitments from all parties followed by diligent work at multiple levels. Even when these ideas have been implemented, the struggle to mitigate and alleviate climate change does not necessarily become easy. However, these ideas are certainly worth trying in order to contribute to the long-term effort to control climate change.

References

Applegate, G., Chokkalingam, U. & Sutanto, S. 2001, The Underlying Causes and Impacts of Fire in Southeast Asia, Center for International Forestry Research (CIFOR), Bogor, Indonesia.

Austin, K.G., Schwantes, A., Gu, Y. & Kasibhatla, P.S. 2019, 'What Causes Deforestation in Indonesia?', *Environmental Research Letters*, Vol. 14, No. 2, pp. 1–9.

Barreiro, V., et al., 2016, The Cost of Conflict in Oil Palm in Indonesia, CRU-IBCSD and DAEMETER, Kadin Indonesia.

Bellman, H.S. & Podziba, S.L. 2012, 'Public Policy Mediation', *Dispute Resolution Magazine*, Winter 2012.

Crowfoot, J.E. & Wondolleck, J.M. 1990, 'Environmental Disputes: Community Involvement in Conflict Resolution', Island Press.

Davis, M. 2002, The Origins of the Third World: Markets, States and Climate, Corner House Briefing 27, UK.

Fisher, R., Ury, W.L., & Patton, B. 2011, Getting to Yes: Negotiating Agreement Without Giving In, The Harvard Negotiation Project.

Forrester, J., 1989, Planning in the Face of Power, University of California Press.

Hardin, G., 1968, The Tragedy of the Commons, *Science*, New Series, Vol. 162, no. 3859, pp. 1243–1248. https://doi.org/10.1126/science.162.3859.1243

Lohmann, L. 2021, Heat, Time and Colonialism, The Corner House, UK.

Moeliono, I., 2005, Uncovering the Hidden Stones: The Use of Participatory Action Research for Conflict Resolution in Natural Resource Management Around Projected Areas in Flores, East Nusa Tenggara, Indonesia. Cornell University, Ithaca, NY, US.

Moore, C. W., 1996, The Mediation Process; Practical Strategies for Resolving Conflict, Jossey Bass Publishers, San Francisco.

Mulyana, A. et al., 2021, Supporting Conflict Sensitive Development: Insights from Mediation Practitioners in Indonesia, FORCLIME and CRU-IBCSD, Jakarta, Indonesia. https://www.conflictresolutionunit.id/wp-content/uploads/2021/02/English-summary-of-Seka-Sengketa-book_8-Dec.pdf

Ostrom, E., 1990, Governing the Commons: The Evolution of Institutions for Collective Action, Cambridge, UK: Cambridge University Press.

Putra, K.I. & Dunphy, M. 2018, The Challenges of Indonesia's Palm Oil Industry: An Overview, ASEAN Studies Centre, Faculty of Social and Political Sciences, University of Gadjah Mada, Yogyakarta, Indonesia.

UNDP Indonesia, 2015, Executive Summary: The 2014 Indonesia Forest Governance Index, Jakarta.

Websites

20th of September 2013, A brief history of climate change, BBC News, Science, https://www.bbc.com/news/science-environment-15874560. Last visit: August 2021.

Dashboard Sebaran Konflik dan Wilayah Kelola (Dashboard of Geographic Distribution of Conflicts and Managed Areas), *Tanahkita.id*, https://www.tanahkita.id/dashboard_portal. Last visit: September 2021.

Timeline (Milestones), The Discovery of Global Warming, https://history.aip.org/climate/timeline.htm. Last visit: August 2021.

Part II

SUCCESSFUL CLIMATE ADAPTATION MODELS AND EARLY WARNING SYSTEMS

Chapter 4
Urban Climate Resilience and Water: Successful Adaptive Planning, and Early Warning Systems

PASCALINE GABORIT PHD

Climate adaptation refers to any adjustment whether passive, reactive, or anticipatory that can respond to anticipated or actual consequences associated with climate change such as droughts, storms, floods, and heat waves (IPCC 2014). In 2050, 20 % of the entire world population could be at risk of floods among other climate disaster events (OECD 2019), while according to the latest IPCC report, half of the world's population is now living in areas that are vulnerable to climate disasters (IPCC 2022). It is implicitly recognized by international organizations and research that future climate changes will occur and must be accommodated in policy, and that adaptation to climate change is a necessity. Specifically, the role of cities and local governments in climate mitigation and adaptation is considered pivotal in these processes (Grimmond 2007, Oleson et al. 2015, Meerow et al. 2016, Rosenzweig et al. 2018). According to the Global Resilient Cities Network, 82 % of cities globally are located in areas that face a high risk of mortality from natural disasters[1]. Building resilience in cities and communities of all sizes is therefore more critical than ever before. Cities are at a pivotal level to understand territories' socio-economic dynamics, to take appropriate decisions, and to protect local populations and ecosystems. They can implement climate disaster prevention and management measures but can also develop adaptive urban planning measures and implement Early Warning Systems towards their communities to strengthen urban resilience.

Urban resilience can be defined by the cities' capacity to respond adapt to and recover from the pressures and crises related to climate change and other changes: subsidence, demographic growth, poor land management, insecurity or attacks, economic downturns, social unrest, unsustainable use of resources, and declining ecosystems (Ziervogel 2017, Gaborit et al. 2021).

---

[1] https://www.rockpa.org/project/global-resilient-cities-network/

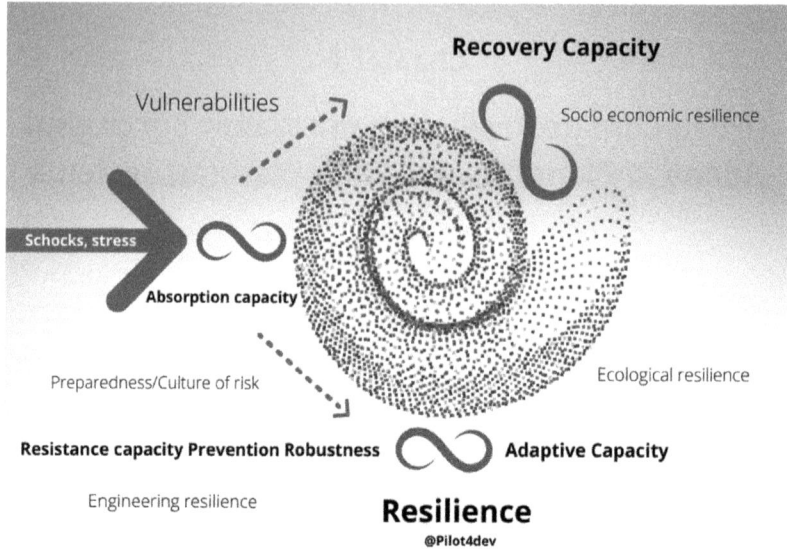

Fig. 1: Urban Resilience (Source: Author)

Resilience is a cross-disciplinary concept referring to the ability to cope, recover and build back from stress, shocks, and crises (Brassett et al. 2013, Brinkmann et al 2017). Originating from the world of physics and sciences where it merely meant 'bounce back', it has been applied to psychology, geography, physics, and ecosystems (Gundersson, Holling, 2003, Folke 2006) and has since been widely used in other areas such as disaster management. In this area, three forms of resilience are distinguished: engineering resilience, ecological resilience, and socioeconomic resilience (Gundersson, Holling, 2003, Folke 2006). See Fig. 1.

Urban resilience can however be mainly considered as a goal, or an objective 'that is not merely responsive to predicted climate impacts, but that also fosters proactive and systemic approaches to preparing for the unexpected and indirect effects of global change' (Moench, 2011: 21). Its very broad scope makes it difficult to operationalize by very clear actions or programs (Meerow et al. 2016).

We assume in this chapter that the challenges ahead of climate adaptation in cities and urban resilience are tremendous, as climate change is having increasing impacts. We argue that we can learn from worldwide successful examples in adaptive urban planning, grey and green solutions, technology in the form of Early Warning Systems (EWS), and people-centered approaches (contingency plans) to find tools for vulnerable areas in different parts of the world including low-lying coastal areas in South Asia, Southeast Asia, Latin America, Europe, and in the Caribbean.

This chapter is therefore presenting possible solutions and tools or examples before we find key features and lessons on how to strengthen urban climate resilience in cities exposed to hazards, including those cities exposed to multiple hazards. We also argue that the most successful examples rely on a good articulation between the local resilience methodologies developed, and national ambitious climate adaptation and contingency plans. We also claim that more attention should be given to medium-sized cities, and that there will be inevitable trade-offs and sacrificed areas and interests in the long run. Finally, the question of engaging with local populations, residents, and communities raises many questions on the possibility of a fair implementation of climate adaptation programs.

The chapter will consider the different trends in climate adaptation, especially with regard to floods, in Part I. Then Part II will focus on the successful examples in Rotterdam and Singapore for creating resilience to floods and water provision, and in Cancún for the resilience to hurricanes and floods. Part III will elaborate on the possibility of implementing Early Warning Systems in order to prevent disasters. Part IV will finally approach several discussion points such as the communication between local and national strategies, the cascading consequences of disasters, the role of insurances, the difficulties in finding criteria for trade-offs and decision makers, and finally the complexities in developing more people-centered and communities-based approaches to climate adaptation.

I. Towards Adaptive Planning Systems

There is no single definition of the resilience concept, but there are necessary boundaries to it. The table below summarizes the main applications, uses, and limitations. In this chapter the resilience concept is considered as an aim for cities to be better prepared to respond to climate change disasters, and to anticipate risks, especially with regard to potential floods.

Tab. 1: Resilience in a nutshell (Source: author)

Resilience	
Possible uses	Disaster management including pandemics or floods, climate adaptation, economic shocks, and external shocks including conflicts or emergency situations.
Criticisms	Buzz word, covering both a process, a trait or an outcome, approach putting the responsibility on the impacted groups, may lead to a confidence excess and a decrease of risk anticipation (Titanic syndrome), may serve the interest of some groups and ignore the questions of power and justice.
Induced by the use of the term	Importance of the concept to deal with uncertainty, shift from understanding resilience to active resilience building, growing emphasis on transformation/changes, and on measuring resilience.
Orientations	Negotiated resilience, economic resilience, socio-economic resilience, resilience to disasters, environmental resilience
Disciplines Covered	Social, environmental, climatic, economic architecture and infrastructure planning, engineering.

The UN Sendai Framework for Disaster Risks Reduction, adopted in 2015, sets out four priority actions to counter this trend: (1) understanding disaster risk; (2) strengthening disaster risk governance to manage disaster risk; (3) investing in disaster risk reduction for resilience; and (4) enhancing disaster preparedness for effective response, as well as to "Build Back Better' in recovery, rehabilitation, and reconstruction. The way in which climate adaptation is implemented by the different strategies at the national and at the local levels show, however, different realities and priorities. This framework that succeeded the 2005 Hyogo declaration is a key text for disasters' management. Other more specific texts on flood prevention and management exist in several regions. In Europe notably, Directive 2007/60/EC on the assessment and management of flood risks entered into force on 26 November 2007. This Directive requires national governments to assess if all water courses and coast lines are at risk from flooding, to map the flood extent and assets and humans at risk in these areas and to take adequate and coordinated measures to reduce this flood risk. Additionally, this directive also reinforces the rights of the public to access this information and to have a say in the planning process. The OECD also published, in 2019, a checklist for action on the prevention of floods, mentioning steps such as the identification of institutional and communication gaps, policy gaps, funding gaps, capacity gaps, accountability gaps, amongst others (OECD 2019: 19).

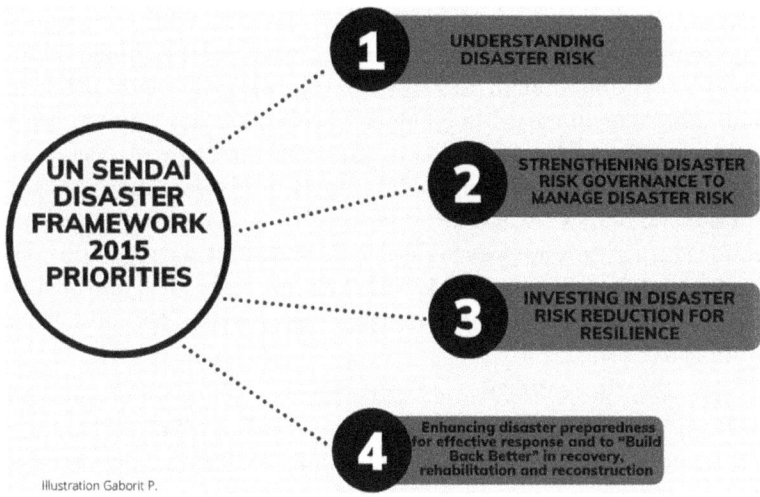

Fig. 2: The UN disaster framework priorities

A focus on security and public health: In the most developed (high income level) countries, there is an increasing focus on the adaptation options linked to the security sector (police, army, defense) to protect the public utilities and critical infrastructure (including metro or underground stations and power plants) from major floods, storms or heat stress. This is the case for instance in Europe. These measures are developed by the national or regional tier level- for instance prefectures in France- in cooperation with the local authorities and the national governments. The security measures can include evacuation plans and routes and the construction of dikes in vulnerable regions in order to reduce damages. Heat waves (such as the ones in Europe in 2003 and 2011) have shown that public health was a major consequence and risk during similar climate events. The protection of the industry sector, such as power plants, refineries, and chemical industries, should be a major security concern to reduce the industrial disaster risk, a potential spillover from the climate risk. The question of whether houses should be constructed in flood-prone areas is a recurrent question after major floods, like the 2021 floods in Liège.

The protection of public utilities: Protecting public utilities is also important in relation and related to security. Power shortages can occur in a possible shortage of cooling water during periods of high temperatures and low precipitations (de Bruin et al. 2009). It is equally important that overhead electricity transmission poles and high tension cables are strong enough to resist extreme weather events (de Bruin et al. 2009). As discussed in Chapter 1, floods are also causing risks to underground water reservoirs and can create problems to secure safe and sufficient drinking water. Public schools, hospitals, public transport, and nursing homes are also key infrastructure necessitating special protection during extreme weather events such as heat waves, storms, and floods. Anticipatory joint work with first aid responders and security forces and awareness raising actions are necessary to develop efficient contingency plans.

Transport: The OECD study (2014) on the influence of flood scenarios in Paris shows that the Paris underground metro could have been entirely flooded in case of a centennial major flood if no action was taken. Tunnels and roads could also be flooded and heavily hamper the traffic. Solutions include the installation of sensors, prevention such as dikes and water catchment areas, and the development of contingency plans, including measures to reduce the inundation of tunnels and underground parking lots. The ports and harbors, especially those located in deltas and the maritime ports, are also playing an increasing role in climate adaptation as they are more exposed to storms and sea level rise (Ecoport, Life Adapt Island project).

Long-term protection also including agriculture and ecosystems: In the long run, the adaptation of the agriculture irrigation to better cope with increasing heat waves and droughts will be necessary, along with precautions and responses to additional floods, storms, and strong winds. Possible solutions include embedding changes into more resilient crop varieties, improvements of water irrigation, and other choices. This also includes an adaptation in the sectors of forestry and fisheries, which are explored in the next chapter.

Finally, the *questions of preparedness, and culture of risks* are also reflected into the adaptation of insurance schemes to cope with future flood disasters. There are some key successful examples in adaptative urban planning to prevent floods. We will go in this chapter through the case studies of Rotterdam, Singapore and Cancún.

II. Successful Examples of Resilient Adaptive Planning for Floods and Water Provision

There are successful examples of urban resilience, especially in flood adaptation. The former 100 resilient cities network from the Rockefeller Foundation, which has split into the Global Resilient Cities Network[2] and the Resilient Cities Catalyst[3] Organization, is developing models and methodologies with worldwide cities to become more resilient, particularly amongst large cities. The local authorities network ICLEI working on sustainability is also increasingly developing models and methodologies for climate disaster risks management (Fig. 6) and urban resilience (Fig. 1).[4] However, the current organizations do not always include an analysis of the interrelation between the local and the national policies.

Successful examples include many cities, but the most integrated ones come from cities exposed to multi-hazard threats and which have adapted their water management systems in anticipation such as Rotterdam and Singapore. Additionally, these cities have benefitted from a strong national policy in the area. Three case studies will be presented to reflect on different approaches and systems, which could be replicated in other cities. The first case study will present Rotterdam's holistic approach to flood prevention. This case study will highlight the steps and tools implemented by the municipality in the Netherlands. It will be followed by the case study of Singapore on integrated water management and flood prevention. Finally, the third case study will present Cancún resilience to hurricanes, strong winds, and floods, and the 'build back better' approach developed after Hurricane Wilma in 2005.

> **Case study 1 Rotterdam/Netherlands holistic approach to floods' prevention**
>
> Located on a delta between the Rhine and the Maas river, Rotterdam is the second largest city in the Netherlands, one of the most densely populated area (over 1.2 million people over the region), flood-prone, home

---

[2] Global Resilient Cities Network – Rockefeller Philanthropy Advisors (rockpa.org). https://www.rockpa.org/project/global-resilient-cities-network/
[3] Resilient Cities Catalyst (rcc.city). https://www.rcc.city/
[4] Home – Resilient Cities 2019 (iclei.org). https://resilientcities2019.iclei.org/

to the largest port in Europe and contains 50 % of the Netherlands' chemical industry. If disaster were to strike, the population would be immediately affected. In particular the territory is highly vulnerable to floods. The infrastructure has been adapted to tackle heavy rainfall, but also prolonged drought and heat waves, groundwater flooding and subsidence. An urban adaptation program of collective design has been set up to accelerate the adaptation measures in public areas as well as on private land. The measures undertaken include tidal banks, water catchment areas, soil stabilization and the creation of a rainwater collection system separated from sewage. At the same time, flexibility is used to adjust the approach based on new forecasts and real time information on floods. The approach focuses mainly on measures at neighborhood level but it also creates dialogue platforms for the involvement of all municipal stakeholders.

Adaptation of the sewage system and stabilization of soils: The Vijverhofstraat, was one of the first streets to become climate-proof when the sewer was replaced[5]. The sewer system in this street has been replaced and a rainwater sewer and a water storage system have been installed simultaneously. The clean rainwater is therefore collected separately from the sewage water. This system, in which the sewage water is separated from the rainwater, is also part of the municipal sewerage plan, and should in the long run cover the entire city. However, this does involve an underground space challenge to have two separate sewer pipes. Since 60 % of the urban land is private, the residents have also been involved in this approach, with measures to maintain foliage in their gardens and facilitate soil absorption and rain collection. In some flood-exposed areas like Agnieseburt, water storage is being installed above the sewerage system, which ensures that the groundwater level remains stable.

Water catchment areas and tidal banks: Tidal banks and water catchment areas have been created along the Maas river (e.g. Nassauhaven) to prevent floods from creating damages to the surrounding housing areas. The tidal banks were also 'greened' with flood-resilient plant species. The Rain Garden is another example of a green area with rain barrels holding rainwater. The rainwater from the gardens and rooftops is collected in a temporary pond. Thanks to the greenery, the water can then slowly sink into the ground. Part of the rain from the Hofbogen roof is also collected in smart rain barrels in the form of the letters ZOHO. The water in the rain barrels is released to the garden with a delay in dry times.

---

[5] A project LIFE urban Adapt (co-funded by the European Union) has been set up by the municipality – Rotterdam is also one of the leaders of the 100 resilient cities network, newly global resilient cities network.

Innovative water Houses: 18 Havenlofts were placed in the Nassauhaven. The floating houses located in the harbor move with the ebb and flow, of 1.5–2 meters twice a day. These houses inconize flood resilient houses with a futuristic representation, but above all pilots the testing and development of new adaptive technologies. The water houses are built sustainably with solar panels and heating via a biomass installation. They also have their own water purification system, and therefore do not rely on the sewerage system[6].

Inclusive governance: The Municipality of Rotterdam, the Delfland Water Board, and Evides Water Company, the Schieland and Krimpenerwaard Water Board, the Hollandse Delta Water Board are involved in the climate-proof Rotterdam master plan that has been extended over the years. There are emergency plans drafted together with the residents in the different districts such as the municipal emergency document *Rotterdams Weerword*. Civil engagement is ensured throughout the entire climate-proof adaptation model process through a 'blueprint approach'.

The link with national spatial policies: National, regional, and local government tiers are primary actors in Dutch Spatial planning and in the floods' prevention. These actors are responsible for flood management and civil protection (Hegger et al. 2017). The Netherlands is indeed an emblematic country of the protection against floods as whole regions, like Flevoland, have been reclaimed from the sea through major public works. Flood protection through dikes, dams, and embankments has been the dominant strategy over the last century (Hegger et al. 2017). Two-thirds of the countries live in dike-protected areas (Pieterse et al. 2009 quoted by Hegger et al. 2017), and the residents living in these areas are entitled to a high safey level (higher than in other countries). National and regional water authorities ensure that the norms are respected. Other residents living in un-embanked areas, however, are not protected by dikes. According to Hegger et al. (2017), they would not be protected by the same security measures and are not necessarily aware of the risks and of the situation. Most responsibilities regarding protection lie primarily at the city's level (Security regions Act of 2010) with the mayor and aldermen (burgermeester and wethouder). If necessary, these responsibilities can be scaled up to the level of one of the 25 security regions,

---

[6] www.urbanadapt.eu

and ultimately to the national level and to the homeland minister or the Minister of the Interior (Minister van binnenlande zaken) (Hegger et al. 2017). As an example, for the Rotterdam-Rijnmond Safety Region (a safety region gathers city boards and operational services), the Regional Emergency Dispatch Center and the Regional Crisis Center coordinate disaster and emergency response and management. The region has 80 disaster contingency plans and is prepared for 29 worst case scenarios[7]. People are informed of disasters and appropriate responses through text alerts or information available on their mobile phones. They also receive information via Dynamic Route Information Panels (DRIPs), which are information screens in public locations such as roads or trains stations. Another successful example of resilient planning to floods and water management is the case study of Singapore. Singapore is indeed the most well-known model of resilience in the area of water management, both in preventing floods and in securing access to clean water.

---

[7] Disaster response management – The Dutch approach – Institute for Environment and Human Security (unu.edu). https://ehs.unu.edu/news/news/disaster-response-management-the-dutch-approach.html

Case study 2 Singapore: transforming water from a threat to an asset

Singapore is often presented as a model for resilience since its independence from Malaysia in 1965. This was mainly due to the country's booming economy (Brinkman 2017), its development model, and its water management (Ying Shan, Wong, 2014). Similarly, to Rotterdam, the city's water management is entirely aligned with the national strategy.

Water Master Planning as a national strategy: In the early 1960s Singapore experienced a demographic boom, leading to the proliferation of informal housing characterized by poor living conditions. High monsoons led to the flooding of low-lying areas. Efforts to improve the public drainage and to create reservoirs for floods were already undertaken during the British colonial times. According to Ying Shan and Wong (2014), one of the valuable engineering legacies implemented by the British was already the separation of the stormwater collection and the sewage systems. This system allowed the drains to capture only rainwater. Soon after its independence, the national leaders prioritized water management in the short- and long-term strategies for the country. Indeed, during the second world war, the water pipes had been used as targets by Imperial Japan, in order to create water shortages and eventually force Singapore to surrender. 'Water was considered as an existential issue by the country, a matter of life and death' Ying Shan and Wong (2014). In 1972, a water master plan was approved with first measures implemented to create water catchment areas. One of the strong steps taken was to stop the activities and industries which caused water pollution. This was later extended to pig farms, which produced a polluted discharge. Between 1977 and 1987, public works were undertaken to clean up the Singapore river and the Kallang Basin, and to relocate the slum dwellers. The master plan also restricted the land use dedicated to urbanization to 34.1 % of the land area. In the 1970s, the Ministry of the Environment worked to enable the necessary infrastructure (separation of the wastewater), before residential and industrial development could spread further. This created many challenges for the planners and engineering teams, as the urbanization could possibly lead to subsidence and further floods. These public works achieved success in the 1990s, with the modernization of the water management, the improvement of public health as Singapore became Malaria free in 1982, and the opening of a mass transit public transportation system (Ying Shan, Wong, 2014).

Innovative water management processes: recycling used water and desalination: A key element in the water self-sufficiency of Singapore was the innovation of the reverse osmosis process to treat wastewater. The idea to recycle used water was indeed considered since the 1970s. However, the technologies were made available only in the 1990s, when it became possible to produce membranes for the water osmosis at lower costs. With more reliable and more affordable technologies, the PUB (Public Utilities Board) stepped up its efforts to make the recycling of used water a reality in the country. In 2000, intensive trials were performed by the PUB to create water reclamation plants. The tensions with the neighboring Malaysia indeed appeared as a driver to secure the water supply. To increase the acceptance of the process by the residents the terms 'used' or 'waste' water have been replaced by the term 'new water' or 'NEWater' to become more widely accepted. Media and public communication have also been necessary to develop public confidence and acceptance in the process. New water is nowadays used for industrial purposes, cooling, industrial estates and retails. It is also regularly checked to meet the international drinking water standards.

In the 1990s, Singapore also started exploring desalination as an innovative method to provide clean water. The first desalination plant was opened in 2005 as a public private partnership process, with the capacity to produce 30 million gallons of water per day. The experience of Singapore in innovative water management resulted in its water technology companies being able to export their technologies, generating billions in investments in Singapore (Ying Shan, Wong, 2014).

Tackling the security of water supply and floods' prevention at the same time: Since the 1970s and the 1980s, drains and canals were quickly constructed to alleviate and prevent floods. In order to prevent floods in low lying areas, the Marina barrage has been a key milestone. It was initially designed as a drainage project from the river to avoid subsequent floods. With the improvement of the membrane technologies, however, and falling costs of operation and maintenance, it soon became possible to treat the raw water from this urban catchment area stretched over one sixth of the total land area and to transform it into freshwater reservoirs. As the cities developed, tests and trials were performed to make sure that the water reservoirs would not deteriorate with the growing urbanization. Initiatives were explored to ensure that the infrastructure and reservoirs would be resilient to aging and further urban developments. Maintenance has become a challenge in the subsequent decades. Finally, climate scenarios such as storms and increased rainfall are regularly tested to optimize the resilience and maintenance of the water system. The 'Source-Pathway-Receptor' approach developed by the public utility board considers more targeted ways to reduce floods.

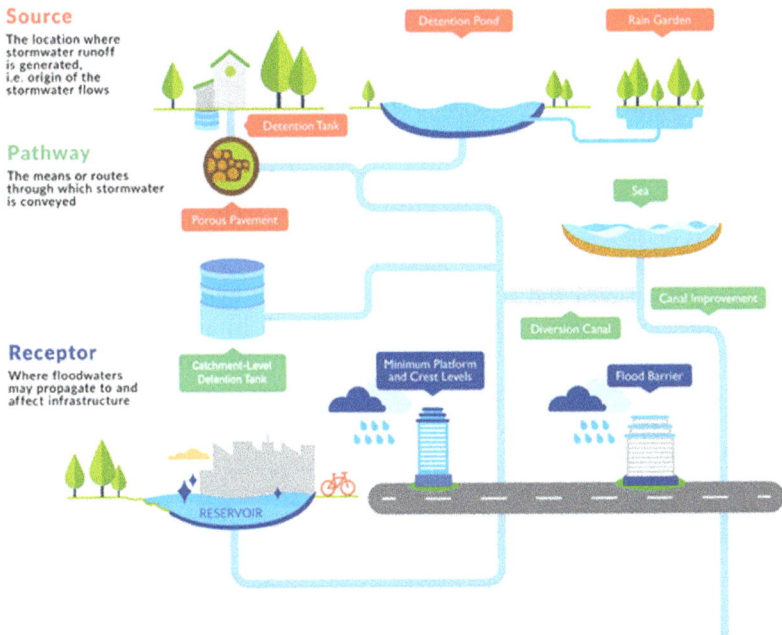

Fig. 3: Source: Pathway receptor methodology to prevent floods in Singapore PUB Public Utilities Board and water national agency. PUB https://www.pub.gov.sg/drainage

The blue infrastructure model: Singapore later used the basins created to develop a 'blue development' model in the city and valorize the secured waterways as assets for the city with water activities such as kayaking and dragon boating. The network of canals created for water drainage were at first utilitarian before being revamped by the Water Bodies Design Panel in the 1990s to improve their design and combine them with beautiful green and blue infrastructure such as increased foliage, mangroves, and water banks. This became possible as the water reclamation plants became more centralized which freed some land for other uses. In 2006, the public utilities board launched the Active, Beautiful, Clean Waters (ABC) program, which aimed at providing new community spaces and integrating waterways and reservoirs within the urban landscape.

Multi-stakeholders' cooperation: The innovative model would not have been possible without the close cooperation of all public entities. New functional lines had to be opened between the Public Utilities Board, which manages gas, electricity, and the water supply, NEWater management agency, the Ministry of the Environment, civil society and the private sector. To streamline these functions, the public utility Board was reconstituted in 2001 to embed a holistic view of all water functions. It later became the national water agency.

Water saving measures and citizens engagement: Since the 1970s, campaigns raising awareness also occurred around the topic 'water is precious' to alert the public of the importance of sustainable water use. The aim was to nurture water-saving habits among citizens and to reduce waste. Tariffs were also adapted and adjusted. Additionally, the water agency strictly controlled the amount of water leakage through its own network and pipes (Ying Shan, Wong, 2014).

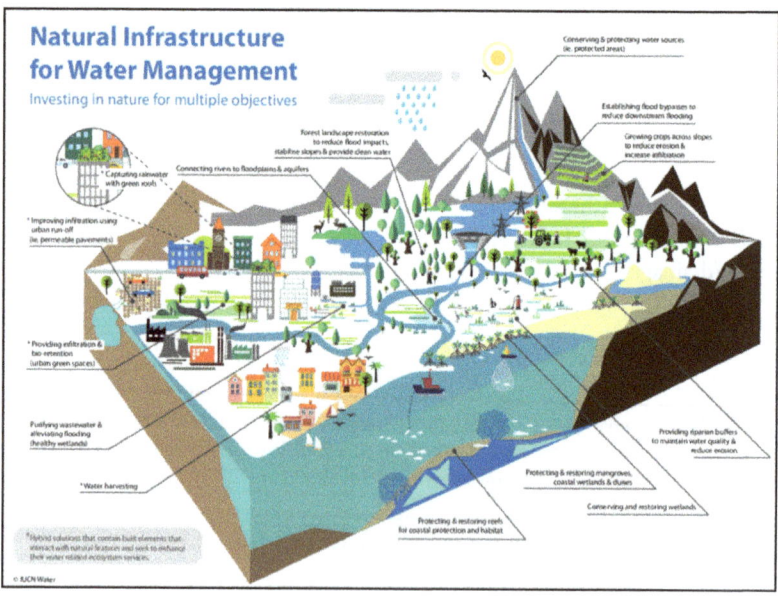

Fig. 4: Natural infrastructure for water management Source: IUCN (as part of 'WISE-UP to Climate' project). See http://www.iucn.org/theme/water/our-work/wise-climate

Both Rotterdam and Singapore have managed to transform water from a threat to an asset, and to develop efficient water management models. Other models are being developed by organizations like the International Union for the Conservation of Nature (See Fig. 4). For smaller cities, the combination of the blue and green infrastructure can also enable a resilience system transforming water from a threat, for example the potential of floods, to an asset by investing in the development of water as water sources, blueprint, waterways, and a natural cooling system. According to Disaster's management cycle (Fig. 5), prevention and preparedness are followed by the important steps of 'response' and 'recovery' or 'build back better'. An increasing number of cities like Cancún, Mexico, have been hit by hurricanes and disastrous events. This has been followed by discussions on how to 'build back better' and increase the prevention and preparedness to future climate events in order to reduce the loss and damage they will cause.

In 2015, the UN Sendai Disaster Management Framework was approved in Japan popularizing the concept 'build back better' which is part of the 'recovery' of the UN Disaster management cycle. The term 'build back better' was first introduced to UN at the United Nations Economic and Social Council (ECOSOC) in July 2005 by former United States President Bill Clinton, as Secretary-General's Special Envoy for Tsunami Recovery[8].

Shinzo Abe, Prime Minister of Japan went a step further at the opening speech of the Third UN World Conference on Disaster Risk Reduction and commented that the use of the concept 'build back better' was common sense to the Japanese people, coming from historical experiences in recovering from disaster and preparing for the future, and it has become an important part of the culture of Japan[9].

During the negotiation period for the Sendai Framework, the concept of 'Build Back Better' was proposed by the Japanese delegation as a holistic concept which states: 'The principle of 'Build Back Better' is generally understood to use the disaster as a trigger to create more resilient countries, areas and societies than before. This was through the implementation of well-balanced disaster risk reduction measures, including physical restoration of infrastructure, revitalization of livelihood and economy/industry, and the restoration of local culture and environment'. The

---

[8] https://www.un.org/press/en/2005/ecosoc6166.doc.htm
[9] https://www.sprep.org/news/third-un-conference-disaster-risk-reduction-opens-welcome-news-pacific-islands

concept therefore added into the Sendai Framework (UNISDR 2017, Hallegate et al. 2018).

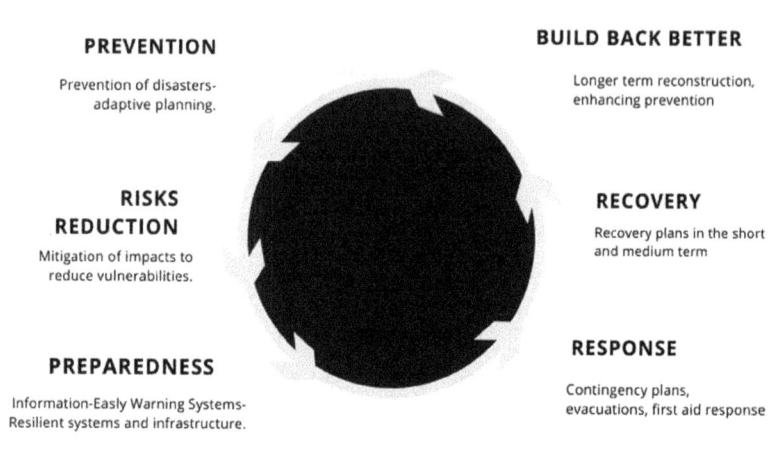

Illustration: Gaborit Pascaline.

Fig. 5: Disaster's management cycle (Illustration: Gaborit P.)

Case study 3 Cancún 'build back better' resilient approach after Hurricane Wilma.

In 2014, the ISOCARP task force gathered to identify recommendations for the 'build back better' options of the city of Cancún. Cancún, of the Yucatán region of Mexico, was hit in 2005 by hurricane Wilma, the third category 5 hurricane of the season. It was recorded as one of the most intense tropical cyclones in the Atlantic Ocean. Although the number of fatalities was not very high, the insured damages to the city were estimated between 5 and 8 billion (USD). It was estimated that among other deteriorations 95 % of the tourism infrastructure was damaged. Some of the beaches were washed away, and many houses were lost in the storm. The task force analyzed the impacts of the different elements to increase future resilience:

(1) Winds: Strong winds can incur damages to buildings because of debris, or because of the structural failure intrinsic to the buildings. During

Wilma, the glass debris were mostly responsible for the damages (Nan et al. 2014). The failure of walls can then lead to floods. Roofs and boarding signs were especially vulnerable to strong winds. Some roofs were torn off, with signs bent and broken. Wind damage was also responsible for most of the electrical shutdown. Trees, wood, and landscape materials became dangerous projectiles. Wooden traditional houses were more vulnerable than concrete modern housing areas. However, even in thick, concrete walled buildings, roofing gravels, glass debris, and other missiles destroyed several windows, propelled by the hurricane's winds. The task force proposed mitigation measures against strong winds such as protection of windows with wood sheets, or the adoption of a building code for resilient buildings similar to the model used in Florida. (2) *Waves*: According to Nan et al. (2014), there are three types of wave damages: storm surge, over wash, and beach transport. *Storm surge* is water pushed by the force of the storm. Wilma created high waves because of storm surge, the waves in the ocean became higher. The winds also contribute to the storm surge effect. As a mitigation solution, the slope of the seabed near the shore greatly influences further damages on the coast. If the seabed approach to the shore is deep, there is less friction, and the waves do not build up. In the case of Cancún, the seabed was shallow, and the storm built into powerful waves. Offshore reefs are important mitigation solutions for this, as they allow the waves to break before reaching the shore. *Overwash* is the term for wave-driven sea water that floods internal areas. Overwash was not a problem in Cancún, as the mainland is located on a bedrock. The overwash fortunately mainly flooded the areas of mangroves. *Beach transport* consists of the erosion and deposition of sand along the shoreline caused by the waves. The wind is also influencing the direction, which explains that in the case of Wilma the northern beaches were mainly affected. In addition to causing damages to beaches, the erosion destroyed soil and construction foundations. (3) Floods. The geology in Cancún is composed of fractured and porous soil and limestone. Water drained slowly in many locations, as the rainfall increased. It was further strengthened by water runoff in urban areas. Flooding was a major problem.

Mitigation solutions included the creation of sea walls, although they have a limited capacity, and the development of engineering studies. One efficient preventive action is also to prohibit constructions on the beachfront. Mangrove swamps also offer interesting water catchment areas to be flooded by the over wash[10], while the restoration of sea dunes can mitigate the impacts of storm surge and soil erosion (beach transport). Resilient construction materials to glass debris have been studied for the

---

[10] See also project LIFE Adapt Island.

> reconstruction and recovery to cope with possible strong winds linked to future events.
>
> The ISOCARP task force also recommended to preserve the mangrove swamps which acted as water reservoirs. Within urban areas, the green infrastructure can, when combined with native plant species, facilitate the water drainage and lessen the floods. Finally, early warning systems seemed equally necessary to further protect the populations and evacuate vulnerable areas. It took millions of dollars and several years to rebuild the tourist area[11]. In areas like Punta Nizuc and Punta Cancún where the hurricane hit the hardest, cement cubes have been placed as a water break to soften the waves before it hits the shore, and also to limit further erosion. More than 6.1 million cubic meters (1.3 billion gallons) of sand was pumped back to Cancún's shorelines using two large dredgers. The sand came from two offshore sand reservoirs so that the seascape was restored as well as the tourist areas. Many of the solutions proposed by the task force, such as the use of resilient materials, the mangrove swamps, and the restriction of urbanization, could, however, not be fully implemented.

The 3 case study examples on floods prevention and disaster management in Rotterdam, Singapore and Cancún, lead to related discussion points in part IV of this chapter to highlight the conditions for urban resilience in terms of governance decision making processes but also stakeholders' involvement. These discussion points seem necessary to understand why urban resilience is difficult to scale up, and why the successful examples cannot always be replicated elsewhere. The discussion points related to Part II and Part III will be presented in Part IV of this chapter. Part III will approach early warning systems as successful examples of **preparedness and response** to climate-related events. Two examples of successful early warning systems will be developed in the two next chapters (with one example taken from India-Tamil Nadu, and another one taken from the Lawu mountains in Indonesia). Part III will briefly present how early warning systems can be fully considered as solutions and parts of urban resilience models.

---

[11] https://www.latimes.com/archives/la-xpm-2005-nov-13-fi-Cancún13-story.html

III. Early Warning Systems as Tools for Disasters' Risks Reduction and Preparedness

Since the Hyoggo framework, and the later adoption of the UN Sendai Disaster Framework, Early Warning Systems have become a major element in disaster risk reduction (Diab 2021, Marchezini 2020, Cumiskey 2015). Early warning for specific hazards like floods can indeed activate preventive action to avoid disasters and fatalities, for example through evacuation and contingency plans. They can prevent loss of life and reduce the economic and material damages linked to disasters. EWS are defined by the UNISDR as 'integrated system of hazard monitoring, forecasting and prediction, disaster risk assessment, communication, and preparedness activities systems and processes that enable individuals, communities, governments, businesses, and others to take timely action to reduce disaster risks in advance of hazardous events' (UNISDR 2017).

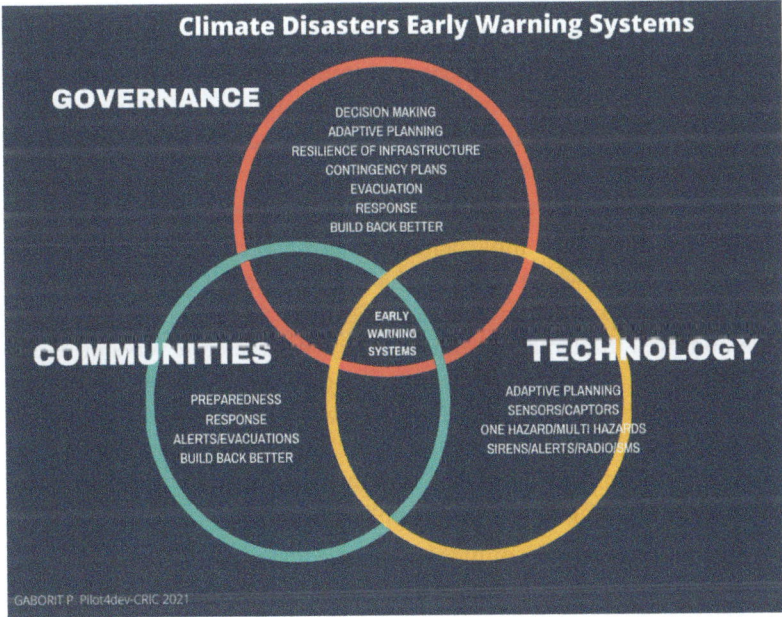

Fig. 6: The interactions of Early Warning Systems

One of the major goals was to 'substantially increase the availability of and access to multi-hazard EWS and disaster risk information and assessments to the people by 2030'. The UN Sendai Disaster Framework details several actions but does not expressly detail the modalities of their implementation (Zia, Wagner 2015, Marchezini 2020).

According to the UNISDR, a people-centered Early Warning system is based on four main pillars: (1) risk knowledge and assessment, (2) monitoring, (3) communication, and (4) response. Risk knowledge and assessment implies data collection and analysis of hazards, threats, and vulnerabilities – physical, social, economic, and environmental, amongst others. Monitoring implies that the resources and capacities for collecting and checking evolutive data are able to make informed decisions based on the analysis of real risks. Communication is the process of sharing data, alerts, and warning situations in real time. Response capability is the preparedness capacity to know how to act, and is dependent on the resources, skills, and networks that people have (Marchezini 2020; Cumiskey 2015).

This is why, as shown in Fig. 6, there is a necessary interaction between technology such as the use of sensors, captors, and real-time communication technology and governance systems like decision making processes, anticipatory measures, the adoption of contingency plans, and evacuation routes, along with the communities for their response and preparedness.

As Marchezini has shown in his sociological analysis and case study (2020), it is difficult to implement integrated and efficient Early Warning Systems from the whole chain to the risk knowledge and assessment to the response capability. He noticed that the risk assessment was sometimes not well implemented and that the EWS therefore was not implemented in the most vulnerable areas (Marchezini 2020). In other cases, he noticed that the communication part was not emphasized enough in the EWS programs and during disasters. Finally, the maintenance costs can have a detrimental effect on the monitoring part (2), especially when risk prevention is no longer on the political agenda.

Fig. 6 illustrates this complexity and the reticulated nature of the different factors. To the contrary of the 4 steps mentioned previously, EWS is an interactive process consisting of necessary complex interactions, including the important coordination between the local, the regional, and the national government tiers, the disaster agencies, the weather agencies, and the communities. Another difficulty is that the Early

Warning Systems need to focus mainly on one single hazard, for instance floods, but cannot integrate all hazards such as floods, earthquakes, landslides, and tropical storms. Additionally, according to Marchezini (2020), floods represent the most frequent hazard in the studies on EWS (almost one third), followed by tsunamis, and droughts (less than 10 %). This illustrates the difficulties in applying the system to multi-risks, while our CRIC project shows that territories and cities are often exposed to more than one single hazard (e.g. the city of Mataram in Java is both exposed to floods and to earthquakes, sometimes simultaneously)[12]. The cascading risks are also not directly visible in EWS: For example, the disruptions of water and flood supply after a storm or during a period of drought.

There are many successful examples of Early Warning Systems. They can be either based on technology, rely on community involvement, or, in the most optimal scenarios, both. Two successful examples include one reflecting the research by Natividad and Mendez in the city of Ilagan, Isabela in the Philippines (Natividad, Mendez 2017) where ultrasonic sensors were installed to prevent floods, a simple technology system, and the other one is based on community warning in Beringin Watershed Semarang in Indonesia, with a community-oriented system. Although there have not been real evaluations on the systems during the time of disasters, these examples appear to be successful in their processes so far. In the Ilagan city, the system created an efficient alarm system using ultrasonic sensors with an automatic relay to communities. In Semarang, at least four community disaster groups have been set up to strengthen preparedness for tidal floods. They involve young people and NGOs. Finally, in the two following chapters of this book, two more complex models and case studies will be detailed further with the case of the SMART early warning system in Tamil Nadu, India (Satyagopal et al.) and the EWS to prevent landslides in the Lawu mountains Indonesia (Kersapati et al.). While we have presented successful models, we also need to thoroughly consider the conditions and factors of their success and highlight probable boundaries. This will be discussed in Part IV of this chapter. The discussion points in Part IV relate to all three previous parts (the theory of adaptive planning and urban resilience, the successful models of urban resilience and adaptive planning, and the early warning systems).

---

[12] Policy Brief Mataram, 2020, accessible on www.resilient-cities.com

IV. Discussion Points

The discussion points approach the first three previous sections of the book on the theory of adaptative planning, the successful examples of urban resilience, and the early warning systems to understand that are the conditions of success and replication (4.1.), and why all successful and innovative solutions are not already widely implemented or the different obstacles (4.2.)

The discussion includes the question of multi-level governance which will be more developed in the conclusion's chapter of this book, and the importance of the articulation between the national and the local strategy, as a condition of success. It discusses the necessary trade-offs, and the proposed methodologies to weigh the different factors before decision making. Finally, it approaches the question of the cascading effects of climate disasters, the complex role of insurances, and the questions related to civic and community engagement.

a. The Factors for Success of Urban Resilient Adaptive Systems

A multi-level governance and articulation between the national and the local levels is needed to have a successful model of climate adaptation. The UN Sendai Disaster Framework declares under Priority 3 'investing in disaster risk reduction for resilience' that both national and local level government agencies must '(f) promote the mainstreaming of disaster risk assessments into land-use policy development and implementation, including urban planning, land degradation assessments and informal and non-permanent housing, and the use of guidelines and follow-up tools informed by anticipated demographic and environmental changes' and '(g) promote the mainstreaming of disaster risk assessment, mapping and management into rural development planning and management of, inter alia, mountains, rivers, coastal flood plain areas, drylands, wetlands and all other areas prone to droughts and flooding, including through the identification of areas that are safe for human settlement and at the same time preserving ecosystem functions that help reduce risks' (UNISDR 2015, p 15).

This need of multi-level governance highlights *the necessary articulation between the local and national strategies in terms of spatial planning and climate adaptation*. In the cases of the Netherlands and Singapore, the national strategy integrates water management and the prevention of floods and access to clean water, etc., as key elements, which enable the cities to develop successful water management solutions. Some authors, such as de Bruin et al. 2009, argue that water management is a key sectoral issue, and that it is 'necessary to check whether current institutions,' like the water boards and local governments, can handle the problems incurred by climate change, and whether they have the necessary means to implement the adaptation actions. Multi-stakeholder cooperation is a requirement and is advocated largely by research to strengthen adaptation strategies. 'Improved coping capability of institutions can be achieved through the cooperation of institutions and stakeholders in new alliances, or through embedding adaptation policies systematically into existing institutions. Problems may however arise, when the urgency of the local and regional institutions differ from the national level' (de Bruin et al. 2009). This shows the necessity for the national government to have a key coordinating role in the area of water management and the protection against floods and other climate-related disasters. But it also shows that local efforts will be vain without a strong, ambitious national adaptation policy and massive investments in the most exposed areas.

b. The Obstacles Towards Urban Adaptive Resilient Systems

The successful examples of urban resilience are mainly found in the literature regarding large cities while ***medium-sized cities are lagging behind***. The successful examples detailed in this chapter are mainly referring to large cities. In many countries, when a climate adaptation strategy is adopted, the focus is principally set on large cities and capitals as they are home to a larger part of the population and they concentrate critical infrastructure, including the economies, industries, and government buildings. Medium-sized cities are given less attention in national spatial policies, unless they are considered a priority or a particularly vulnerable target. In medium-sized cities, the question of funding for successful climate adaptation models is especially pertinent, but so is the access to smart technologies (such as Early Warning Systems). In terms of population awareness in France, for instance, campaigns against heat waves or possible floods are more exemplary in large capital cities (Paris, for instance) than in medium-sized or smaller cities exposed to hazards such as floods.

Another issue is **trade-offs, or whether and under what criteria we weigh adaptation options**: As we have seen in Chapter 1, the local and national authorities will be increasingly confronted with trade-offs and will have to choose between different options as climate disasters become a threat. The example of Cancún shows clearly that priorities need to be set. For example, there should be a prohibition of constructions on the beachfront versus tourist economic developments, protections of informal housing areas instead of protections of modern concrete buildings from glass debris, and the creation of sea walls versus mangrove swamps. It also raises the question of who is bearing the costs of future adaptation needs. Several authors, for example,e de Bruin et al. 2009, have proposed a system to score the adaptation options with criteria using a multi-criteria analysis. This scoring still seems very relevant and could be used by national and local governments to develop efficient adaptation systems. Efficient systems include four sections: (1) the importance of the option in terms of gross benefits, (2) the urgency of the option, reflecting the need to act sooner and not later, (3) the no-regret characteristic of the option (the option is worthwhile whether the disaster occurs or not), and (4) the co-benefits to other sectors and domains, typically public health or biodiversity, and the effect on climate change mitigation, for instance with changes in land use. However, the results of their study (poll) ranked first very general ideas such as 'integrated nature and water management' before costly options like 'abandoning the low-lying parts of the Netherlands', or specific options like the 'selection of agricultural crops'. This reduced the implementation and replication potential of the proposed options to many cities. However, the criteria and similar methods could be used for more specific solutions. One of the main trade-offs can be, for instance, the balance between security and mitigation: Not all proposed climate adaptation measures reinforce climate mitigation. The city's massive use of air conditioning to cool critical infrastructure during heat waves is also reinforcing climate change and indirectly exposing more vulnerable groups. Finally, the question of multiple hazards is also raised: there are numerous examples and illustrations of successful climate adaptation strategies for floods, but fewer 'strategies for storms, strong winds, fires, and droughts without including other hazards which could be linked to climate change in terms of public health, pandemics, biodiversity, food supply, and crops, amongst others (See also the IPCC report 2022).

When disasters cannot be prevented, the other steps of the disaster management cycle (preparedness/response and recovery) come into play (see Fig. 5). However, the consequences of disasters and especially the spiraling consequences have an impact, which can be cascading or compound, and these consequences can create disruptions in the longer term, raising questions like the role of insurances.

Cascading Consequences of Disasters: The questions of power and of disruption of community livelihoods are also important in the disaster management process, as is the prevention of fatalities. In 2005 hurricane Katrina in New Orleans did not lead only to fatalities, but also to regional migrations. The storm caused a wave of refugees who had lost their houses into surrounding areas and states. In the case of floods in the region of Liège in July 2021, the first aid responders raised complaints that they could not find enough boats for rescue operations, and when boats were available, they were not equipped with enough lights to continue overnight[13]. Disasters create a moment of disruption and disorganization with possible additional losses during the rescue operations. Long-term disruptions can prevail in entire areas if humanitarian and recovery aid does not quickly follow. In 2022, Hurricane Batsirai, followed by Emnati in Madagascar also demonstrates that the number of fatalities (around 120), does not entirely reflect the whole humanitarian crisis. The hurricane also caused the destruction of an entire city (Mananjary), supply chain and quality problems in the water and food supply, the destruction of agricultural crops, salinization of water reservoirs, and disrupted access to medicine. The number of indirect victims and the impacts of disasters on the disruption of livelihoods is often underestimated in the whole Disaster Cycle Management process (see Fig. 5). This leads us to consider what actually constitutes preparedness, and the culture and different perceptions of risk.

The role of insurance is overlooked: The literature and research are putting increasing emphasis on insurances as a way to reduce disaster risk (in the preparedness and culture of risks, in case EWS would not be sufficient). Hegger et al. insist on the role of insurances to restrict the 'loss' of residents in case of damages, in the disaster risk prevention, preparedness,

---

[13] Documentary Investigations: inondations le temps des sinistrés – La nuit du déluge: December, RTBF La une 2021. https://www.rtbf.be/auvio/detail_investigation?id=2820945

and response (see Fig. 5). Nevertheless, the example of the floods in the Liège Province in 2021 shows that the question of insurances is very complex (CRIC interview Partagence November 2021)[14]. The victims who lost a house were very dependent on the reactivity of the insurance and the expertise of the loss. New inequalities appeared between the groups of people who were covered by the insurances and the others. As climate change will create more uncertainties, the development of insurance schemes to cover the communities against climate disaster-related loss seems like a rather logical evolution. The role of insurance, however, will not compensate for human loss, the disruption of livelihoods, or the spiraling consequences of the disasters such as the outbreak of diseases and subsequent industrial disaster. Caution should therefore be exercised when integrating insurances as a holistic solution for climate adaptation. A strong involvement of residents and communities can support better resilient adaptive systems in a way that goes beyond infrastructure and technologically feasible solutions.

c. People's and Residents' Engagement in the Disaster Management Cycle

People's and residents' engagement in the disaster management cycle: Diverse authors insist on the necessity of better participatory approaches and civic engagement in climate adaptation and in Early Warning Systems (Hegger et al 2017, Marchezini 2020). Hegger et al show that local governments in the Netherlands have been active in involving residents in stormwater management, and in the prevention of heat stress. It was also the case in the city of Paris, after the heat wave in 2003, which led to over 25,000 casualties (in France), and after the implementation of Early Warning Systems to prevent floods on the river 'La Seine' (Landau, Diab 2016). The most active examples referred to in several cities involve the creation of solidarity networks in case of crisis, or the construction of artisanal dikes with sand banks along the rivers (CRIC project database). Hegger et al. show that community engagement and awareness were, however, very low among the local population both for heat stress and floods (Hegger et al. 2017). This is also exemplified by the floods which occurred in the Liège region in july 2021. These floods showed that the citizens and the local governments were not

---

[14] The interviews are available on the blog. https://www.pilot4dev.com/blog

entirely prepared for such a disaster. Solidarity networks have emerged in the 'post disaster' phase, especially with regard to the distribution of food, but they were considered scattered and not entirely well-planned (interviews). Some scams have also been noticed: food vendors had exaggerated prices, and insurance intermediaries proposed abusive contracts[15]. This shows that resilience can only happen if different sectors work together to be prepared: The institutions in charge, the first-aid responders, and the communities must work in a complementary way. However, the preparedness also necessitates a certain level of risk acceptance.

The recognized difficulty also often concerns how to communicate about risks (Hegger et al. 2014). It is indeed difficult to engage residents in a context of uncertain risks in terms of time, such as when the disaster will occur if it occurs, and scale, like where the disaster is likely to happen if it happens. If there is no 'preparedness' or 'culture of risks,' the residents will not see the need to be involved in contingency actions. Here again, some interesting examples of risk communication and EWS have been developed punctually. For example, the LIFE Franca project (which ended in 2019) has set up interesting guidelines in this area to communicate on the risks of floods in the Alps mountains in Europe[16]. The project has identified the need to work in particular with schools and young people to inform them of climate disasters and preparedness without the 'fear effect.' Young people are indeed identified as an important target group in community's engagement by several experts (Cumiskey 2015). Social contexts where citizenship and participation exist are very diverse, unstable, and complex as are the different systems of governance (See Fig. 6 in the case of EWS). The possibilities of engagement, when they exist, are not equally distributed between the different social groups and can exclude entire vulnerable communities. More worryingly, the vulnerable, poorer, and sometimes more densely populated neighborhoods can be 'sacrificed' in climate disaster management, leading to an unequal exposure to hazards between social groups. On the contrary, the successful examples of Rotterdam, Singapore, and Cancún show that inclusive and participatory approaches are necessary within a strong national and local cooperation.

---

[15] Documentary Investigations: inondations le temps des sinistrés – La nuit du déluge: December, RTBF La une 2021. https://www.rtbf.be/auvio/detail_investigation?id=2820945

[16] https://www.lifefranca.eu/en/

Conclusion

The examples of Rotterdam, Singapore and Cancún show that a successful urban adaptive planning can transform water from a threat, with floods, sea level rise, and water pollution, to an asset when creating a clean water supply, useful waterways, and blue and green jobs. Water management can create many opportunities to develop innovation and valorize the cities with a blueprint model ensuring security from disasters and sustainability of water by reducing its pressure, pollution, and smart water use. This shows that solutions do exist. However, these success stories show that many conditions are required, including a strong national strategy and articulation, a political will, strong monitoring, maintenance, and continuous improvement of the systems and used technologies. These factors are complemented by a necessary strong risk analysis, multi-stakeholders' cooperation, and citizens' or residents' engagement. A good complementarity between top-down and communities' initiatives can be successful if there is enough knowledge and understanding about the risks (in a culture where these risks are accepted). In a context of cities increasingly exposed to hazards, early warning systems can become powerful tools, but cannot replace a whole strategy of risk analysis, monitoring, communication, and response, as promoted by the UN Sendai Framework Disasters Framework. The interrelation between governance systems, technologies, and communities will become central (Fig. 6). Climate adaptation and integrated approaches are necessary for the creation of resilient territories. The question of implementation regarding the inclusion of medium-sized cities and poorer neighborhoods, the anticipation of the cascading and spiraling effects of disasters, the strategies against multi-hazards, especially for droughts and storms, and the costs for recovery programs remain unsolved and inaccessible for a large part of the world's hazard-exposed areas.

References

Amri M., Jamalianuri, R.D. 2020, 'Urban Analysis Report Samarinda', https://www.resilient-cities.com/en/?preview=1&option=com_dropfiles&format=&task=frontfile.download&catid=41&id=34&Itemid=1000000000000, last accessed 31.03.2021

Brassett J., Croft S., & Vaughan-Williams N. 2013, 'Introduction: An agenda for resilience research in politics and international relations', *Politics* 33(4), 221–228

Brinkmann H., Harendt C., Heineman F., & Nover J. 2017, *Economic Resilience: a new concept for policy making?* Bertelsmann Stiftung research paper

de Bruin K., Dellink R.B., Ruijs A., Bolwidt L., van Buurn A., Graveland J., de Groot R.S., Kuikman P.J., Reinhard S., Roetter R.P., Tassone V.C., Verhagen A., & van Ierland E.C. 2009, 'Adapting to climate change in the Netherlands: an inventory of climate adaptation options and ranking alternatives', *Journal Climatic Change* 95, 23–45. https://doi.org/10.1007/s10584-009-9576-4, last accessed 10.02.2022

Cumskey L. 2015, 'Case Study: Flood Early Warning System', UN-Water Annual International Zaragoza Conference

Diab Y. 2021, 'Note on the CRIC tools' CRIC project common drive

Dwitama P. 2021, 'Policy Brief', https://www.resilient-cities.com/id/?preview=1&option=com_dropfiles&format=&task=frontfile.download&catid=45&id=70&Itemid=1000000000000, last accessed 31.03. 2021

Folke C. 2006, 'Resilience: The emergence of a perspective for social-ecological systems analysis', *Global Environmental Change*, 16, 253–267

Grimmond S. 2007, 'Urbanization and global environmental change: Local effects of urban warming'. *Geography Journal* 173, 83–88.

Gunderson L.H., & Holling C.S. 2003, *Panarchy, understanding transformations in human and natural systems*, Washington, Island Press

Hallegatte S., Rentschler J., Walsh B 2018, 'Building back better: Achieving resilience through stronger, faster, and more inclusive post-disaster reconstruction' World Bank/GFDRR https://www.gfdrr.org/sites/default/files/publication/Building%20Back%20Better.pdf

Hegger D.L.T., Mees H.L.P., Driessen P.J., & Runhaar A.C. 2017, 'The roles of residents in Climate Adaptation: A systematic Review in the case of the Netherlands', *Environmental Policy and Governance*. https://doi.org/10.1002/eet.1766

IPCC (Intergovernmental Panel on Climate Change) 2014, 'Climate Change 2014: Impacts, adaptation, and vulnerability' IPCC Working Group II Contribution to AR5, IPCC Cambridge UK and New York USA

Intergovernmental Panel of Climate Change (IPCC) 2022, Climate Change 2022: Impacts, adaptation and vulnerability, Summary for Policy Makers

Landau B. & Diab Y. (Eds.) 2016, 'Résilience, vulnérabilité des territoires et génie urbain', Presse des Ponts

Marchezini V. 2020. 'What is a sociologist doing here?: an unconventional People centered approach to improve warning implementation in the Sendai Framework for Disaster Risk Reduction', International Journal of Disaster Risk Science 11, 218–229, https://doi.org/10.1007/s13753-020-00262-1, last accessed 12.02. 2022

Meerow S. & Newell J.P. 2016, 'Defining urban resilience a review', in Landscape and urban planning, March 2016. https://doi.org/10.1016/j.landurbplan.2015.11.011

Nan S., Reilly J., & Klass F. 2014, 'Planning to mitigate hurricane damage and to ensure the continued growth of Cancún and its Region', ISOCARP Cancún urban Task Force Team, https://www.isocarp-institute.org/knowledge-base/planning-to-mitigate-hurricane-damage-and-to-insure-the-continued-growth-of-Cancún-and-its-region/, last accessed 11.02.2022

Natividad J.G., & Mendez J.M. 2018, 'Flood monitoring and Early Warning System using Ultrasonic Sensor', *IOP Conference Series: Materials Science and Engineering* 325, 012020

OECD 2019, 'Applying the OECD principles on water governance to floods: a checklist for action', https://doi.org/10.787/d5098392.en

Oleson K.W., Monaghan A., Wilhelmi O., Barlage M., Brunsell N., Feddema J., Hu L., Steinhoff D.F. 2015, 'Interactions between urbanization, heat stress, and climate change', *Climatic Change* 129, 525–541

Pisano U. 2012, 'Resilience and sustainable development: theory of resilience, systems thinking and adaptive governance', European Sustainable Development Network ESDN Quaterly report n°26

RESIN Supporting decision-making for resilient cities 2015 report 'state of the art Report: resilience adaptation and disaster risk reduction', www.resin-cities.eu

Reveuny R. 2007, 'Climate Change Induced migrations and violent conflicts', *Political Geography* 26, https://doi.org/10.1016/j.polgeo.2007.05.001, last accessed 06.02.2021

Rosenzweig C., Solecki W., Romero-Lankao P., Mehrotra S., Dhakal S., & Ali Ibrahim S. (Eds.) 2018, Climate Change and cities: Second assessment report of the urban climate change research network, Cambridge University Press, New York

Salim W., & Hudalah D. 2020, 'Urban governance challenges and reforms in Indonesia: Towards a new urban agenda', in Dahiya B. & Das. (Eds.), *New Urban Agenda in Asia-Pacific*, Springer Nature Singapore Pte Ltd.

UCLG ASPAC 'Early Warning Systems Assessment on the situational information for Bandar Lampung City', December 2021 CRIC project drive

UNISDR (United Nations International Strategy for Disaster Reduction) 2017a, 'Build back better in recovery, rehabilitation and reconstruction' in support of the UN Sendai Framework for disaster reduction, https://www.unisdr.org/files/53213_bbb.pdf

UNISDR (United Nations International Strategy for Disaster Reduction) 2017b, 'Terminology: Basic terms of disaster risk reduction', https://www.unisdr.org/we/inform/terminology#letter-e

Worowirasmi T.S., Waluyo M.E., Rachmawati Y., & Hidayati Y. 2015, 'The Community-based flood disaster risk reduction in Beringing watershed in Semarang city', *Jurnal Wilayah dan Linkungan*, 131–150, http://dx.doi.org/10.14710/jwl.3.2.131-150, last accessed 12.02.2022

Ying Shan W., & Wong M. 2014, 'Water and cities: the Singapore story', in ISOCARP Review n° 10, p 243, https://www.isocarp-institute.org/knowledge-base/water-and-the-city-the-singapore-story/, last accessed 11.02.2022

Zia A., & Wagner C.H. 2015, 'Mainstreaming early warning systems in development and planning processes: Multilevel implementation of sendai framework in Indus and Sahel', *International Journal of Disaster Risk Science*, 6, 189–199, https://doi.org/10.1007/s13753-015-0048-3, last accessed 12.02.2022

Ziervogel G., Pelling M., Cartwright A.A., Chu E., Deshpande T., Harris L, ..., Zweig P. 2017, 'Inserting rights and justice into urban resilience: A focus on everyday risk', *Environment and Urbanization*, 29(1), 123–138

Chapter 5

System for Multi-hazard Potential Impact Assessment, Alert, Emergency Response Planning and Tracking (SMART) in Tamil Nadu India

KORLAPATI SATYAGOPAL, PhD, ITESH DASH, PhD,
JOTHIGANESH SHANMUGASUNDARAM, PhD, RAMRAJ
NARASIMHAN, PhD, SUBBIAH ARJUNAPERMAL, PhD

Introduction

Advances in weather prediction have increased the lead times of weather forecasts, for a suite of products from 3 to 5-days weather forecasts to 5–10-days medium-range forecasts, 20–25-days extended-range forecast, monthly to seasonal climate outlooks, and climate projections years and decades into the future. Highly accurate weather forecasts of up to 3 days lead time have contributed to the reduction of casualties associated with extreme weather events, through early warning[1]. As forecast lead time increases, however, uncertainty also increases[2]. Although with reduced reliability, probabilistic forecasts are useful when applied in a seamless manner within a risk management framework. For disaster managers, seasonal climate outlook is useful for disaster risk management planning, 20–25-days forecast for logistics planning, and 5–10 days

---

[1] World Meteorological Organization (2021). WMO Atlas Of Mortality And Economic Losses from Weather, Climate and Water Extremes (1970–2019). WMO-No. 1267.

[2] Slingo J., & Palmer T. 2011, Uncertainty in weather and climate prediction. Philosophical Transactions of the Royal Society A: Mathematical Physical and Engineering Science 2011 Dec 13, 369(1956): 4751–4767. https://doi.org/10.1098/rsta.2011.0161. PMCID: PMC3270390.

forecast for hazard impact mitigation, to complement the 3–5 days forecast for securing lives and property[3].

In most countries, National Meteorological and Hydrological Services (NMHSs) provide forecasts of various time-scales ranging from daily, weekly forecasts, 10-days for various geographical and administrative regions using Numerical Weather Prediction (NWP) models. The National Disaster Management Organizations (NDMOs) could make use of these forecasts to evaluate potential impacts that extreme weather, such as heavy rains, could bring to a locality, and guide the preparation of options for mitigating potential impacts. For example, heavy rains could exceed the threshold for a specific location, and result in landslides or flooding. Areas at risk could be identified using a risk map, prepared from layers of information on geographic features and land use, administrative boundaries, population, livelihoods, infrastructure and facilities, historical hazards and impacts, hazard threshold, and capacities (e.g. practices, resources) to absorb losses or recover from hazard impacts. Once areas at risk are identified, targeted watch/alert/warning could then be issued, evacuation and response teams could be readied, and resources for response could be prepared.

Most NDMOs do not have such systems to make use of the NMHSs' medium-range forecasts for assessing potential impacts and impact management options to guide their operations. It was in this context that the Regional Integrated Multi-Hazard Early Warning System for Africa and Asia (RIMES) developed a proposal in 2012 to collaborate with the Department of Disaster Prevention and Mitigation (DDPM) of the Government of Thailand in the development of a multi-hazard potential impact assessment and management system, or the SMART, which stood for System for Multi-hazard potential impact Assessment and emergency Response Tracking. SMART was intended as a web-based system for assessing potential impacts of a hazard using weather forecast information, and the evaluation, generation, and dissemination of impact management options. The system also was to act as a data management system for managing and processing weather, disaster risk, and emergency response resources data.

---

[3] Hellmuth M.E., Mason S.J., Vaughan C., van Aalst M.K., & Choularton R. (Eds). 2011, A better climate for disaster risk management. International Research Institute for Climate and Society (IRI), Columbia University, New York, USA.

SMART in Thailand was designed to comprise the following components: (i) Hydro-meteorological observation data management system; (ii) *Weather* forecast data management system; (iii) Hazard and impact data management system; (iv) Physical, social, and economic data management system; (v) Emergency response options system; (vi) Dissemination system. SAHANA EDEN, an open-source disaster management system, was to be used in developing the Emergency Response Options and Dissemination subsystems. While the SMART development in Thailand did not move forward, majorly due to non-availability of exposure and vulnerability data, a high-level delegation from the Government of Tamil Nadu, India during its visit to RIMES in Thailand recognized the potential for adopting this methodology to address disaster risks in Tamil Nadu, and showed keen interest in its co-development.

Tamil Nadu, India: Tamil Nadu, a well-developed State of India, ranking high on most socio-economic indicators, is spread over 130,058 km^2 in the southern peninsular India. It has a long coastline of over 1076 km and a densely populated coastal area that is home to over half of its 72 million population. About half of the state's fisher population live within 2 km of the coast, while more than 40 % reside within 1 km. An average of 1–2 cyclonic events and depressions annually affect these coastal populations during the Northeast Monsoon period, while several low-pressure/depressions in the adjoining Bay of Bengal result in intense spells of heavy rains over 3–4 days leading to wide-spread floods and inundation[4].

Tamil Nadu is also exposed to various other seasonal hydrometeorological hazards such as heavy rainfall, floods, landslides, storm surges and droughts, which affect several thousands of people and result in casualties of people and livestock annually, in addition to other non-seasonal hazards such as coastal erosion, forest fires and tsunamis. Heavy rains, storm surge and local flooding damage over 5000 houses annually, across the coastal districts of Chennai, Kancheepuram, Tiruvallur, Cuddalore, Thanjavur, Nagapattinam, Thiruvarur, Pudukottai, and Tiruchirappalli, where over 0.45 million hectares of land is estimated as prone to floods[5]. Landslides affect predominantly the hills of the Nilgiri range, as well

---

[4] Tamil Nadu State Disaster Management Authority. Physical Vulnerability. TNSDMA. https://tnsdma.tn.gov.in/pages/view/physical-vulnerability

[5] Tamil Nadu State Disaster Management Authority. Hydro-Meteorological. TNSDMA. https://tnsdma.tn.gov.in/pages/view/Hydro-Meteorological

as other hills of Salem, Erode, Coimbatore, Vellore and Dindugal districts. The state having only 3 % of water resources of the country is severely affected by droughts due to erratic rainfall during Southwest and Northeast monsoon which has an adverse impact on hundreds of thousands of families directly or indirectly dependent on agriculture, especially in the districts of Dharmapuri, Madurai, Coimbatore, Ramanathapuram, Salem, Tiruchirapalli, Tirunelveli and Kanyakumari[6].

The socio-economic impacts of these climate-related disasters pose a huge burden on the state and its people[7], threatening hard-won development gains. In the last 30 years (since 1990), the economic damages due to 25 large-scale hydrometeorological disasters, that are captured in the global database (EM-DAT[8]) itself amounts to over USD 2.2 billion, though the actual damages are likely to be much larger, as it does not include the localized/smaller-scale disasters that are far more numerous, cumulatively significant and affecting the most vulnerable[9]. The more official data base Desinventar records over 8,500 incidences of hydrometeorological/climatic disasters between 1968 and 2011[10].

Urban flooding has been severe in recent times, especially in the capital city, Chennai, affecting the poorer households disproportionately. Patnaik et al, 2019, have assessed the economic impact of the 2015 Chennai floods and have highlighted the greater hardship faced by the poor households and those in weaker houses structures, reflected in only a marginal increase in their consumption expenditure, compared to over 32 % increase overall; and further identified the lingering impacts extend beyond a year, resulting in reduced savings, assets purchase[11]. Patankar,

---

[6] Geethalakshmi V., & Ganesan D. 2008, Impact of climate change on agriculture over Tamil Nadu. Climate Change and Agriculture Over India, 80–93.
[7] United Nations Development Programme (UNDP), Bureau for Crisis Prevention and Recovery (BCPR). 2013. Climate Risk Management for Agriculture Sector in Tamil Nadu State of India. New York, NY: UNDP BCPR
[8] EM-DAT, The International Disaster Database, Centre for Research on Epidemiology of Disasters – CRED. EM-DAT, CRED/UCLouvain, Brussels, Belgium – www.emdat.be
[9] UNDRR. Understanding Disaster Risks. PreventionWeb. https://www.preventionweb.net/understanding-disaster-risk/disaster-losses-and-statistics
[10] UNDRR. Tamil Nadu (India)-033. Desinventar Sendai. https://www.desinventar.net/DesInventar/country_profile.jsp?countrycode=033&lang=EN
[11] Patnaik I., Sane R., & Shah A. 2019, Chennai 2015: A novel approach to measuring the impact of a natural disaster. NIPFP Working Paper No. 285.

2019 estimated average household losses during the 2015 Chennai floods, ranging from INR 7,500 (*97,5 Dollars*) to INR 125,000 (*1625 Dollars*) in addition to loss of identification, bank documents and certificates causing severe disruptions to their lives beyond the direct losses[12].

The Government of Tamil Nadu is cognizant of these impacts recognizing in its 12th Five Year Plan (2012–2017)[13] the direct and indirect paths for transmission of weather risks across primary, secondary and tertiary sectors, and the potential for reducing the direct and indirect losses through the incorporation of a 5–10 days forecast system into decision making. A comprehensive and proactive decision support system becomes a vital aid to strengthen policy making, enhance operational efficiencies and in empowering the communities to address short-, medium-, and long-term risks.

Objectives: The objective of this paper is to shed light on the development of the Tamil Nadu System for Multi-Hazard Potential Impact Assessment and Emergency Response Tracking (TN-SMART) as an operational tool for the Government of Tamil Nadu, and its implementation in Tamil Nadu for managing and processing weather, disaster risks and emergency response resources data for use during the different phases of disasters. The paper also describes the quantum leap leading to development of other improved decision support systems in other states in India and elaborates the potential for Decision Support System (DSS) to reduce and manage disaster risks in other countries.

TN-SMART was developed by the Commissionerate of Revenue Administration and Disaster Management, the Department of Revenue Administration, the Disaster Management and Mitigation of Government of Tamil Nadu with the assistance of Regional Integrated Multi-hazard Early Warning System for Africa and Asia (RIMES) and funded by the World Bank through its Coastal Disaster Risk Reduction project, to support with advanced predictive and prescriptive data analytical capabilities to archive, analyze, model, and communicate disaster risk information.

---

[12] Patankar A. 2019, Impacts of natural disasters on households and small businesses in India, ADB Economics Working Paper Series No. 603.
[13] State Planning Commission, Chennai 2012, Twelfth Five Year Plan (2012–2017). Government of Tamil Nadu.

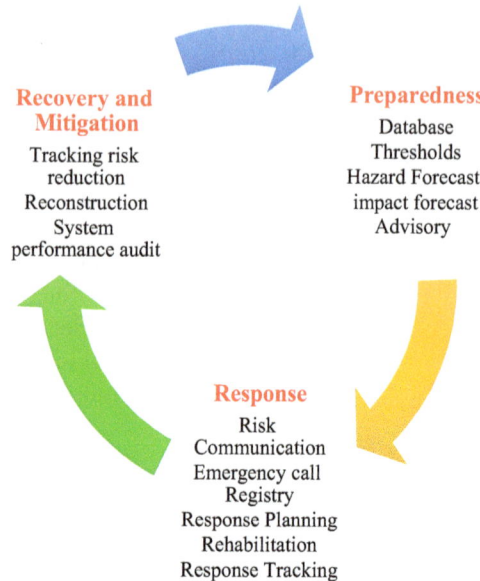

Fig. 1: Role of TN-SMART in all phases of disaster management

This operational decision support system is aimed at catering to the differential risk information and product needs of policy makers, operational users and communities for their informed decision making during all the phases of disaster management (Fig. 1). It especially:

- Informs policy decisions on preparedness, response, and mitigation measures for effective risk reduction.
- Enables emergency managers to have better understanding of risk; communicate risk; assess impact management options; deploy personnel, material, and equipment; plan relief measures; respond to emergency calls from public; track field response.
- Communicates alerts and disseminates advisories in understandable form to users for timely evacuation to prevent loss of lives and safety of movable assets and enhanced preparedness.

Data and Methodology

Data: The TN-SMART is built around a robust data architecture drawing from multiple sources within the Government of Tamil Nadu and the Government of India as tabulated (Tab. 1) below:

System for Multi-hazard Potential Impact Assessment

Tab. 1: Data architecture of TN-SMART

Layers	Description	Source
Administrative boundaries	GIS layers of Administrative boundaries (State, District, Block, Taluk, Revenue Village)	Directorate of Survey and Settlement, Government of Tamil Nadu (GOTN)
Land use	GIS shapefile of Tamil Nadu Land use	Institute of Remote Sensing, Anna University
Vulnerability profile	Flood vulnerable locations (4133) due to incessant rains/cyclones/storm surges over Tamil Nadu. Locations are categorized into four flood-risk levels based on inundation depth: Very High (>5 feet), High (3–5 feet), Medium (2–3 feet) and Low (<2 feet).	Commissionerate of Revenue Administration and Disaster Management, GOTN.
Rainfall observation	Daily rainfall data for the period 2011 to present (488 stations)	India Meteorological Department, GOI Commissionerate of Revenue Administration and Disaster Management, GOTN
Water level	Water level data from 20 major reservoirs (daily), tank status by district (daily during North east monsoon season) – 2018-present	Public Works Department, GOTN
Weather Forecast	10 days forecast for rainfall, maximum temperature, minimum temperature and humidity	India Meteorological Department, GOI
Response database	Inventory of the response equipment (power saws, JCBs, suction pumps, generators, etc.) Details of First responders Relief Centres details Emergency contacts	Commissionerate of Revenue Administration and Disaster Management, GOTN

Methodology: Understanding disaster risks and being prepared to anticipate and manage these risks can provide adequate lead-time to effectively respond to disasters. Establishing dynamic risk assessment tools in place and providing risk watches through innovative technologies can enhance disaster preparedness capabilities to manage disaster risks. In this line, the Commissionerate of Revenue Administration and Disaster Management (CRADM) and the Regional Integrated Multi-Hazard Early Warning System (RIMES) co-developed Tamil Nadu System for Multi-hazard Potential Impact Assessment and Emergency Response Tracking (TN-SMART) to support Tamil Nadu's State Emergency Operations with essential technical capacities for effective disaster risk management in Tamil Nadu. The system design considered addressing the requirements of policy makers, operational users, and the community during all phases of a disaster encompassing preparedness, response, relief, reconstruction, and mitigation measures; as well as covering multiple hazards such as flood, cyclone and tsunami, and lightning in the State.

TN-SMART engine is built with big-data analytics that incorporates traditional descriptive analytics to understand the risk, predictive analytics to model the risk based on hazard forecast, and prescriptive analytics to provide impact management options (Fig. 2). TN-SMART also integrates legacy data for vulnerability mapping and other static data with dynamic data of rainfall/storm surge forecast products to assess the potential flood scenarios (predictive analytics) based on the understanding from past impacts (descriptive analytics). It can then evaluate and disseminate forecast-based alerts and prescribe the impact management options to ensure better preparedness for managing risk (prescriptive analytics)[14].

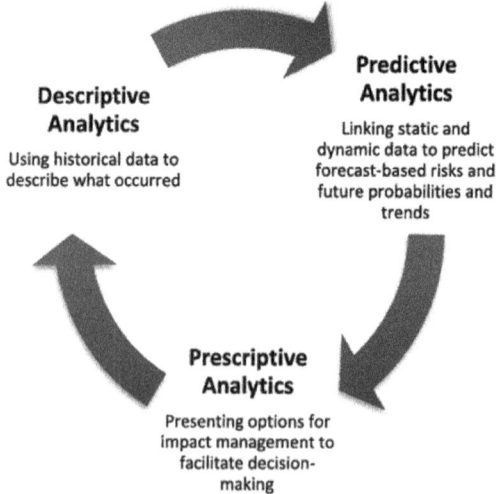

Fig. 2: Big data analytics in TNSMART

TN-SMART is designed in modular and easily customizable web-based interfaces, using open-source tools (PHP, MySQL, JavaScript, C, C++, Open Layers, Google Earth), generic mapping tools, and an open-source disaster management system. The system integrates artificial intelligence-based machine learning algorithm for automatic evaluations of

---

[14] Dr. K. Satyagopal IAS., Dr. Itesh Dash and Dr. Jothiganesh Shanmugasundaram (2019). 2019, TNSMART – A web-GIS based decision support system for strengthening disaster management. RIMES, Thailand

potential impacts and impact management options. A modular development approach is adopted for TN-SMART in the title User-needs based Design of TN-SMART and in the sentence 'adopted for TN-SMART (Fig. 3), so that any future requirements in terms of functions/analytics could be easily adopted. This module-based approach also allows differential levels of access for users to data layers within the system, ensuring that sensitive data is accessible only to authorized users.

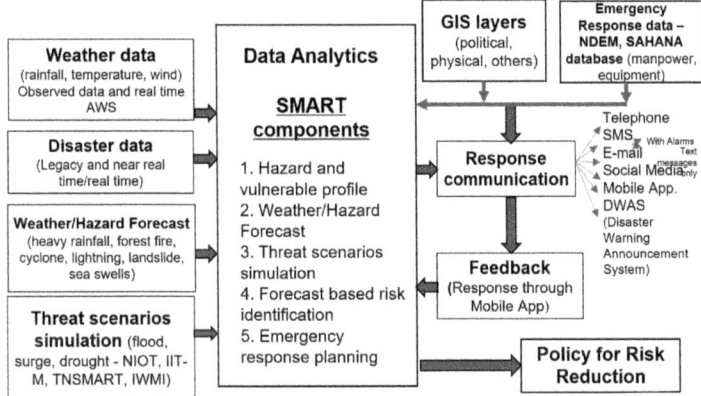

Fig. 3: SMART system architecture

Results and Discussions

User-needs-based Design of TN-SMART

For disaster managers, generic forecasts are not very helpful in identifying specific areas likely to be impacted by disasters. Further, disaster managers lack systems that could superimpose rainfall forecast data over vulnerable locations to assess inundation levels and disaster vulnerability. Thus, warnings and notifications to stakeholders and communities at risk become a challenge[15].

To address these challenges, a web-GIS application with 11 modules has been developed, as tabulated in Tab. 2. This aims to address the varied needs of policy makers, operational users, and communities during all the phases of disaster risk management – early warning, preparedness, response, relief, recovery, and mitigation.

---

[15] Commissionerate of Revenue Administration and Disaster Management, Government of Tamil Nadu and Regional Integrated Multi-hazard Early Warning System, 2019. TNSMART – A web-GIS based decision support system for strengthening disaster management. Pathumthani, Thailand: RIMES.

Tab. 2: TN-SMART modules – Functions, features and application for policy makers, operational users and communities

Modules	Function	Features	Application		
			Policy makers	Operational Users	Community users
1 Database	Database management system with GIS layers, meteorological and hydrological data, vulnerability profile, disaster damage data for understanding disaster risks	• Rainfall, Water Level, Wind speed • Disaster Loss and Damage • Vulnerability Profile • Emergency response database (SDRF; Relief centres, First responders, Inter. Dep. Team) • Automatic data flow in real or near real time data flow (rainfall, water level) • Near real time disaster Loss and damage data	Prioritizing DRR for vulnerable locations	Understanding risk patterns, characterizing hazard thresholds, modeling potential impacts and plan response	Avoid risk and build resilience

	Modules	Function	Features	Application	
2	Thresholds	Threshold values of hazards that have manifested into disasters	Rainfall thresholds triggered flood. Water level in water bodies that triggered flood Wind speed that caused damage Surge heights that impacted the coastal belt Inundation depth that decides vulnerability class	Review thresholds for policy changes (sector specific and region specific) e.g., Fishermen warning	Serves as base for Risk forecasting. The trigger that turns hazard into disaster over a location.	–
3	Hazard Forecast	Integration of various weather forecasts from short term to seasonal forecast (e.g., from IMD) from various sources and tsunami alerts from INCOIS	IMD 3 days forecast ECMWF 10 days forecast Ensemble forecast for 3 days (IMD WRF, RIMES WRF, ECMWF)	Policy based on seasonal forecast	Identify the risk locations.	Avoid risk and build resilience
4	Impact Forecast	Forecast values exceeding thresholds the level of risk will be indicated and anticipated impacts of hazards will be modeled.	Forecasted hazard values exceeding the identified thresholds to forecast the risk Locations (villages) and population likely to be affected	Policy based on potential impacts for the seasonal forecast	Identify the risk locations. To understand the likely regions to be affected and how many people are living in the risk zones.	Avoid risk and build resilience

(Continued)

Tab. 2: Continued

	Modules	Function	Features	Application	
5	Advisory	Bulletin generation for different administrative and operational level officers	Magnitude of the hazard and its likely impacts Likely affected villages and population Response resource inventory available with districts	Requisition additional response teams including Armed Forces based on need	To guide the operational officers on the likely risk and response options available	Avoid risk by leaving risk prone areas
7	Emergency call Registry	Register the emergency calls/information from community, forward to concerned focal point for action Response planning – Prepare strategies to respond to the anticipated risks in short, medium, long term and mitigation measures	An icon is designed to flash red in the TNSMART whenever emergency calls or messages are registered	Make policies based on number and types of emergency calls received	Respond to emergency calls from the communities, expand reach and provide timely response	Communicate distress calls regarding disasters to Incident Response Systems

	Modules	Function	Features		Application	
8	Risk communication & Alerts through Emergency Alarms	Communicate risk through Email, SMS, App notifications and message with alarms, via social media, Website		Make policy for improved risk communication	Communicate risk to Incident Response System for better preparedness, pre-emptively deploy personnel and equipment, prevent loss of life through advanced evacuation.	Evacuate from vulnerable areas to safe grounds or relief shelters; Avoid going to sea for fishing or return to shore if on High Seas
9	Response Tracking	Track action taken on distress calls and alert calls through feedback mechanism	Tracking responders and their responses through Feedback mechanism	Monitoring the performance of officers in action, as well as TN-SMART system	Respond to emergency messages/calls from community, expand reach and provide timely response	Communicate distress calls/ messages to Incident Response System
10	Tracking Risk Reduction	Review vulnerable locations annually based on mitigation measures implemented to guide policy level decisions for risk reduction		Policy to continue the mitigation measure or revise the mitigation strategies	feedback on usefulness of the mitigation measures helps in deployment strategies.	Benefit from improved mitigation measures
11	Performance Audit	Carry out annual performance audit for all the modules of TN-SMART to improve the system		Make better policy for disaster management	Strengthen preparedness, response, and 'build back strategies'	

(Continued)

TN-SMART Relevance with Global Frameworks for DRR

The TN-SMART is designed to assist the government to monitor the implementation strategies for compliance with the four priority areas identified in the Sendai Framework[16] (which are also similar with priorities set forth in the Tamil Nadu State Disaster Management Perspective Plan 2018–2030), viz.[17]:

i. Understanding disaster risk
ii. Strengthening disaster risk governance
iii. Investing in disaster risk reduction for resilience
iv. Enhancing disaster preparedness for effective response, and to 'Build Back Better' in recovery, rehabilitation, and reconstruction.

TN-SMART's 11 operational modules enable the Tamil Nadu State Disaster Management Authority (TNSDMA) to effectively address the priorities of the Sendai Framework for DRR, as analyzed in Tab. 3.

Tab. 3: Relevance of TN-SMART with global frameworks – SFDRR

TN-SMART Modules	Relevance to Sendai Framework for DRR Priorities
Module 1. Database	Data management of static and dynamic data; integration of forecast information from IMD, INCOIS, and other global sources (for reference); forecast analysis against hazard thresholds; evaluation of the hazard's potential impacts; assessment of risks all contribute to *'Priority 1: Understanding Disaster Risk'*.
Module 2. Thresholds	
Module 3. Hazard Forecasting	
Module 4. Impact Forecasting	
Module 5. Response Planning	Dissemination of forecast and risk information through unique alarms (alarms are triggered even when the mobile is on silent mode) enables communication of severe risks to all stakeholders within a short time ensuring safety of at-risk communities; advisories for emergency response and planning, Do's and Don'ts address *'Priority 3: Enhancing disaster preparedness for effective response, and to "Build Back Better" in recovery, rehabilitation and reconstruction'*
Module 6. Advisory	
Module 7. Risk Communication	

---

[16] UNDRR. 2015. Sendai Framework for Disaster Risk Reduction 2015–2030.
[17] Dr. K. Satyagopal IAS., Dr. Itesh Dash and Dr. Jothiganesh Shanmugasundaram (2019). 2019, TNSMART – A web-GIS based decision support system for strengthening disaster management. RIMES, Thailand.

Tab. 3: Continued

TN-SMART Modules	Relevance to Sendai Framework for DRR Priorities
Module 8. Emergency Call Registry	Registration of emergency calls from affected areas; tracking of response actions to these calls address *'Priority 2: Strengthening disaster risk governance to manage disaster risk'*
Module 9. Response Tracking	
Module 10. Tracking Risk Reduction	Module on mitigation and risk reduction actions undertaken in risk prone areas, addresses *'Priority 4: Investing in disaster reduction for resilience'*
Module 11. Performance Audit	TN-SMART performance audit module to guide system improvement addresses *'Priority 2: Strengthening disaster risk governance to manage disaster risk'*

TN-SMART is also aligned with the World Meteorological Organization's Global Framework for Climate Services and provides climate-forecast-based impacts for making climate-smart decisions to minimize the negative impacts of climate shocks.

The Tamil Nadu State Government adopted TN-SMART as a tool for dynamic risk assessment, in its State Disaster Management Perspective Plan 2018–2030. The TN-SMART also strengthens the ability of operational users to deploy personnel and materials at appropriate locations for effective emergency response to evacuate quickly and avoid loss of lives and protect moveable properties, addressing the major targets of Sendai Framework for Disaster Risk Reduction: *'Reducing disaster mortality and reducing economic losses and loss of livelihood opportunities for the vulnerable population'.*

Cases on Operational Performance of TN-SMART

a. During 2018 Gaja Cyclone, 2020 Nivar and Burevi Cyclones

The TN-SMART functioned as designed and effectively, from genesis of Gaja Cyclone on 10 November 2018 until its landfall in Tamil Nadu on 16th November 2018, and ever since during subsequent events, including the 2020 Nivar and Burevi cyclones. Using TN-SMART's analysis of potential impacts from the cyclone based on IMD forecast, potential impact information and risk-based warnings were communicated to districts at risk, and resources for emergency response were pre-positioned days before the cyclone's landfall. During the emergency, TN-SMART facilitated tracking of TNSDMA response action to the distress messages received from affected areas, contributing to accountable and effective risk governance.

b. Evidence-based Impacts' Analysis of TN-SMART During 2021 Floods

In November 2021, Chennai received over 915 mm of rainfall (exceeded only in November 2015, when it received 1049 mm), which resulted in wide-spread flooding across the city. Including the preceding month of October 2021, Chennai received over 1130 mm of rainfall which was 83 % higher than average. This huge deluge led to waterlogging in streets that had to be evacuated using heavy-duty pumps from flooded streets into sewerage pipes- an estimated 920 million liters per day (mld) was being pumped out round the clock. instead of the typical 585 mld, an increase by almost 60 %[18].

However, the impacts were not as severe as in 2015, due to several reasons including the significant investments on improving drainage systems and the deliberate monitoring of the water bodies surrounding Chennai. The TN-SMART too played a role in supporting operational decision-making within the TNSDMA, through: (i) regular and user-friendly depiction of water levels in reservoirs of the state, its variance from past year(s); (ii) the daily (and seasonal) observed rainfall and its variance from past; (iii) overlay of forecasted rainfall and (iv) generation of potential flood outlooks and communication through user-friendly advisories, as detailed in Tab. 4.

Tab. 4: TN-SMART during November 2021 heavy rainfall

Duration/ Time period	Alerts generated and disseminated via TN-SMART	Actions taken
7 Nov– 11 Nov 2021	7 Flood alerts 54 Fisher alerts, 157 heavy rainfall alerts	49,927 people in vulnerable, low-lying areas evacuated in advance 263 relief camps with food, safe drinking water and basic amenities set up
15 Nov– 19 Nov 2021	15 Flood alerts 140 Fisher alerts, 335 heavy rainfall alerts	112,707 people in vulnerable, low-lying areas evacuated 1764 relief camps set up

---

[18] Chandrababu, D. (2021, Dec 01). Deep dive: Why does Chennai flood every time it rains heavily?. The Hindustan Times: India Edition. https://www.hindustantimes.com/india-news/deep-dive-why-does-chennai-flood-every-time-it-rains-heavily-101638297634345.html

Flood risk advisories generated by TN-SMART identified potential flooding areas in each district that enabled district authorities to take proactive actions to reduce adverse impacts. The complaint registry module of the TN-SMART registered 603 complaints during the 2021 monsoon season of which 185 were raised via the TN-SMART mobile app, 213 via the web and 204 via Whatsapp. Of these on 227 complaints action was taken – another 307 were forwarded for actions, including 27 to concerned departments for resolution in a time-bound manner. It has not only expedited the grievance redressal processes through automated escalation of complaints to next levels if not addressed within the stipulated time, but through responsive administrative mechanisms for resolving potential problems and issues, these also resulted in improved preparedness and enhanced risk governance.

c. Performance of TN-SMART During 2020 Monsoon Season

To ensure that TN-SMART is relevant to the needs and demands of the users and to constantly improve the system, an annual performance audit is an integral feature of TN-SMART. This performance audit conducted by the RIMES in collaboration with the TNSDMA assessed performance of each of its modules during 2020, to identify strengths and weaknesses as well as opportunities for improvement during the next season (2021). Key findings relating to the relevance and usefulness of the TN-SMART across the various categories of users is tabulated (Tab. 5)

Tab. 5: Performance of TN-SMART modules during 2020/21 season

Modules	Performance	Remarks
Module 1 Database	Rainfall data (488 stations) and water level data (20 reservoirs), tank status (37 district) updated on daily basis 4133 flood vulnerable locations in 37 districts were identified and made available in public domain.	
Module 2 Thresholds	Flood triggering thresholds were used for assessing likelihood of the flood during BUREVI cyclone,	
Module 3 Hazard Forecasting	Weather forecast for NIVAR and BUREVI cyclone from IMD, visualized with reference to districts and taluks in respect of hazards associated with cyclones	
Module 4 Disaster Risk Forecasting	Potential flood scenarios for the vulnerable locations in 37 districts were identified, based on NIVAR and BUREVI Cyclone forecasts.	The forecast-based potential flood vulnerable locations for BUREVI cyclone were disseminated to District Administrators for taking necessary precautionary measures.
Module 5 Response Planning	Details of 4,696 relief shelters (place name, building name, capacity, facility, monitoring officers) in Tamil Nadu updated. 61,974 first responders' details updated.	The relief centre details and first responder details were used by the State and District level decision makers during the preparedness and response phases of NIVAR and BUREVI Cyclones.
Module 6 Advisory & Module 7 Risk Communication	193 alerts, sourced from IMD and CWC, disseminated using TN-SMART during the 2020 Northeast monsoon season 88 alerts issued through TN-SMART mobile application during NIVAR cyclone 2020 (21–27 November 2020) 96 alerts disseminated during BUREVI cyclone (29 Nov–07 Dec) A voice over alarm sounded 12 hours before the landfall of cyclone onwards to caution people to take action	More than 10,000 new users registered and received cyclone related alerts, since the first cyclone watch was issued for NIVAR Cyclone 2020 on 24 November 2020. Alerts communicated both in English and Tamil language

(Continued)

Tab. 5: Continued

Modules	Performance	Remarks
Module 8 and 9 Emergency Call Registry and Response Tracking	During 2020, 190 emergency messages were registered through TN-SMART App, out of which, 119 messages were forwarded to officials for taking action and 25 action taken report was recorded, and duplicate/irrelevant messages were disregarded	
Module 10 Tracking Risk Reduction	Permanent Mitigation activities such as construction of walls, percolation pits, pipe culverts to box culverts and check dams etc. were updated Temporary Mitigation activities such as desiltation carried out under various departments were updated State level decision makers checked this data before the NE Monsoon season to evaluate the mitigation activities at district level.	
Module 11 Performance Audit	RIMES On-Site team conducted the performance audit with RIMES Program Unit.	

d. Significant Increase in TN-SMART User-Base

CRADM's continuous efforts to enhance effectiveness of the TN-SMART and its outreach led to an increase in mobile application users from 13000+ users during GAJA Cyclone 2018 to 245000+ users during NIVAR Cyclone 2020. Currently there are over 400,000 users, including over 250,000 registered users who can receive location-specific alerts[19].

TN-SMART disseminated 193 alerts during the 2020 Northeast monsoon season (95 cyclone alerts, 37 fishermen alerts, 44 heavy rainfall alerts, 17 flood alerts). TN-SMART has all details of first responders (127,349 disaggregated by gender), relief shelters (5,106) capable of accommodating 1.39 million people, and a database related to response planning for guiding CRADM on forecast-based response planning.

---

[19] Tamil Nadu State Disaster Management Authority, Government of Tamil Nadu. TNSMART Web Portal. https://beta-tnsmart.rimes.int/index.php/TNSMART_App_version_details/tnsmart_app_registration_report

TN-SMART Mobile App[20] has emergency contact details for the general public to reach during emergencies, and Do's and Don'ts to guide them for safeguarding themselves from disasters. CRADM tracks the temporary and permanent mitigation measures through TN-SMART for monitoring Disaster Risk Reduction in the State.

e. Indian Ocean Tsunami and Other Rapid Onset Hazards

While no major tsunamis have impacted the Tamil Nadu coasts since the 2004 Indian Ocean tsunami, the TN-SMART incorporated certain critical features recognizing the specific characteristics of a tsunami, such as the unique alarm, which rings even when the phone is on silent-mode. A tsunami may occur at any time of the day or night, and unlike a cyclone which is forecast with a lead-time of several days, a tsunami affecting Tamil Nadu (due to an undersea earthquake off the Andaman-Sumatra subduction arc) might only have a lead-time of 2–3 hours at best. If such an event were to occur in night-time when at-risk population is asleep, it will pose a huge challenge to communicate the tsunami warning to the vulnerable population using conventional methods, or even advanced cell broadcasts especially if phones are in silent mode. Hence the TN-SMART incorporates this unique alarm which ensures that the phone will ring even if on silent mode, until the tsunami warning message has been read.

The same feature can also be leveraged for alerting specific population at risk in case of gas leaks from industrial facilities, if it occurs during night-time, or any other time, as lead-time from onset to impact is similarly quite short for industrial gas leaks.

f. Decision Support System (DSS) for Disaster Management in Operation in the Region and Beyond

The authors are not aware of any other DSS for disaster management in operation in the South Asia region. However, in Southeast Asia, the ASEAN Coordinating Centre for Humanitarian Assistance on disaster

---

[20] Tamil Nadu State Disaster Management Authority, Government of Tamil Nadu. TN SMART (TamilNadu System for Multi-hazard potential impact Assessment and emergency Response Tracking) Mobile Application. https://play.google.com/store/apps/details?id=int_.rimes.tnsmart&hl=en_IN&gl=US

management (AHA Centre) is using the Disaster Aware platform of the Pacific Disaster Center. AHA Centre is an intergovernmental organization, established by the ten ASEAN Member States with the aim to facilitate cooperation and coordination of disaster management amongst ASEAN Member States[21]. The Disaster Monitoring and Response System (DMRS) of the AHA Centre is powered by the Disaster Aware platform and is used by AHA Centre since 2012 to monitor hazard and to provide early warning notifications for pending disaster events.

The Disaster AWARE Enterprise™(DAE) is a proprietary, commercial software providing sophisticated risk intelligence and resilience for both the public and enterprises ensuring the safety of assets and business continuity. It has a smart alerting feature that informs of incoming hazards that could adversely affect safety, predictive impact models built on layers of Vulnerability Index, Resilience Index, Impacts and Losses aiding decision making by Disaster Risk Managers. It supports monitoring, preparedness and response to threats and hazards[22].

The TNSMART, however, was conceived jointly by the TNSDMA of the Government of Tamil Nadu and RIMES and continues to evolve organically to address the felt needs and operational demands of TNSDMA in managing disaster risks in Tamil Nadu.

g. Future Enhancements in TN-SMART and Its Sustainability

TN-SMART performance audits carried out as an integral part of its design enables feedback from all relevant stakeholders on the real-world performance of each of the modules and identifies specific improvements that can enhance its utility as a decision support tool for the operational users and decision-makers. Based on the audit conducted in 2020, a work plan for 2021 was drawn up in consultation with the CRADM to address key opportunities for enhancing effectiveness of the TN-SMART and is currently under implementation.

---

[21] AHA Centre. ASEAN Coordinating Centre for Humanitarian Assistance on disaster management. https://ahacentre.org/about-us/

[22] DisasterAware Enterprise. The State of Situational Risk Intelligence- The Importance of Digital Risk Tools in the Modern Risk Management Era. Whitepaper. https://disasteraware.com/wp-content/uploads/2021/02/DAE-Whitepaper-v1.0.pdf

The TN-SMART can be further updated to incorporate new developments in disaster-risk modeling to ensure its relevance and enhance its application for the State's decision-making. The system will accumulate enormous data over time, encompassing emergency calls/messages, weather observations and weather forecasts and their impacts. Therefore, a machine learning algorithm will be integrated into TN-SMART to uncover patterns in the data dynamically and improve the understanding and forecasting of risk, and will be introduced for the following areas to begin with[23]:

i. *Vulnerability characterization*: The emergency call registry database reflects the ground-truth, which could serve as a proxy to understand the vulnerability of a region corresponding to the hazard.

ii. *Impact assessment and advisories*: Weather forecast data from various models as well as observational weather data are archived in TNSMART which is utilized to dynamically assess the performance of weather forecasts spatially. Thus, the system can be further enhanced to incorporate this forecast reliability score by locations and the new patterns of vulnerability uncovered for improving location-specific impact assessments and advisories

TN-SMART's sustainability is ensured through an innovative institutional mechanism wherein the CRADM collaborates with RIMES even beyond the project development phase to provide back-up for operations as well as to continuously refine the system to ensure it stays relevant to the needs of the State Government. Furthermore, the importance of TN-SMART as highlighted in government policy documents and the State Disaster Management Perspective Plan 2018–2030 will provide policy support for sustaining TNSMART. Linkages of historical and forecast databases for generating tailor-made advisories and continuous updating of data will enable the system to be relevant all the time and this will ensure sustainability. Empowerment of policy-makers, operational-users, and communities by providing necessary early warning information to avoid and reduce disaster risks will ensure sustainability.

---

[23] Dr. K. Satyagopal IAS., Dr. Itesh Dash and Dr. Jothiganesh Shanmugasundaram (2019). 2019, TNSMART – A web-GIS based decision support system for strengthening disaster management. RIMES, Thailand.

h. System for Tracking and Alerting Disaster Risk Information based on Dynamic Risk Knowledge (SATARK) in Odisha State, India

Recognizing the successes and considerable potential of decision support systems such as the TN-SMART for effective management of disaster risks, facilitating timely and actionable warning and efficient emergency response, RIMES shared details on the TN-SMART with the Government of Odisha State. The Government of Odisha has since then taken a quantum leap, creating with RIMES, a sophisticated decision support system – the System for Tracking and Alerting Disaster Risk Information based on Dynamic Risk Knowledge (SATARK)[24] – currently in advanced development stage and under pilot implementation.

SATARK is being developed with advanced infographics and visualization support for exposure data layers for population, livestock, agriculture lands, infrastructure, alerting potential threats through SMS, mobile App, and prescribing preparedness measures, for extreme events such as the dangerous thunderstorms, heat wave conditions and stress levels, flood inundation, likelihood of drought situation. SATARK also integrates the emergency contact details of Government Officials from State to Gram Panchayat for emergency management.

i. Relevance of SMART Decision Support System (DSS) for Disaster-prone Countries

Analysis of disaster incidents, damages and impacts indicates that while the technological advances and enhanced accuracy in forecasting and early warning have enabled countries to reduce the number of lives lost, the economic damages are on the other hand increasing manifold. Analysis of data from the global database (EM-DAT[25]) undertaken for the World Bank's South Asia Hydromet Forum shows that in South Asia Region alone, a total of 799 weather/climate-related disasters were reported in the last two decades affecting 3 out of 5 people (1.22 billion)

---

[24] OSDMA. Odisha State Disaster Management Authority, Government of Odisha. System for Tracking and Alerting Disaster Risk Information based on Dynamic Risk Knowledge (SATARK). https://satark.rimes.int/Satarklogin/login_satark

[25] EM-DAT, The International Disaster Database, Centre for Research on Epidemiology of Disasters. CRED, EM-DAT, CRED/UCLouvain, Brussels, Belgium – www.emdat.be, accessed on 28 Dec 2021.

and resulted in 76,350 deaths and damaged assets worth over US$144 billion.

Fig. 4 shows that the number of lives lost due to disasters in South Asia region in the decade of 2010 has reduced by 12 % (to 35,721 compared to 40,630 in the previous decade), while damages due to the disasters on the other hand has almost tripled to USD107.5 billion from USD 37 billion.

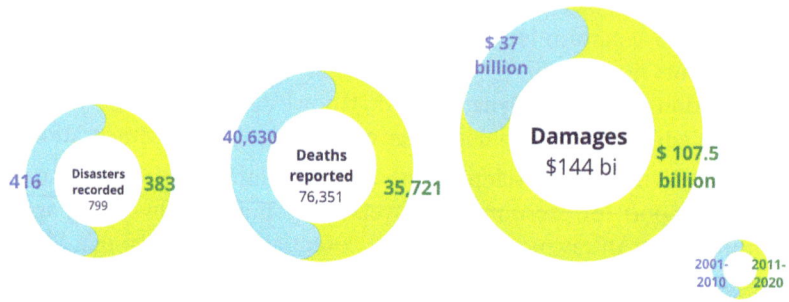

Fig. 4: Increasing disaster damages, decreasing deaths in South Asia

These increasing impacts on societal systems reinforce the need for a Decision Support System like TNSMART or SATARK for disaster managers to undertake proactive risk management through generation of potential impacts of disasters specific to sectors and communities built on a thorough understanding of their vulnerabilities and exposure, and its dissemination leading to risk reduction, as well as provision of guidance to policy-makers on undertaking necessary mitigation measures to reduce or prevent impacts.

Conclusion

The TN-SMART has been recognized as an innovation in the region and featured as a showpiece during the Second South Asia Hydromet Forum organized by the World Bank in 2019 in Kathmandu, Nepal where the eight South Asian countries along with Myanmar participated[26].

---

[26] World Bank. 2019. South Asia Hydromet Forum. https://www.worldbank.org/en/events/2019/11/19/south-asia-hydromet-forum-ii

The successful experience of TNSDMA in implementing the TN-SMART as a decision support system for operational disaster risk management has led to collaboration of RIMES and the India Meteorological Department (IMD) for creating a unit at IMD, New Delhi dedicated to support replication of similar decision support systems (DSS) across all States and Union Territories in India through the joint efforts of State Disaster Management Authorities (SDMAs) and IMD, considering their shared responsibility for generating and applying actionable early warning information on a sustained basis.

As the SATARK experience in Odisha illustrates, though grounded initially in severe/extreme events, being an integrated system bringing together science-based information providers, with sectoral agencies and societal applications, the scope will gradually broaden and deepen to address medium-, and longer-term multi-hazard risks including from a climate change perspective, contributing to risk informed development.

The comprehensive scope and pragmatic design of the SMART in Tamil Nadu and in Odisha (in form of SATARK though in its nascent stages) has already attracted international attention and interest to replicate the initiative through RIMES across 48 Member regions and Collaborating States in Asia, Pacific and in Africa. Specifically, the development of DSS (decision support system) based on SMART is under active consideration in Nepal and in several other South Asian countries[27], and in many other countries including in Africa through other RIMES sub-regional hubs.

The System for Multi-hazard Potential Impact Assessment, Alert, Emergency Response Planning and Tracking (SMART) offers a demonstrated solution of early warning system at any administrative level, that can be further customized for anticipating and managing risks emanating from natural hazards, as well as from anthropogenic hazards.

References

Chandrababu, D. 1st of December 2021, 'Deep dive: Why does Chennai flood every time it rains heavily?', *The Hindustan Times: India Edition*,

---

[27] As well as Pacific Island countries through the RIMES Sub-Regional hub in Papua New Guinea.

https://www.hindustantimes.com/india-news/deep-dive-why-does-chennai-flood-every-time-it-rains-heavily-101638297634345.html

Commissionerate of Revenue Administration and Disaster Management, Government of Tamil Nadu and Regional Integrated Multi-hazard Early Warning System, 2019, TNSMART – A web-GIS based decision support system for strengthening disaster management, Pathumthani, Thailand: RIMES

Disaster Aware Enterprise 2021, The State of Situational Risk Intelligence – The Importance of Digital Risk Tools in the Modern Risk Management Era, Whitepaper, https://disasteraware.com/wp-content/uploads/2021/02/DAE-Whitepaper-v1.0.pdf

EM-DAT, The International Disaster Database, Centre for Research on Epidemiology of Disasters – CRED. EM-DAT, CRED/UCLouvain, Brussels, Belgium, www.emdat.be

Geethalakshmi V., & Ganesan D. 2008, 'Impact of climate change on agriculture over Tamil Nadu', Chap IV, *Climate Change and Agriculture Over India*, CRIDA, Hyderabad, pp. 80–93

Hellmuth M.E., et al. 2011, A Better Climate for Disaster Risk Management. International Research Institute for Climate and Society (IRI), Columbia University, New York, USA

Patankar A. 2019, 'Impacts of Natural Disasters on Households and Small Businesses in India', Asian Development Bank Economics Working Paper Series No. 603

Patnaik I., Sane R., & Shah A. 2019, 'Chennai 2015: A novel approach to measuring the impact of a natural disaster', NIPFP Working Paper No. 285

State Planning Commission, Chennai 2012, Twelfth Five Year Plan (2012–2017), Government of Tamil Nadu

Slingo J., & Palmer T. 2011, 'Uncertainty in weather and climate prediction', Philosophical Transactions of the Royal Society Vol. 369, pp. 4751–4767. https://doi.org/10.1098/rsta.2011.0161. PMCID: PMC3270390

Tamil Nadu State Disaster Management Authority, Government of Tamil Nadu, TNSMART Web Portal, https://betatnsmart.rimes.int/index.php/TNSMART_App_version_details/tnsmart_app_registration_report

Tamil Nadu State Disaster Management Authority, Government of Tamil Nadu, TN SMART (TamilNadu System for Multi-hazard potential impact Assessment and emergency Response Tracking) Mobile Application, https://play.google.com/store/apps/details?id=int_.rimes.tnsmart&hl=en_IN&gl=US

Tamil Nadu State Disaster Management Authority, Government of Tamil Nadu, Physical Vulnerability, TNSDMA, https://tnsdma.tn.gov.in/pages/view/physical-vulnerability

Tamil Nadu State Disaster Management Authority, Government of Tamil Nadu, Hydro-Meteorological, TNSDMA, https://tnsdma.tn.gov.in/pages/view/Hydro-Meteorological

UNDRR, Understanding Disaster Risks, PreventionWeb, Disaster losses & statistics, https://www.preventionweb.net/understanding-disaster-risk/disaster-losses-and-statistics

UNDRR, Tamil Nadu (India)-033, Desinventar Sendai, https://www.desinventar.net/DesInventar/country_profile.jsp?countrycode=033&lang=EN

United Nations Development Programme (UNDP), Bureau for Crisis Prevention and Recovery (BCPR), 2013, Climate Risk Management for Agriculture Sector in Tamil Nadu State of India, New York

World Bank, 2019, South Asia Hydromet Forum, https://www.worldbank.org/en/events/2019/11/19/south-asia-hydromet-forum-ii

World Meteorological Organization, 2021, WMO Atlas Of Mortality And Economic Losses from Weather, Climate and Water Extremes (1970–2019), WMO-No. 1267

Chapter 6

Landslide Vulnerability, Risk and Resilience Management of Cultural Heritage Sites in the Western Slope of Lawu Mountain, Indonesia

Muhamad Iko Kersapati, Muhammad Attorik Falensky, Gina Fitri, Heri Purwanto

Introduction

Climate change and global warming are the most discussed environmental issues today. Human activities since the industrial era capitalize coal, oil, and other fossil sources for combustion have produced major greenhouse gases whose concentrations continue to increase drastically. Changing the concentration of these gases (such as CO_2) in the atmosphere leads to higher temperatures and changes in precipitation patterns. Implicatively, climate change associated with a particular geographical characteristic can lead to another environmental degradation such as a decline in organic matter, soil biodiversity loss, salinization, droughts, and another solemn catastrophe: a landslide (European Commission, 2021). On the other hand, changing weather patterns also contribute to excessive rainfall accompanying earthquakes disturbing the stability of natural and engineered slopes (Gariano & Guzzetti, 2016).

Indonesia with its geographical condition and as an archipelagic country is very vulnerable to climate impacts. As one of the disaster-prone areas in the world, it is that part of the meeting point of three dynamic plate tectonics, i.e., the Pacific plate, the Indo-Australian plate, and the Eurasian plate, known as the Ring of Fire. Due to those conditions, earthquakes and volcanic eruptions often occur. A crucial factor that directly affects slope stability is the amount of water within the slope and is concerned with variables such as evapotranspiration, infiltration rate, conductivity, and hydraulics of slope materials (Crozier, 2010). On the top of this, Indonesia is positioned in a tropical zone with a six-month wet season. The increase in the earth's surface temperature will affect the hydrological cycle, as a result, there will be an increase in the intensity of rainfall and escalation of the frequency of landslide events.

The National Coordinating Agency for Disaster Management (BAKORNAS PB) in February 2007 stated that the highest rainfall intensity occurred after 30 years and 1998 was the longest year with the highest temperature in history. It is predicted that Indonesia will continue to experience disasters related to climate change in the following years. During 2003–2005, 1,429 disaster cases occurred, about 53.3 % of which were related to hydrometeorology. Floods were the most frequent disasters (34 %) followed by landslides (16 %). Christianto et al. (2008) reported that 49 landslide events were occurring every year during the period 1981–2007 (Cepeda et al., 2010). Inventory for Indonesia covering the period 1998–2009 reported 890 landslide events that killed 1280 people. According to the recorded data from 2006–2008, the average number of landslide events occurred in Java (52 %), Sulawesi (24 %), and Sumatra (18 %) (Indonesian Geological Agency, 2008).

The highland of Mount Lawu accounts for one of the prone areas to landslide catastrophic events. The physical conditions in Karanganyar specifically slope and steepness increase the landslide potential where its constituent rocks are young volcanic deposits products from Mount Lawu. The land experienced quite thick weathering and rainfall intensity was relatively high. As anthropogenic forces, land use in Karanganyar generally consists of agricultural activities such as rice crops, horticultural gardens, forests, and settlements. Some massive plantations conducted in areas with steep topography become prone to erosion and landslide (Naryanto, 2011). The Western Slope of Lawu Mountain is home to important archaeological relics from the Majapahit dynasty: Sukuh Temple, Cetho, Planggatan, and many others. Sukuh Temple and Cetho Temple are known as the center of Hindu religious sites on the island of Java.

Other types of heritage tourism such as parks, tropical forests, and water heritage represent both natural and cultural that are classified into different values such as historical, economic, social, and scientific. With an integrated framework of heritage conservation and disaster management, this paper focuses on risk assessment and resilience management of heritage objects in Karanganyar as valuable national assets. The objectives of this research will be fulfilled by answering the main questions: How is the risk assessment in the observed region? How is the priority scale set for heritage tourism based on landslide risk assessment? How is resilience management in the observed region? This paper is expected to be a new reference and recommendation of risks and resilience management in the future.

I. Methodology

1. Conceptual Framework

Heritage conservation framework is typically complex defined in any relationship to other disciplines. Leblanc (2006) proposed six elements of project management in the heritage field comprising a lifecycle with initiating the process, planning, selecting the heritage objects based on their types, characteristics, and values. Meanwhile, monitoring and evaluating were implemented as the ultimate phase at the end of the cycle (Leblanc, 2006). This paper focuses on the disaster field as a specific integrated perspective to monitoring and evaluating processes. Crichton (1999) introduced the risk triangle as a comprehensive approach in assessing the risk of one type of disaster that relies on several elements: exposure, vulnerability, and hazard. However, there are possibilities of change to this approach based on the disaster characteristics (Crichton, 1999). The different interpretations of each institution in the decision-making process resulted in no exact definition of risk.

The key points are understanding 'how' the catastrophe events threaten a specific community of Mount Lawu, 'where' and 'when' it will happen. In this case, the landslide is identified as the element of the risk. **Exposure**: the humans are involved (represented by the number of visitors), **vulnerability**: lack of resistance to destructive forces (represented by slope, land use, lithology, distance to stream, and distance to geological structure), **hazard**: occurrence of natural events (represented by rainfall pattern, and landslide history) (Genovese et al., 2007). The three of these components are analyzed by weighted overlay to produce the landslide risk assessment (explained in more detail in the next section). The following figure represents the integrated framework between heritage conservation and disaster risk and resilience management in this research.

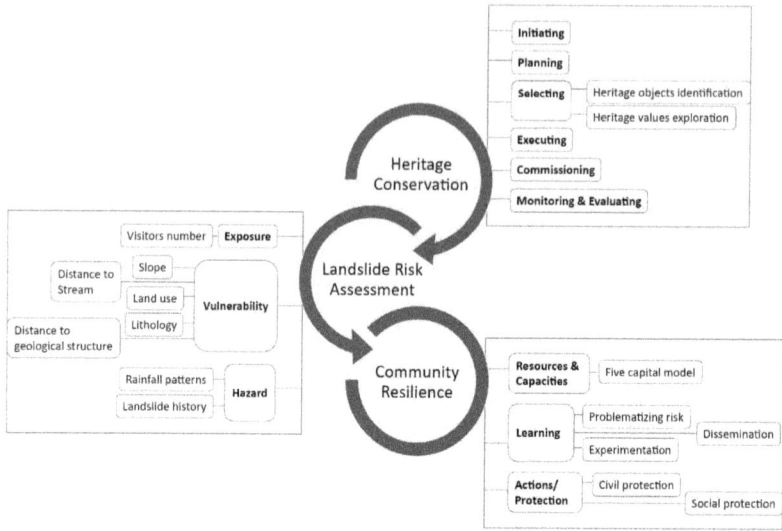

Fig. 1: Conceptual framework

The last section of the framework is community resilience identified by three aspects: resources and capacities (the five-capital model), learning (problematizing risk, experimentation, and dissemination), and actions (civil and social protection). Kruse et al. (2017) suggested a multidimensional approach in investigating resilience on different scales and multiple actors that complements the preventive strategy. Concerning the risk triangle, the community resilience approach aims to build knowledge of how to manage the potential issues before, during, and after the disaster event by reducing the exposure and increasing the capacity of the folks. Community resilience emphasizes the role of stakeholders (local people and institutions, regional, national governments, and international organizations) in a wide range of actions as a manifestation of collaboration in disaster management (Kruse et al., 2017).

2. Data and Resources

The required data are obtained from various sources to assess the landslide risk in the established area such as statistics, images, news, and spatial data that complexify the analysis. The data of visitors are collected from the Office of Education, Youth, and Sports Karanganyar Regency,

Central Java Province, while the rainfall records for the last 10 years are acquired from the Ngaringan station, Grobogan Regency, Central Java Province. The data from other sources such as news websites, social media, and anonymous records of disaster history are collected. The spatial data for vulnerability assessment is collected from several sources. The following table reveals the acquired data for the analytical purpose of this research.

Tab. 1: Data requirements

Variable	Sub-variable	Type	Source
Exposure	Number of visitors	Statistics	Education, Youth, and Sports Office (2019–2020)
Vulnerability	Slope	Spatial	NASA (ASTER-GDEM V03)
	Land use	Spatial	Google Earth Engine (Landsat 8 OLI)
	Lithology	Spatial	Geological Agency (Geological Map of Indonesia 1:250,000)
	Distance to lineament	Spatial	Geological Agency (Geological Map of Indonesia 1:250,000)
	Distance to stream	Spatial	Geospatial Information Agency (Topographical Map of Indonesia 1:50,000)
Hazard	Rainfall pattern	Statistics	Meteorology Climatology and Geophysics Council (2011–2020)
	Landslide history	Statistics	Various online sources (2011–2020)

3. Research Analysis

The assessment of landslide risk involves three different analyses: (1) weighted overlay to acquire the vulnerability index; (2) scoring for concluding the spatial distribution of the risk triangle (vulnerability, exposure, and hazard); while (3) time series management is utilized to obtain the temporal distribution of the landslide events potential based on the rainfall pattern. The weighted overlay integrates the vulnerability factors by its arithmetical feature of raster data layers utilizing ArcGIS software. The vulnerability factors are classified consecutively utilizing AHP (Analytical Hierarchy Process) for each vulnerability factor on a scale of 0–9. The higher weight composes the more susceptible landslide probability on the area and vice versa (Banuzaki & Ayu, 2021). The next step is integrating the calculation components: the causative factors percentage and the weight of classes to generate Landslide Potential Index (LPI) as the outcome of the vulnerability assessment.

Meanwhile, the rainfall pattern analysis is performed using the time series management tools package in R (Hydro TSM) to analyze and interpolate the hydrological modeling by plotting the time series indicators (date, month, and year). In the next step, the vulnerability index results are combined with the exposure variable (tourists' number) and hazard (landslide history) to produce the risk probability to describe uncertainties associated with the random occurrence of landslide events (Lee, 2009). The risk assessment comprises modified probability determination from Lee's version (2009) by transforming the LPI value into Potential Ratio (PR) and visitors number into Visitors Ratio (VR) by comparing the value of these two variables for each heritage object to the total value in the region. Meanwhile, the Landslide Frequency (LF) was acquired from the number of recorded landslides per 10 years.

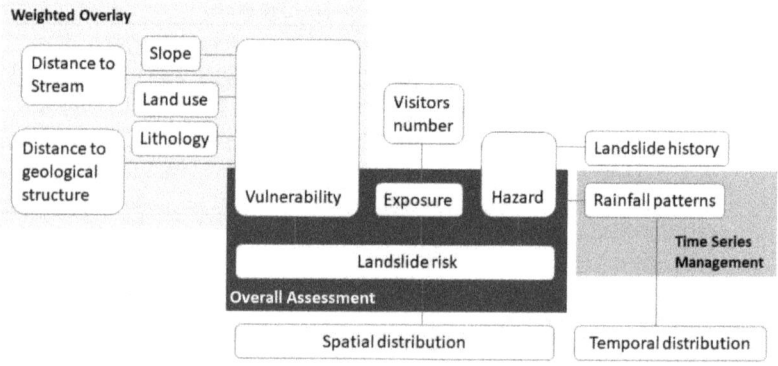

Fig. 2: Research analysis

4. Limitations and Disclaimers

These analyses are limited to the location of tourist destinations that are considered heritage objects (natural and cultural). However, there are still many other tourist destinations that are not included in this analysis, both heritage and non-heritage tourism. This is related to tourist data as an indicator of exposure aspects that affect the results of risk analysis. In terms of data availability, the rainfall data is limited and only collected from the closest observation station (Ngaringan, Grobogan Regency, Central Java Province). Hence, more data is needed to produce a more in-depth analysis in the future. The absence of spatial data of landslide history as a part of hazard indicator that manifests the specific location of landslide events with certain coordinates reduces the accuracy level of data in the past 10 years.

II. Region of Interest

1. Administrative Units and Regional Profile

Lawu Mountain is an inactive volcanic mountain with a height of 3,265 m above sea level or 10,712 ft which is the 76th highest mountain in the world, and The third highest mountain in Java. Located in the southern border of Central Java and East Java. Most of the western slope belongs to the Karanganyar Regency, Central Java Province in the District of Metasih, Karangpandan, Jenawi, Tawangmangu, and Ngargoyoso. On the other side, some small areas are in Plaosan and Poncol, Magetan Regency, and Ngawi Regency East Java Province. This mountain extends from the north to the south, separated by the highway connecting the provinces of Central Java and East Java, with Cemoro Sewu as the highest hamlet. Covering a total area of 15,000 ha, this region is very distinctive from the topographical perspective. The flow of wet southeast wind contributes to the rainy conditions that are relatively fertile with dense vegetation, even in the dry season.

The intensity of rainfall in the Lawu Mountain is divided between >2000–3000 mm/year and >3000–4000 mm/year (Nita et al., 2019). Based on hydrological conditions, this region has various water sources. There are 42 rivers grouped into six sub-watersheds: Kedung sub-watershed, Jlantah Walikan sub-watershed, Samin sub-watershed, Pepe sub-watershed, Mungkung and Kenatan watersheds. From a geomorphological perspective, this mountain constitutes a depression filled with young volcanic sediment. The activity of Mount Lawu produces rocks of andesite to basalt composition. The morphology of this area is dominated by wavy, steep, and very steep. The western slope of Lawu Mountain has a significant role in the environment such as the source of water for Karanganyar, Sukoharjo, Wonogiri, Sragen, Magetan, Ponorogo, and Madiun administrative areas. The other advantages are the vivid ecosystem for typical flora and fauna (Setiawan, 2001).

2. Heritage Objects and Values

Heritage as a part of the tourism sector recently has had a significant increase in visitors. Swarbrooke (1994) explains that heritage tourism pursues the people's curiosity in experiencing heritage (Nguyen & Cheung, 2013). However, various ways in defining the heritage objects based on their values take the attention of heritage experts for specific evidence related to the indigenous attributes. Indonesia's heritage is classified into the category of cultural (immovable and movable) heritage (including cultural property) and natural heritage, divided into five subcategories (UNESCO, 2010).

Archaeological Remains

As an element of resources, archaeological remains hold their new role in tourism activity chiefly in improving the socio-economic conditions of the local communities (Comer, 2012). 38 identified points of these relics are retaining the historical value of Hindu civilization in Indonesia, particularly on the island of Java. Three main Temples: Cetho, Sukuh, and Kethek represent religious buildings that are inhabited and used by sages and ascetics to perform hermitages on mountain slopes, hills, riverbanks, and in the forest to study Hinduism (Purwanto & Titasari, 2018). In 1995, the Temple of Sukuh was registered to UNESCO as a part of the world's cultural heritage and became famous for its similarity in shapes to the Mayan pyramids. At the top of Mount Lawu, 10 terraced buildings are believed to have a certain spiritual meaning. As the development Islamic belief, visiting this mountain remains as the cultural practice of the local people every new Hijri year (1 *Suro*) since the establishment of the Hijriyah calendar by Sultan Agung Hanyokrokusumo in the 16th century (Purwanto et al., 2018).

Agriculture

Agrotourism in terms of rural development emphasizes the harmonious relationship between the economic sphere, natural resources, and specific traditions of local agriculture (Ammirato & Felicetti, 2014). Kemuning Tea Plantation offers memorable experiences such as walking around the tea plantation, enjoying the natural scent of green tea and the fresh atmosphere. Meanwhile, Amanah Agrotourism offers agricultural activities by involving the local communities participate in a sustainable framework.

Fig. 3. Some tourism heritage destinations. Source: https://puromangkunega ran.com/perkebunan-teh-kemuning-mangkunegaran/https://travelspromo. com/htm-wisata/agrowisata-amanah-tawangmangu/

Parks

Parks have a significant increase in relevance to sustainable development as a manifestation of cultural values (González, 2013). This is related to the visitors' positive behavioral intentions to the environment as one of the main goals of educational value in a national park context (Buonincontri et al., 2017). Balekambang Park is one of the alternative tourism destinations that gives experience traveling the world through miniature landmarks and Puri Taman Saraswati represents a place of worship as a part of Cetho Temple, established by the Regents of Karanganyar and Gianyar Bali in 2007 that contains cultural and religious values for the Hindu local community.

Tropical Forests

Tropical (rain) forests are a settling place for the oldest living cultures of aboriginal people (Weber et al., 2021). Taman Hutan Raya Karanganyar constitutes a majestic heritage from the reign of Mangkunagoro I and provides shelter to more than thirty types of fauna and is a home for hundreds of flora notably bougainvillaea flowers. Tenggir Park and Gunung Campground (also known as Sekipan Hill) offer natural views of the forest with a very beautiful line of pine trees.

Water Heritage

Water as the earliest resource for human life became a crucial part of development in heritage science as well as the discoveries of prehistoric creatures that settled down along rivers, lakes, and coastlines (Hein,

2020). Waterfalls as a part of geocultural heritage involve several multidimensional frameworks to be analyzed. For instance, in terms of environmental assessment, they should be viewed based on the relationships between geomorphology and cultural aspects such as archaeological, historical, architectural, and anthropological (Mesmin & Etoga, 2014). Pringgodani is the highest waterfall in Karanganyar and has two levels with a height of more than 100 m. The other exotic natural heritage destination is Madirda Lake with its relation to historical and belief values of *Cupu Manik Astagina*'s heirlooms (local sacred figure) (Rudi, 2021).

III. Results and Discussions

1. Risk of Landslide

Impacts and cataclysms in the aftermath of climate change have risen as research topics, particularly in heritage sites affected by physical conditions (Guzman et al., 2020). The main tasks of their works are identifying related aspects between the disastrous events' patterns, cause factors, and their threats to the heritage objects by documenting the constraints, assessing the vulnerability, giving some recommendations to reduce the risks (Seekamp & Jo, 2020). This section consists of four parts of the discussion in exposure, vulnerability, hazard, and overall assessment derived from the risk triangle.

2. Exposure to Heritage Tourism

The line graph below indicates that during 2019–2020, there were some significant fluctuations in the number of tourists who visit the Karanganyar region. It should be mentioned that this graph is representing the general visitors' number for all of the tourism destinations (the selected destinations are included). In 2019, the graph reached the highest peak (720,000) in June before it fell dramatically to the lowest point (380,000) in August. Several months later, it recovered and increased moderately to the second-highest point (530,000) in December. The two highest trends were indicating the peak season of summer holidays (around June) and new year holidays (December).

Landslide Vulnerability, Risk and Resilience Management

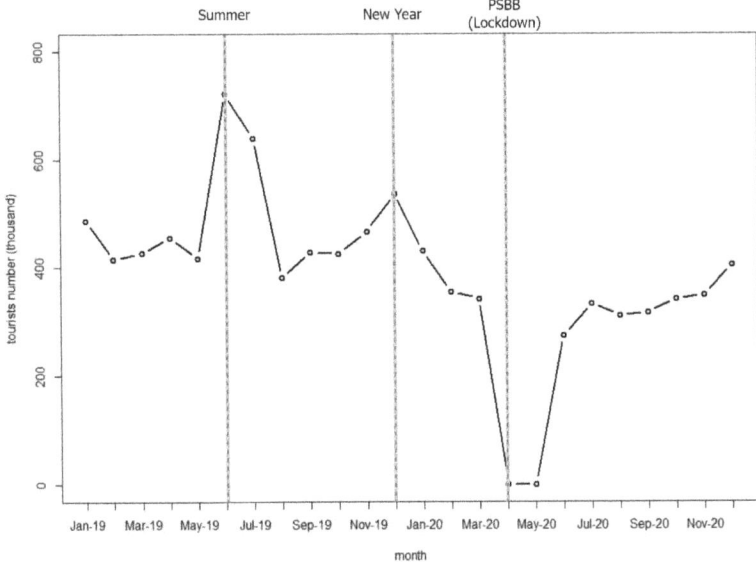

Fig. 4: Number of Visitors (2019–2020)

The next year (2020) had a very different trend. The Covid-19 pandemic punched the graph down since most of all destinations were closed in April-May. It contributed to none of the visitors during the PSBB policy (Indonesian policy for lockdown) being enforced. Several months later, when the summer holidays came and some of the destinations reopened for tourists, it was restored to the amount before the lockdown. For the focused heritage tourism destinations, the visitors' number is collected from accumulated numbers during 2020 (Jan–Dec). The spatial distribution of the tourists specifies that Kemuning Tea Plantation, Grojogan Sewu Waterfall, and Madirda Lake are the top three destinations based on the number of visitors (see Fig. 5).

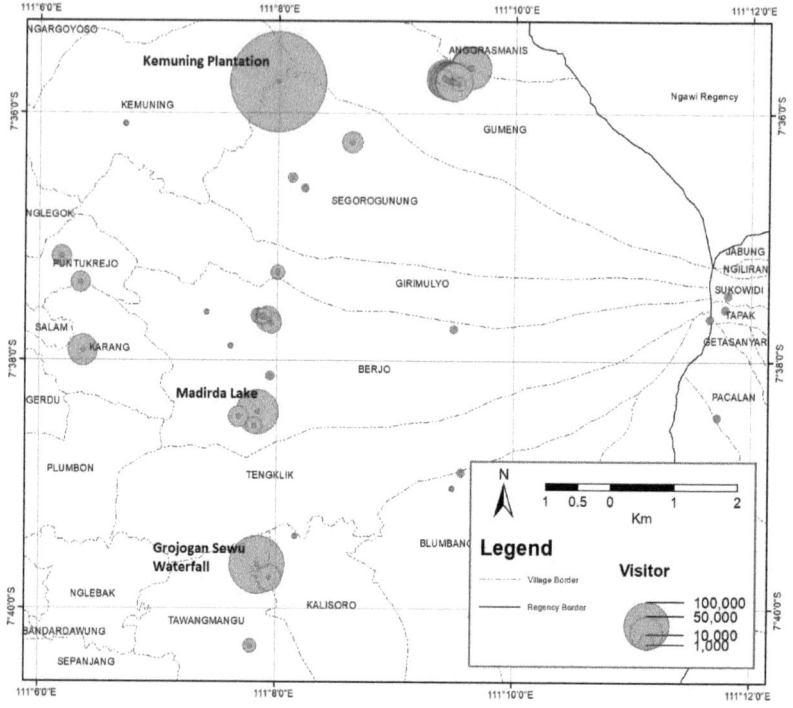

Fig. 5: Visitors distribution map

3. Environmental Indicators of Vulnerability

Three classes of the slope were determined with a dominant area of gentle slope (0°–10°) in the west, moderate slope (10°–25°) covering the foot of the mountain area, and steep slope (>25°) dominating the peak. The land use was dominated by forest, while the other types were identified as agricultural land, built-up area, and barren land. In terms of lithology, the entire research area was covered by younger volcanic rock which was carved up into six categories: Candradimuka Lava, Jobolarangan Breccia, Jobolarangan Tuff, Lawu Lahar, Lawu Volcanic Rock, and Sidoramping Lava.

From a geological perspective, distance to lineament was classified into three classes: 400 m, 700 m, and 1000 m. The distance to these geological features may affect the stability of the land by fragmenting the rock body into plates. This condition will increase the possibility of weathering, which implies the landslide potential elevation. As well as the lineament, streams were the other prominent factor of slope instability. The rising volume of water during the rainy season contributed to rock erosion and flow formation on the land. The distance of slope to the stream was classified into three classes, i.e., close (10 m), moderate (30 m), and distant (50 m).

The amalgamation of the environmental indicators was processed into a map of vulnerability levels as shown in the following figure. Medium category area was predominant with covering percentage 57.14 %, while low and high vulnerability area was 21.59 % and 21.28 % consecutively.

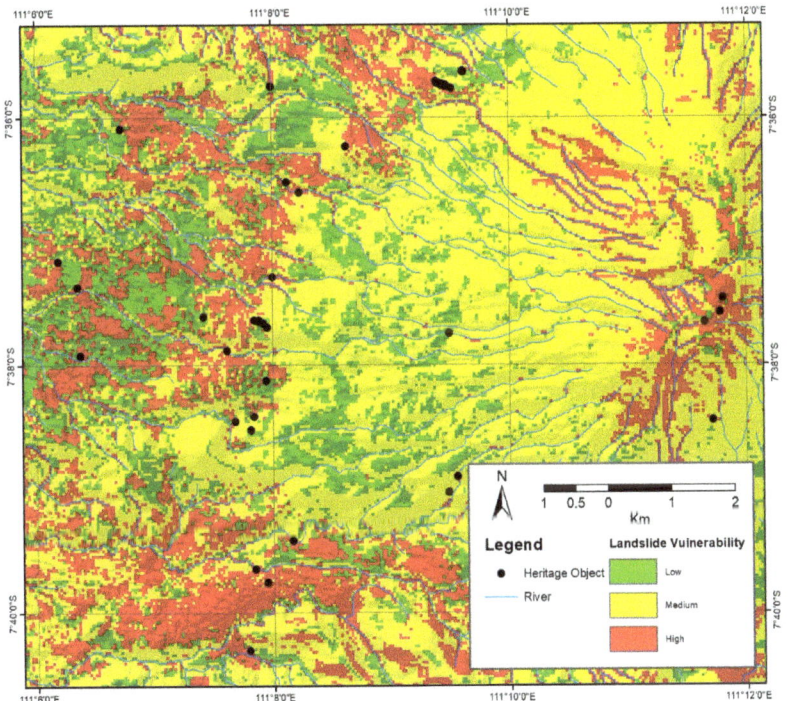

Fig. 6: Vulnerability levels map

The major finding has been highlighted on high vulnerability whereby half of the total heritage tourism destinations were placed in a vulnerable area (48.1 %). Cetho Temple and Sukuh Temple were in the top ranking. By comparison, the moderate and low rates of vulnerability stood in the same percentage around a quarter of the total locations (25.9 %). Segoro Gunung Waterfall and Segoro Site were in the zenith position for the moderate category, while Kethek Temple and Cemoro Bulus Site were the prominent attractions at the low vulnerability level.

4. Hazard: Rainfall Patterns and Landslide History

Results of rainfall pattern analysis using HydroTSM convey that the values for daily rainfall varied throughout time.

For monthly patterns, the time-series graph shows that the rainfall tended to reach the highest point at the beginning (Jan–Mar) and the end of the year (Nov–Dec), while the lowest amount is in the middle (Jul–Aug) every year. This is confirmed by the boxplot that appears the different distributional characteristics of the months were resembling a horse's saddle. Meanwhile, the given histogram indicates 0–50 mm/month as the value that had the highest probability (>0.005) to occur within ten years (2011–2020). Based on geographical distribution, Karanganyar is part of the monsoon type area. It's characterized by the existence of a significant difference between the rainy season and the dry season in a year where the highest rainfall tends to occur during Jan–Apr and Nov–Dec (wet months), while the lowest rainfall tends to occur during May–Oct (dry months) (Tukidi, 2010). These wet months are considered to have the highest probability of climate-related phenomena, including landslide events related to disaster risk probability.

Landslide Vulnerability, Risk and Resilience Management

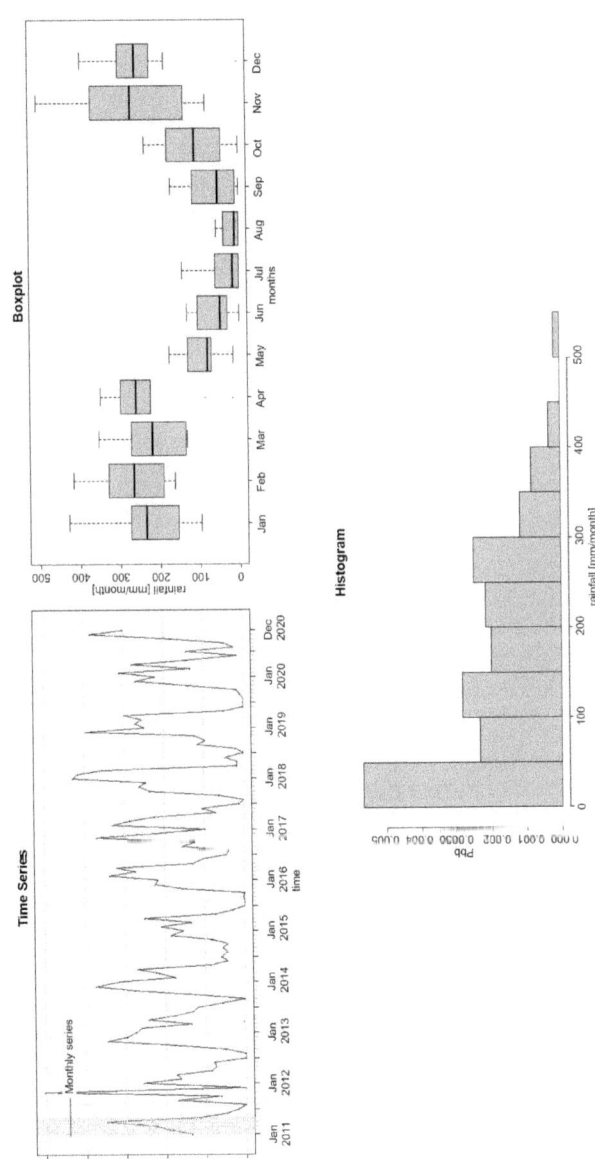

Fig. 7: Monthly rainfall pattern

For annual patterns, the result reveals that 2015 was the year with the lowest rainfall, while 2016 had the highest level of rainfall (2390 mm/year). Meanwhile, the landslide historical records proclaim that the highest number of landslide incidents occurred between January and February for several years (2011, 2012, 2014, and 2017). However, there was an outlier recorded in August 2021 indicating that there was an anomaly in landslide phenomena during the year that was identified as a dry month. Several sources indicate that the probability of landslide events can occur during the alternation between dry and wet months, when extreme increases in rainfall intensity submerge the initially dry soil and enter through cracks on the surface, causing rapid lateral movements (Rahman & Samsurizal, 2018).

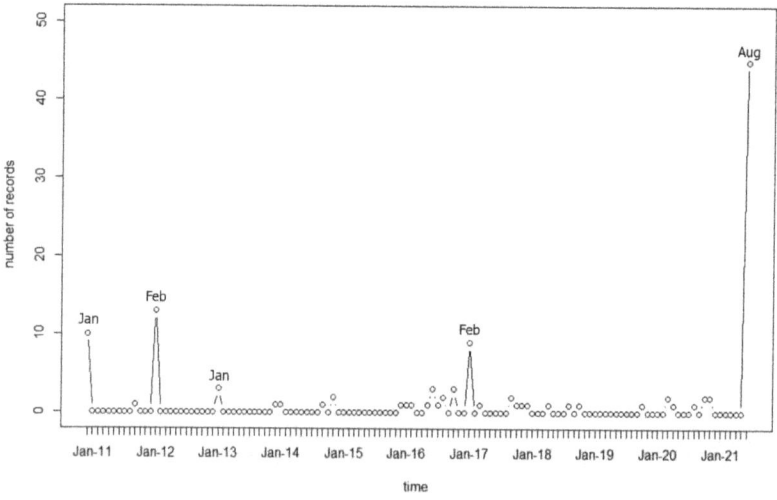

Fig. 8: Landslide history

The ground movements can be divided into several categories (Cruden & Varnes, 1996). Soil crawl is the slowest ground movement, potentially occurring on the coarse and fine grain surface. Although almost undetectable, a prolonged type of landslide can cause telephone poles, trees or houses to tilt downward. Meanwhile, debris flow is the fastest type of landslide that occurs when the moving soil mass is driven by water. The flow rate depends on the slope, volume, pressure, and the type of dissolved material that can cause heavy casualties. The most potential area along the valley with hundreds to thousands of meters to the watersheds around the volcano.

5. Overall Assessment

The results show that the maximum probability of landslide occurrence was 0.35 at Balekambang Park, while the lowest value was the Planggatan Site with a value of 0.08. Some heritage objects which have similar probability were clustered to abbreviate the result report. These probability values indicated the chance of landslide events every year. For example, the annual probability of landslide occurrence at Balekambang Park was 0.35 (35 in 100 chance), the annual probability of no event would be 0.65 (65 in 100 chance). The detailed risk probability and the cluster of similar valued heritage objects are expressed in the figure below.

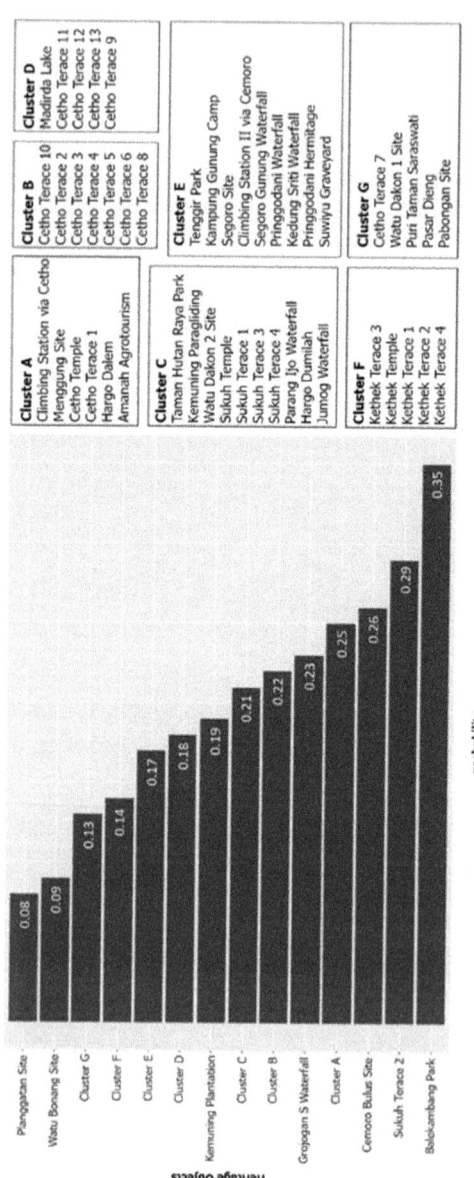

Fig. 9: Risk probability of landslide events for each heritage objects

IV. Resilience Management of Community

The notion of resilience in heritage tourism essentially accounts for diverse models. However, not all resilience models can provide proper answers in addressing questions or problems in a certain region with a specific characteristic of human-environment interactions. This section focuses on the discussion of the resilience applications and implementation in particular for the heritage tourism sector in the region of Karanganyar.

1. Previous Resilience Application, Practical Issues and Challenges in Heritage Tourism Sector

The concept of resilience today is predominantly sourced from Holling (1973) who gestated the 'ecological resilience' that focused on the ability to absorb disturbance and accentuate persistence (Holling, 1973). However, the environmental-oriented dominant approach in this concept failed to convey the more complex issues in resilience management. This was driven by the research conducted by subsequent experts. Walker et al. (2004) proposed an expansion of the ecological resilience concept to the socio-ecological realm to solve questions related to disturbances impact in regime shifting and reorganization (Walker et al., 2004).

Changes in socio-economic conditions due to a catastrophic event as it is currently taking place also clarify the need for an integrated resilience framework. For example, in this era of the Covid-19 pandemic, the travel and tourism sector has unquestionably been one of the hardest hits (Cambra-Fierro et al., 2022). UNWTO Secretary-General presents Five Priorities for Tourism's Restart (Costa, 2021). These five priorities highlight diverse scopes:

1. Mitigate socio-economic impacts on livelihoods,
2. Boost competitiveness and build resilience, including through economic diversification,
3. Advance innovation and digital transformation of tourism,
4. Foster sustainability and green growth,
5. Coordination and partnerships to restart and transform the sector towards achieving SDGs.

The threshold to determine if a heritage object is in danger is currently not well-defined as it cannot be unilaterally applied to the diverse types of heritage: the concept of transformation is required. Thus, adaptation can be planned when monitoring projects that a heritage object will be negatively affected (e.g., climate change) (Seekamp & Jo, 2020).

However, although adopting the social-ecological systems approaches can broaden the scope to the political or economic factors, they tend to focus on structures and functional features of institutions and neglect the wider 'political, historical and cultural sense'. Consequently, the post-disturbances transformation of heritage sites requires integration and collaboration among the stakeholders. Kruse (2017) offered a new approach to enhance the socio-ecological system in resilience concept: community resilience. Through the embRACE framework, conceptualizing community resilience convolutes specific parties in a three-dimensional framework.

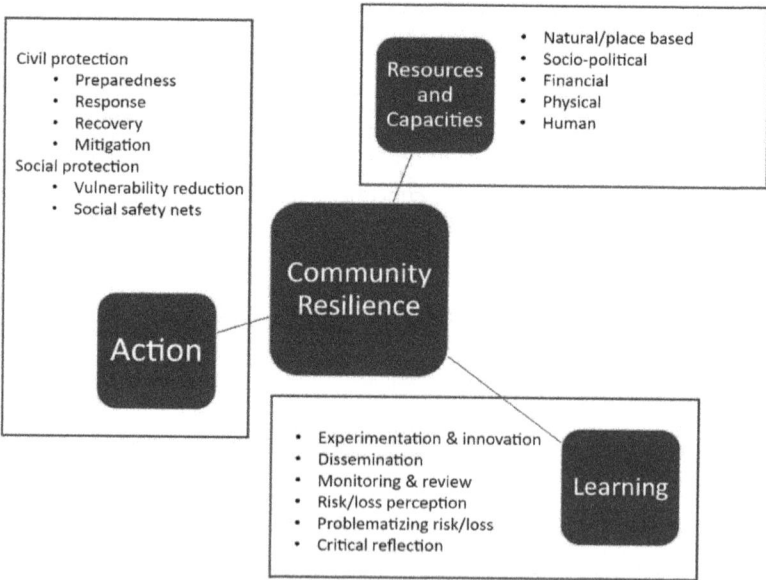

Fig. 10: embRACE framework of community resilience (Reprocessed from Kruse et al., 2017)

2. Community Resilience Review of Heritage Tourism in Karanganyar

This section focuses on heritage tourism in Karanganyar in terms of resources and capacities, actions and learning as the main components of community resilience. Karanganyar represents abundant natural and physical potential with a very beautiful and enchanting beauty in the Mount Lawu area. In terms of resources and capacities, the development of natural resources is not limited to natural and cultural tourism but also includes agricultural and livestock production (Toni, 2012). The folks' habits and behavior are manifesting the relationship between humans and land (nature) as a balanced relation through existing local cultural values such as the myth of the *Sabuk Janur* (Lestari et al., 2018). The farmers have developed many years of experience and knowledge in managing the sustainability of water resources as a consequence of the physical conditions of mountainous areas of their settlements. These capitals contributed to 5.93 economic growth in 2019 although it fell −1.87 in 2020 due to the pandemic, while the economic structure of Karanganyar is dominated by the manufacturing industry (47 %) and agriculture, forestry, and fishing (13 %).

In the aspect of action, the local regulation rigorously set the expedient coordination for preparedness, response, recovery, and mitigation between the Regional Disaster Management Agency (BPBD), the district and provincial governments, the tourism service providers, and the volunteers. Entering the rainy season, BPBD Karanganyar collected and conducted training for volunteers as part of the emergency response (Ludiyanto, 2021). The Office of Education, Youth, and Sports Karanganyar urged tourism service providers such as hotel and restaurant managers to anticipate natural disasters that could threaten the safety of visitors, regarding the estimation of the surge of visitors during the year-end holidays. Mitigation also includes data collection on vulnerable assets, environmental conditions and early detection of damage, mainly in tubing areas that utilize rivers and springs. These processes are supported by a strong network and regulation for facile coordination (Handayani, 2020).

Global disaster management has undergone a paradigm shift from responsive to preventive, from sectoral to multisectoral, from government responsibility to shared responsibility, from centralization to decentralization, and from emergency response to disaster risk reduction. Referring to the analysis results of risk probability in the previous section, the priority scale is set as the basis for decision making both before, during, and after the disturbance. As a part of the learning process, it can be identified that a consciousness between the local people and other related stakeholders had been built which is divided into four stages:

1. Emergency response carry out immediately when a disaster occurs to save human victims;
2. Rehabilitation after a disaster to restore the basic functions of the community's lives;
3. Reconstruction to restore conditions as before; and
4. Disaster prevention is aimed at reducing the risk of disasters that may repeat in the future.

The strength of resilience recline on several main aspects such as community-based expertise, the network of volunteers, independence and neutrality (IFRC, 2020). However, the consciousness of the disaster threats is not followed by proper implementations. Indonesia's heritage conservation is established by the Law of the Republic of Indonesia Number 11 of 2010 Concerning Cultural Conservation, while disaster management rests on the Law of the Republic of Indonesia Number 24 of 2007 Concerning Disaster management executed by the National Agency Disaster Management (BNPB). This regulation covers funding issues in the field of disaster as well. Referring to BNPB Regulation Number 5 of 2018 concerning Conditions and Procedures for Implementing Disaster Management in Certain Circumstances article 11, funding includes:

1. Mobilization of human resources;
2. Deployment of equipment;
3. Logistics deployment;
4. Immigration, excise and quarantine;
5. Licensing;
6. Procurement of goods/services;
7. Management and accountability of money and/or goods; and
8. Command rescue to coordinate sectors/agencies.

Although this set of laws is updated by Presidential Regulation Number 87 of 2020 Concerning Disaster Management Master Plan Year 2020–2044, the implementation of this regulation is still being developed. From this point of view, a policy regarding the general definition and related terms to detrimental events had been established, but remained practically immature. Social dilemmas often arise due to the misalignment of individual goals (maximizing profit) and community goals (reducing social impact) (DPIS IPB, 2021).

For example, the National Agency mentioned a problem in financing the action in particular a limited national budget. This may be affected by a lack of integrity exacerbated by a very vague financing framework. Another issue is coming from the technical realm such as the availability of data and database management systems for disasters. The maps of disaster-prone areas which are associated with the existence of cultural heritage are unattainable including those associated with the threat of natural, human-affected and multi-hazard disasters. Even though the map of hazard risks is produced, the other concern is the level of scale detail where the available maps are only can be analyzed on the level of the province. The InaRISK platform recently provides the scale of 1: 250.000 only, while the more specific areas on the regional levels (scale of 1: 50.000 and 1: 25.000) are still under development and require cross-sectoral or institutional collaboration (Rahardjo, 2021).

Meanwhile, in 2006, initial attempts to transfer data, software and ownership of DesInventar (a project was initiated by LA RED) to the Government of Indonesia were met with reluctance. This reluctance was partly due to the existence of a similar inventory database of individuals at the ministry level and the ineffective marketing of the utilized methodology. This condition was exacerbated by the absence of government acknowledgment of data validation and the absence of laws and regulations regarding the development of the database (*BNPB*, 2008). Therefore, an integrated management approach needs to be continuously improved as well as the issues on the limited human and funding resources require coordination between professionals and institutions in charge of sites, volunteers, governmental roles, and other related institutions. To address the identified problems in disaster resilience management for heritage tourism, some recommendations were given to improve the quality of work in the future.

Tab. 2: Recommendations

Aspect	Review	Priority to Address
Law, policy, and regulation	General law and policy are available but the particular regulation need to be set up	(1) Development planning lies on disaster risk reduction principles (2) Integrated designation of governor and commander for the emergency task force (3) Designing specific regulations for technical aspects of disaster risk management
Financial management	Need to be improved	(1) Establishing a financial framework (2) Expanding budget rate for preventive attempts, emergency response, and mitigation
Massive involvement of academics and experts	Some of the academic improvements are accomplished but other aspects need to be enhanced	(1) Improvement in heritage values and disaster education (2) Development of data science for heritage and disaster risk reduction (3) 'Penta Helix' collaboration by involving the government, business institutions, academics, the community, and the participation of the media
Database management systems	An appropriate decision and a massive development are needed	(1) Establishing national standards and guidelines for statistics, spatial, and other types of heritage and disaster data (2) Development of an integrated Early Warning System (EWS) (3) Quality improvement of available mobile application
Collaboration and networking	Some of the aspects are running well but enhancements are still necessary	(1) More massive local society involvement (2) Expanding collaboration and networks with local, national, and international heritage and disaster-related institutions (3) Simulation of disaster management exercises on a regular and continuous basis

Conclusion

The geographical background of Indonesia as an archipelagic country constitutes a vulnerable area for many types of hazards, in particular caused by climate change. Landslide, as the climate-originated disaster threatens a wide range of human aspects, one of them is heritage tourism activity. Lawu mountain represents a vast area with diverse heritage tourism destinations such as archaeological remains, parks, tropical forests, and water heritage representing both natural and cultural heritage as valuable national assets. The general analysis of tourists' numbers as the exposure element indicates significant shifts during 2019–2020 with the lowest point caused by the policy during the Covid-19 pandemic in April 2020. The spatial distribution of visitors shows on the one hand, that Kemuning Tea Plantation, Grojogan Sewu Waterfall, and Madirda Lake are the top three destinations in Karanganyar.

On the other hand, the environmental indicators reveal that 48.1 % of the destinations were located in a vulnerable area, while 25.9 % of them are had moderate and low rates of vulnerability. The analysis of rainfall patterns as the element of hazard gives different trends for each time unit during 10 years (2011–2020). The general course of the daily pattern tended to be by 0 (zero), only some days in January–February 2017 gave the highest rainfall by 140 mm/day as the outliers. For monthly patterns, the highest point is reached at the beginning (Jan–Mar) and the end of the year (Nov–Dec), while the lowest amount is in the middle (Jul–Aug) every year. In annual patterns, 2015 was the year with the lowest rainfall, while 2016 had the highest level of rainfall (2390 mm/year). Meanwhile, the risk probability that had been produced from the overall assessment proclaimed that Balekambang Park constituted the maximum probability (0.35), while Planggatan Site contributed the lowest probability (0.08).

The community resilience framework review to the Karanganyar region provides ample evidence on how the interconnected stakeholders are taking account to manage the landslide disaster through a set of principles in resources and capacities, action, and learning. Some main activities such as volunteers training, data collection on vulnerable assets, environmental conditions, early detection of damage, and funding are implemented as part of the preparedness, response, recovery, and mitigation. However, the quality should continue to improve to provide more robust and comprehensive outputs in the future. Some recommendations and solutions are suggested to address the identified issues including law, financial management, the role of academics and experts, database management systems, collaboration and networking.

References

Alexander, D. E. 2013, 'Resilience and Disaster Risk Reduction: An Etymological Journey', *Natural Hazards and Earth System Sciences*, Vol. *13*, Issue 11, pp. 2707–2716. https://doi.org/10.5194/nhess-13-2707-2013

Ammirato, S., & Felicetti, A. M. 2014, 'The Agritourism as a Means of Sustainable Development for Rural Communities: A Research from the Field', *The International Journal of Interdisciplinary Environmental Studies*, Vol. 8, Issue 1, pp. 17–29.

Andreu, M. D. 2017, 'Heritage Values and the Public', *Journal of Community Archaeology & Heritage*, Vol. 4, Issue 1, pp. 2–6. https://doi.org/10.1080/20518196.2016.1228213

Banuzaki, A. S., & Ayu, A. K. 2021, 'Landslide Vulnerability Assessment Using Gis and Remote Sensing Techniques: A Case Study from Garut – Tasikmalaya road', *IOP Conference Series: Earth and Environmental Science*, 622. https://doi.org/10.1088/1755-1315/622/1/012005

Buonincontri, P., Marasco, A., & Ramkissoon, H. 2017, 'Visitors' Experience, Place Attachment and Sustainable Behaviour at Cultural Heritage Sites: A Conceptual Framework', *Sustainability*, Vol. 9, Issue 7, pp. 1–19. https://doi.org/10.3390/su9071112

Cambra-Fierro, J. J., Fuentes-Blasco, M., Huerta-Álvarez, R., & Olavarría-Jaraba, A. 2022, 'Destination Recovery During COVID-19 in an Emerging Economy: Insights from Peru', *European Research on Management and Business Economics*, Vol. 28, Issue 3, pp. 1–9. https://doi.org/10.1016/j.iedeen.2021.100188

Central Bureau of Statistics, 2021, *Kabupaten Karanganyar dalam Angka 2021*, https://karanganyarkab.bps.go.id/publication/2021/09/24/2461f686aa12a81b564d6abd/kecamatan-kebakkramat-dalam-angka-2021.html

Cepeda, J., Smebye, H., Vangelsten, B., Nadim, F., & Muslim, D. 2010, 'Landslide Risk in Indonesia', *Global Assessment Report on Disaster Risk Reduction*, pp. 20.

Comer, D. C. 2012, Tourism and Archaeological Heritage Management at Petra Driver to Development or Destruction, New York, Springer.

Costa, J. 2021, 'Sustainability as a Measure of Tourism Success: The Portuguese Promotional Tourism Boards' View', *Worldwide Hospitality and Tourism Themes*, Vol. 14, Issue 1, pp. 65–71. https://doi.org/10.1108/WHATT-10-2021-0131

Crichton, D. 1999, 'The Risk Triangle', *Natural Disaster Management*, Vol. 3, pp. 102–103.

Crozier, M. J. 2010, 'Deciphering the Effect of Climate Change on Landslide Activity: A Review', *Geomorphology*, Vol. 124, Issue 3–4, pp. 260–267. https://doi.org/10.1016/j.geomorph.2010.04.009

Cruden, D. M. & Varnes, D. J. 1996, 'Landslide Types and Processes', *Transportation Research Board, U.S. National Academy of Sciences, Special Report*, Vol. *247*, pp. 36–75.

DPIS IPB, 2021, *Menyoroti Bencana di Indonesia: Dampak, Penanggulangan, dan Pencegahan*. Direktorat Publikasi Ilmiah dan Informasi Strategis, https://dpis.ipb.ac.id/menyoroti-bencana-di-indonesia-dampak-penanggulangan-dan-pencegahan/

European Commission, Environment challenges, Climate Action, retrieved 5th of November 2021, from https://ec.europa.eu/clima/eu-action/adaptation-climate-change/how-will-we-be-affected/environment-challenges_en

European Comission, 2007, 'An Assessment of Weather-Related Risks in Europe: Maps of Flood and Drought Risks'.

Gariano, S. L., & Guzzetti, F. 2016, 'Landslides in a Changing Climate', *Earth-Science Reviews*, Vol. *162*, pp. 227–252. https://doi.org/10.1016/j.earscirev.2016.08.011

González, P. A. 2013, 'Cultural Parks and National Heritage Areas: Assembling Cultural Heritage, Development and Spatial Planning' (1st ed.), Newcastle on Tyne, UK, Cambridge Scholars Publishing.

Guzman, P., Fatoric, S., & Ishizawa, M. 2020, 'Monitoring Climate Change in World Heritage Properties: Evaluating Landscape-Based Approach in the State of Conservation System', *Climate*, Vol. *8*, Issue 3, pp. 1–19. https://doi.org/10.3390/cli8030039

Handayani, S. S. *Waspada Bencana, Pengelola Wisata di Karanganyar Diminta Lakukan Mitigasi*. Solopos, retrieved the 16th of January 2022, from https://www.solopos.com/waspada-bencana-pengelola-wisata-di-karanganyar-diminta-lakukan-mitigasi-1095516

Hein, C. 2020, Adaptive Strategies for Water Heritage, Springer Nature.

Holling, C. S. 1973, 'Resilience and Stability of Ecological Systems', *Annual Review of Ecology and Systematics*, Vol. *4*, pp. 1–23. https://doi.org/10.1146/annurev.es.04.110173.000245

Indonesian Geological Agency, 2008, Yearly report of the Geological Agency of Indonesia.

International Federation of Red Cross and Red Crescent Societies (IFRC), 2020, Strengthening Law and Disaster Risk Reduction (DRR) in Indonesia.

Kreator: Mei Norm. kompasiana. Retrieved January 16, 2022, from https://www.kompasiana.com/meinorma/5519cb53813311dd7a9de0e7/karanganyar-dan-sejuta-pesonanya

Kruse, S., et al. 2017, 'Conceptualizing Community Resilience to Natural Hazards – the embRACE Framework', *Natural Hazards and Earth System Sciences*, Vol. *17*, Issue 12, pp. 2321–2333. https://doi.org/10.5194/nhess-17-2321-2017

Leblanc, F. 2006, The Built Heritage Conservation Process. http://ip51.icomos.org/~fleblanc/publications/pub_2007_conservation_process.html

Lee, E. M. 2009, 'Landslide Risk Assessment: The Challenge of Estimating the Probability of Landsliding', *Quarterly Journal of Engineering Geology and Hydrogeology*, Vol. *42*, Issue 4, pp. 445–458. https://doi.org/10.1144/1470-9236/08-007

Lestari, et al. 2018,. 'Sabuk Janur: Tools to Move Community Participation in Reducing Natural Disasters and Environment (Case Study at Lawu Mount Slope in Indonesia)', *IOP Conference Series: Earth and Environmental Science*, Vol. *142*, Issue 1, pp. 1–8.

Ludiyanto, A. 11th of October 2021, Musim Hujan Tiba, BPBD Karanganyar Siapkan Sukarelawan Tanggap Bencana. Solopos, retrieved 16th of January 2022, from https://www.solopos.com/musim-hujan-tiba-bpbd-karanganyar-siapkan-sukarelawan-tanggap-bencana-1171299

Mesmin, T., & Etoga, M. H. 2014, 'The Lobé Waterfall, an Exceptional Geocultural Heritage on the Coast of Cameroon between Sustainable Tourism and the Conservation of Cultural Identities', *A Tourism Review*, Vol. *4*, Issue 5, pp. 1–18. https://doi.org/10.4000/viatourism.964

Muhammadi, R., et al. 2019, 'Penerapan Sistem Informasi Geografi Dalam Pendugaan Sebaran Daerah Rawan Longsor Di Kecamatan Ngargoyoso, Kabupaten Karanganyar', *Jurnal Tanah dan Sumberdaya Lahan*, Vol 6, Issue 1, pp. 1083–1092. https://doi.org/10.21776/ub.jtsl.2019.006.1.7

Naryanto, H. S. 2011, 'Analisis risiko bencana tanah longsor di Kabupaten Karanganyar, Provinsi Jawa Tengah', *Jurnal Dialog dan Penanggulangan Bencana*, Vol. *2, Issue 1*, pp. 21–32.

Nguyen, T. H. H., & Cheung, C. 2013, 'The Classification of Heritage Tourists: A Case of Hue City, Vietnam', *Journal of Heritage Tourism*, Vol. *9*, Issue 1, pp. 35–50. https://doi.org/10.1080/1743873X.2013.818677

Purwanto, H., & Titasari, C. P. 2018, 'The Worship of Parwatarajadewa in Mount Lawu', *Kapata Arkeologi Scientific Journal of Archaeology and*

Cultural Studies, Vol. *14*, Issue 1, pp. 37–48. http://dx.doi.org/10.24832/kapata.v14i1.472

Purwanto, M. R., et al. 2018, 'Sultan Agung's Thought of Javanis Islamic Calender and its Implementation for Javanis Moslem', *International Journal of Emerging Trends in Social Sciences*, Vol. *4*, Issue 1, pp. 9–14. https://doi.org/10.20448/2001.41.9.14

Rahardjo, S. 2021, Managing Heritage Sites in Disaster-Prone Zone Is Indonesia Ready? *Prosiding Balai Arkeologi Jawa Barat*, Vol. *4*, pp. 283–303.

Rahman, K., & Samsurizal, D. 6th of December 2018, Mitigasi Bencana Tanah Longsor, http://bpbd.jogjaprov.go.id/berita/mitigasi-bencana-tanah-longsor-1

Rudi, A. 29th of September 2021, Telaga Madirda, Lokasi Camping Seru Di Kaki Gunung Lawu, *Native Indonesia*, https://www.nativeindonesia.com/telaga-madirda/

Seekamp, E., & Jo, E. 2020, 'Resilience and Transformation of Heritage Sites to Accommodate for Loss and Learning in a Changing Climate', *Climatic Change*, Vol. *162*, pp. 41–55. https://doi.org/10.1007/s10584-020-02812-4

Setiawan, A. 2001, 'Potensi Gunung Lawu Sebagai Taman Nasional', *Biodiversitas*, Vol. *2*, Issue 2, pp. 163–168.

Toni, S. 22nd of November 2012, Karanganyar dan Sejuta Pesonanya Konten ini telah tayang di Kompasiana.com dengan judul 'Karanganyar dan Sejuta Pesonanya', Klik untuk baca: https://www.kompasiana.com/meinorma/5519cb53813311dd7a9de0e7/karanganyar-dan-sejuta-pesonanya

Tukidi, 2010, 'Karakteristik Curah Hujan di Indonesia', *Jurnal Geografi FIS UNNES*, Vol. *7*, Issue 2, pp. 136–140.

UNESCO, 2010, Indonesia Laws, UNESCO Database of National Cultural Heritage Laws UNESCO/CLT/Natlaws, https://whc.unesco.org/en/statesparties/id/laws/

United Nations Development Programme, 2008, Data dan Informasi Bencana Indonesia, https://bnpb.go.id/uploads/migration/pubs/446.pdf

Walker, B., et al. 2004, 'Resilience, Adaptability and Transformability in Social– ecological Systems', *Ecology and Society*, Vol. *9*, Issue 2, pp. 1–9. https://www.ecologyandsociety.org/vol9/iss2/art5/

Weber, E. T., et al. 2021, 'Managing a World Heritage Site in the Case Study of the Wet Tropics in Northern Queensland Face of Climate Change', *Earth*, Vol. *2*, Issue 2, pp. 248–271. https://doi.org/10.3390/earth2020015

Part III

NATURE BASED SOLUTIONS AND THE CIRCULAR ECONOMY

Chapter 7

Nature-Based Solutions in Climate Adaptation: A Shift from Specific, Isolated Tools to Large-scale Global Conservation

Pascaline Gaborit PhD and Zoé Thouvenot

Nature-based Solutions (NbS) are defined by the IUCN as '*actions to protect, sustainably manage, and restore natural or modified ecosystems, that address societal challenges effectively and adaptively, simultaneously, providing human well-being and biodiversity benefits*' (Cohen- Shacham et al. IUCN 2016). They embrace different areas and sectors such as tree planting, restoration and maintenance of ecosystems, riverbed or wetland restoration to prevent coastal erosion, and green roofs and gardens to 'cool' the city during heat waves. Nature based solutions can therefore play a large role in climate adaptation, as well as in climate mitigation. '*The benefits and opportunities achievable using nature-based solutions to address global and societal challenges have never been more relevant, important or urgently needed than now*' (Wild et al. 2020). NbS is an 'umbrella concept' for other established 'nature-based' approaches such as ecosystem-based adaptation (EbA) and ecosystem-based mitigation, eco-disaster risk reduction, and green infrastructure (Seddon et al. 2020).

Several authors argue that a focus on nature-based solutions may overlook the impacts of climate change on the ecosystems themselves, as these ones experience cascading effects (Malhi et al. 2019). The same authors underscore that the choice of nature-based solutions needs to be based on a sound preliminary analysis on both ecosystems and its impacts. We have realized, for instance, that the deforestation caused by the production of palm oil-based biofuel, could not be integrated into 'nature-based sustainable solutions.' We argue in this chapter that the strong emphasis on nature-based solutions is indeed hiding diverse realities ranging from small-scale city infrastructure to the conservation of large ecosystems. This is a very broad concept, creating some concerns for

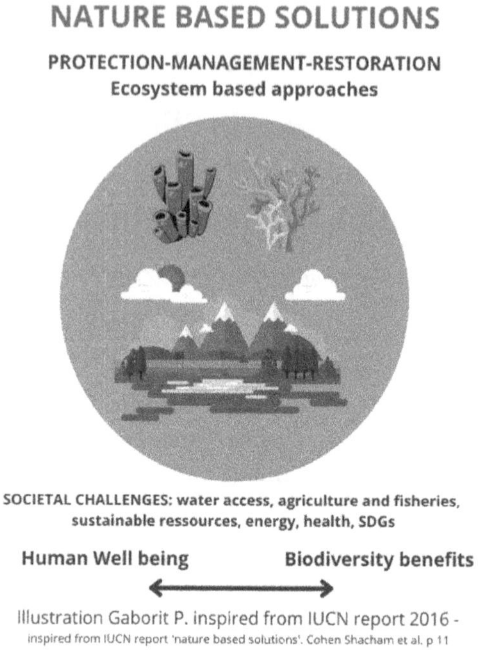

Fig. 1: Nature based solutions (Source: Gaborit P., illustration inspired by Cohen – Shacham et al. IUCN 2016, p.11)

possible misuses[1]. Our analysis is based on a thorough literature review and project findings (*such as our LIFE Adapt Island* project). Many uncertainties are still associated with interventionist approaches to land and marine environmental management operations due to spatial and temporal variations in hydrology, currents, weather, and other unpredictable disturbances, making it almost impossible to predict the trajectory of ecological processes after one or more interventions. As a result, many authors have recently advocated the need to view interventionist environmental management methods as experiments and to adopt a 'learning-by-doing approach', which can be described as an adaptive approach to environmental management. This requires careful monitoring and knowledge of ongoing ecosystem dynamics to learn and adapt based on observed results. This is the approach developed among others 'by the

---

[1] https://www.nature.com/articles/541133b.pdf

LIFE Adapt Island project and other projects. Our findings show that the conservation and the protection of ecosystems should be prioritized in terms of both climate adaptation and climate mitigation, including in cities, but most of all with the protection of marine ecosystems and forests. Small-scale measures such as green infrastructure in cities can still be beneficial at the neighborhood level but will not massively counteract the impacts of floods, storms, and heat waves or dry spells. There has been an initial optimism in nature-based solutions, which led to overestimations of the impacts of the 'small scale actions' before the first evaluation reports could be used. We then argue the importance of ecosystem protection, particularly with regard to marine ecosystems, in both climate mitigation and climate adaptation. In this chapter we will approach the green infrastructure in cities as a case study (Part I), the broader question of deforestation and forest degradation (Part II), the protection and restoration of marine ecosystems including innovative renovation actions (Part III) and the limits and boundaries of nature-based solutions with the example of the Great Green Wall in Sahel (Part IV). This last example shows not only the needs of nature-based solutions in regions affected by drought, such as Sahel, but also the difficulties and challenges in implementing large-scale programs.

I. Green Infrastructure in Cities '*renaturing cities*'

Directly confronted with climate-related hazards like floods and heat waves, cities have been at the forefront of the development of nature-based solutions as one of the solutions for climate adaptation. The development of green roofs and emphasis on urban green spaces had indeed been widely accepted as a 'model' to cool the city against heat waves in many European and Asian cities. Other cities have created water catchment areas and tidal banks against regular and seasonal floods. Local authorities have a key role to play not only in terms of incorporating NbS into local land use planning and master planning but also for adapting the solutions to the local context (OECD 2021). It is estimated that in 2050 two-thirds of the world's population will be living in urban areas and that half of them will be based in Asia. Cities will be nexuses for climate impacts and adaptation (Malhi et al. 2020). Some of them experience subsidence, floods, and storms, but also stronger heat waves as they experience heat island effects and increasing demands for freshwater resources. They are also more exposed areas to climate hazards as they contain dense populations, critical urban systems, and infrastructure. Spatial plans (land use or urban planning) shape the built environment and human activity. They define the future medium- and long-term urban development planning of cities, and the share among different land use (new housing constructions, green areas, and protected nature areas) in coordination with the national and regional land use agencies and spatial plans. They clearly define what is authorized in terms of future development (number of dwelling units and grey and green infrastructure) based on certain risk assumption (or analysis of the vulnerability to floods, landslides, or earthquakes), geology, and the socioeconomic forecast. The growing urbanization and the urban expansion are often key parameters in the medium-term planning of cities.

According to the OECD (2021), the national planning policies have also played a key role in promoting green and nature-based solutions in

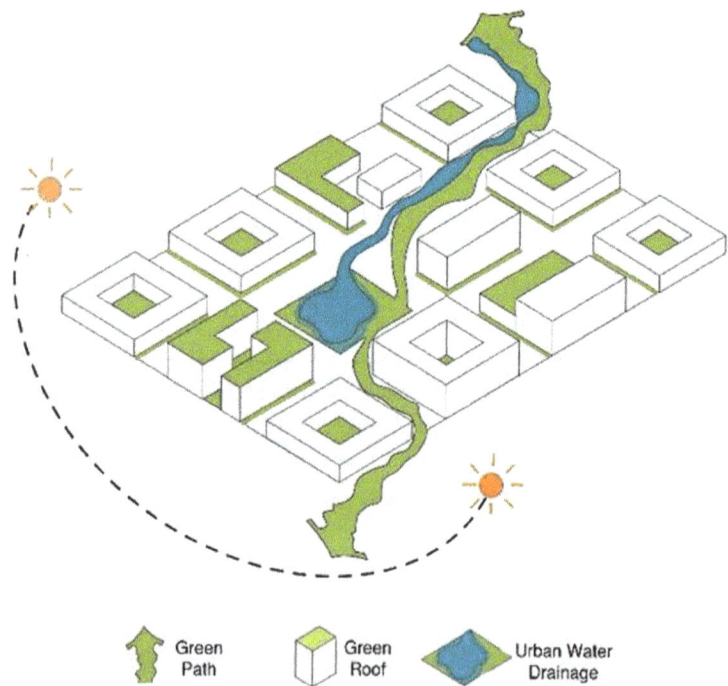

Fig. 2: Source urban form and layout (Source: Urban Climate Lab, Graduate Program in Urban & Regional Design, New York Institute of Technology, 2016)

cities. The UK national policy framework encourages local authorities to maintain and enhance green infrastructure. Norway and some other countries also require both counties and municipalities to consider the use of NbS in planning processes before they utilize alternatives such as grey infrastructure in the prevention of floods or for the stabilization of soils.

The regulations in many countries require new buildings to be equipped with green roofs, while a minimum percentage of green spaces is required per ha or per population in other countries (minimum 20% of land use in Indonesian cities for instance). In other cities, like Paris and Brussels, the green roofs have been developed through the cities' calls for proposals to 'green' specific areas. In 2009, Toronto, Canada, became the first city on the North American continent to adopt a 'green roof bylaw' for which new developments over 2000 m² require green roofs (OECD 2021). Green roofs have been required on the majority of all new buildings in Copenhagen, Denmark since 2010 and have since been spread to cities like Paris and Barcelona. The city of London (UK) integrates green roofs as a solution in its 2016 London plan, saying the 'major development proposals should be designed to include…green roofs and walls where feasible' to deliver climate change adaptations and sustainable urban drainage benefits.

Fig. 3: Nature based solutions and climate adaptation in cities (Source: Gaborit P.)

In terms of urban planning, the question of integrating the green (and the blue) infrastructure is not new: there were developments of garden cities in the UK in the 1920s, and green cities with vegetable gardens symbolized urban development in former socialist countries (Gaborit 2014, Gaborit 2015). But at that time, green infrastructure was considered in terms of 'wellbeing' and 'countryside's ideal' brought into the city's grey environment, rather than a solution to 'cool' cities faced with 'heat waves'. The development of the 2010s was the use of these 'greening' infrastructures as a solution for climate adaptation. The new developments also needed to rely on more extensive research in order to understand which species would have the best cooling effect on buildings and which ones were less water reliant, as the initial garden cities relied on maintenance and irrigation, which could sometimes default. Studies also had to show which green infrastructure was needed to develop corridors for biodiversity and which ones facilitated soil stabilization and water drainage, amongst other issues. However, as of 2022 we can confirm that urban and spatial planners are very much acquainted with the use of green infrastructure in climate adaptation (ISOCARP 2020). International and regional networks have been created around the topic of 'nature-based cities,' 'nature in cities,' 'green cities,' and 'biophilic cities,' among other names.

The reports produced by the OECD (2021) and the European Commission report on nature-based solutions note that the difficulty of the 'greening cities movement' is that the results have been extremely 'scattered' and 'ad hoc': 'the majority of NbS initiatives have been implemented as one-off projects and in an ad hoc way, often on a pilot basis and on a small scale' (OECD 2021). Challenges such as the difficulty to uptake and scale up the good practices, the lack of performance evaluation, and the gaps in the regulatory frameworks and incentives hamper their potential roles as key elements for climate adaptation. Succinctly, although the greening of cities' infrastructure with nature-based solutions has proven to have a positive impact on cooling certain buildings and neighborhoods (0.3–3°C according to Wilde et al. 2020), and in mitigating the flood's impacts with the water catchment fields, these solutions alone will not be sufficient to entirely prevent heat waves and to reduce the floods' impacts in the long term. According to the OECD (2021), the difficulty of collecting robust, quantified data makes it difficult for NbS to be considered on an equal field with grey solutions in terms of adaptation to flood hazards. Planners and decision-makers perceive the performance of nature-based solutions as more uncertain than grey solutions.

However, nature-based solutions are not only the small-scale projects used to reduce the heat waves and flood hazards in certain city neighborhoods; they also encompass broader solutions against forest degradation and reduce other climate risks such as heatwaves and wildfires.

II. Deforestation and Forests' Degradation

It is increasingly acknowledged that forests are central actors of climate mitigation and adaptation. According to the International Panel on Climate Change (IPCC) much of the mitigation potential from terrestrial ecosystems stems from the restoration and management of forests and from curbing deforestation (Wild et al. 2020). Following UNEP's reports (2014), forests are among green infrastructure solutions with the greatest environmental and socioeconomic co-benefits. In addition to the immediate benefits that they have in regulating water quantity and quality, they can also function as carbon sinks, increase pollination for nearby agricultural fields, improve air quality, regulate local climate (including cooling), and help preserve biodiversity. They absorb 2.4 billion tons of carbon dioxide per year on average according to the IUCN (2021), which represents one third of the annual carbon dioxide released from burning fossil fuels. Deforestation and forest degradation not only reduce the potential for the absorption of carbon, but these would-be carbon sinks also release large amounts of carbon into the atmosphere when they are burnt. Additionally, forests help to reduce risks from disasters like coastal flooding and regulate water flows and microclimates (IUCN, 2021)[2]. Forests sustain livelihoods and societies' wellbeing. They contribute to clean air, water drainage and management, and healthy soils on which agriculture is dependent. In addition to this, the FAO report (2020) states that 80 % of the world's land diversity is located in forests. Deforestation and forest degradation therefore are a direct threat to the survival of a large variety of species. Additionally, many communities worldwide directly rely on forests for their living. 'Over half of the tropical forests worldwide have disappeared since the 1960s and every second more than one hectare of tropical forests is destroyed or drastically

---

[2] IUCNhttps://www.iucn.org/resources/issues-briefs/deforestation-and-forest-Degradation

degraded' (IUCN, 2021). The resilience of forests is very dependent on the history and recent evolutions, in a context where 'land clearing' and 'burning' are still widely used (Malhi 2019). According to the WWF, over 43 million hectares of forests were lost on 24 deforestation fronts in South America, Sub-Saharan Africa, Southeast Asia, and Oceania between 2004 and 2017 (Pacheco et al., 2021).

The actions to protect and restore forests are therefore recognized as nature-based solutions by the IUCN as they are defined as 'actions to protect, sustainably manage, and restore natural or modified ecosystems, that address societal challenges effectively and adaptively, simultaneously providing human well-being and biodiversity benefits' (Cohen- Shacham et al. IUCN 2016).

The pace of deforestation, enhanced by increasing forest fires, is concerning even though countries have committed to action during the 2021 COP 26 in Glasgow. At this event more than 100 countries, which hold up to 85 % of the world's forests, have committed to halt deforestation by 2030. This commitment was, however, not recognized as sufficient by environmental organizations, as more global changes and follow-ups are needed. Indonesia has undertaken public action to protect the forests in its different islands, especially in Kalimantan, Borneo. The Indonesian forests were indeed at the heart of land use stakes linked to agriculture (palm oil plantations) and urbanization. The identified threats stemmed not only from the loss of forests but also from the destruction of the peatlands beneath the forests, one of the most important land carbon sinks[3]. The country reduced forest loss by 75 % between 2019 and 2020, due to the efforts of the Ministry of the Environment and forests to more widely protect forest land use, the reduction in forest fires, and the reduction of land use conversions linked to palm oil[4]. Even with these efforts, the country lost 115,459 hectares (285,300 acres) of forest cover in 2020, representing the equivalent of the territory of Los Angeles[5].

There is a common distinction between deforestation and forests' degradation: deforestation refers to the suppression of the tree coverage of a land in order to use it for another purpose, like agriculture or

---

[3] https://edition.cnn.com/interactive/2019/11/asia/borneo-climate-bomb-intl-hnk/

[4] https://news.mongabay.com/2021/03/2021-deforestation-in-indonesia-hits-record-low-but-experts-fear-a-rebound/

[5] https://news.mongabay.com/2021/03/2021-deforestation-in-indonesia-hits-record-low-but-experts-fear-a-rebound

infrastructures' construction. Forests' degradation refers to the negative impact of activities on forests in the services they produce and the ecosystems they harbor.

The WWF (2020) identifies various direct and indirect causes of deforestation and forests' degradation. Agriculture, extractive activities, and infrastructure development are well known direct drivers of deforestation – agriculture dominates as the main driver of deforestation (Pacheco et al., 2021). According to 'The State of World's Forests' FAO report (2020), large scale commercial agriculture represented 40 % of tropical deforestation between 2000 and 2010 and local subsistence agriculture represented 33 % of tropical deforestation on the same period (due, for instance, to burnt land clearing for small scale farmers). These factors vary depending on more indirect factors like demographic growth, changes in consumption patterns, political regulations, economic changes, and climate and topographic related reasons (FAO, 2020, WWF, 2020). The causes of deforestation and forest degradation largely vary depending on the regions and even the countries - people's use of a resource depends on the economic, historical, cultural, and local contexts (FAO, 2020).

Various area or sector-based responses have proven to be efficient to tackle deforestation and forests' degradation at the national, regional, and local levels. Conservation responses protecting biodiversity-rich forest areas from human activities have been broadly acknowledged as a priority by national, regional and by local governments, and also by NGOs and civil society organizations.

The conservation of forests also means protection against wildfires, which is increasingly considered an important element of climate adaptation. Indeed, forests fires jointly deteriorate the climate mitigation potential of forests and also the forests' climate adaptation potential with regard to soil erosion and water drainage against floods. According to several studies, one single forest fire can lead to many cascading effects on entire ecosystems (Malhi et al. 2020). The resilience and vulnerability of forest species composition to changing fire regimes depends on a variety of local factors, including climate, soil conditions, and communities' historical legacies. In some cases, extreme events combined with biophysical feedbacks can cause ecosystems made up of long-lived species to completely shift their composition in response to a single fire event (Iglesias, Whitlock 2020, Malhi et al. 2020).

The European 'Land Based Wildfire Prevention' (2020) report states that actions can be taken to prevent and/or mitigate the impact of wildfires. A few essential actions are: (1) an efficient and clear governance-distribution of responsibility and power in leadership – it appears essential to have a clear idea of who is in charge of what to act efficiently in response to fires, (2) a work on anticipation of potential drivers and risks-Such risks include seasonal changes and periodic winds that drive fires-, (3) an intelligent forest management system. This system should rely on a deep knowledge of ecosystems to manage land and potential restorations as efficiently as possible, (4) a strong investment in prevention and potential operations, notably the enforcement of rigorous rules regarding human activities around forests, or investment in firefighter's teams and equipment.

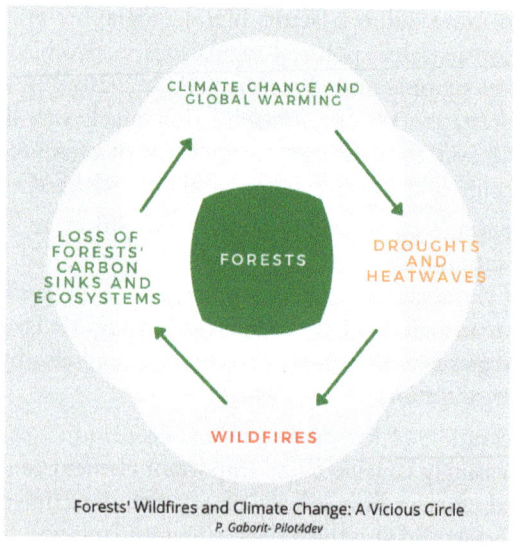

Fig. 4: The prevention of wild fires as a NBS

The OECD (2021) report states that nature-based solutions cannot always be evaluated positively as they are often implemented as 'ad hoc' and 'small-scale solutions.' However, by taking a broader overview of the definition of 'nature-based solutions' and by including the actions created to protect nature in the forms of forests, soils, oceans, and marine ecosystems, nature-based solutions could become a cornerstone in both climate adaptation and climate mitigation.

III. Marine Ecosystems: Restoration or Protection?

While the oceans are often presented in the press as a threat due to sea level rise, or as an opportunity to extract resources or energy as part of the blue economy, they play a central role in climate mitigation. Meanwhile, marine ecosystems such as mangroves and coral reefs play an important role in climate adaptation. Nature-based solutions are often analyzed through the angle of land ecosystems as opposed to marine ecosystems (Malhi et al. 2020). In this third part of the chapter, we briefly approach the natural and anthropic pressures faced by marine ecosystems. Then, we will present the potential and limits of 'blue' and 'marine' nature-based solutions – coral and seagrass restoration, artificial reefs implementation, marine protected areas, regulations on human activities, waste management, synergies and partnerships, and, lastly, the importance of research, education, and sensitization.

1. *Oceans playing a key role in climate mitigation, adaptation, and biodiversity*: According to the IPCC (2014) the Ocean plays a central role in Earth's climate and has absorbed 93 % of the extra energy from the enhanced greenhouse effect and approximately 30 % of anthropogenic carbon dioxide (CO_2) from the atmosphere (Hoegh-Guldberg et al. 2014). The oceans, as ecosystems, are the most important, although often overlooked, carbon sinks of the planet[6]. The role of phytoplankton in particular has been acknowledged as playing an important role both in climate regulation and in the oceans' marine food chain.

Marine ecosystems shelter habitats of a multitude of organisms, and they are important reservoirs for biodiversity. They also benefit societies that rely on them by providing food and materials for research and medicine and attract tourism. Marine ecosystems also protect societies and terrestrial environments against hazards. They control pollution and some ecosystems like corals and mangroves protect coastal communities from storms, floods, and shoreline destabilization and erosion. '50 % of salt marshes, 35 % of mangroves, 30 % of coral reefs, and 29 % of seagrasses have already been lost or degraded worldwide over several decades' (Barbier 2017).

---

[6] https://www.pilot4dev.com/knowledge/185-policy-brief-n-4-the-preservation-of-marine-ecosystems-a-comparative-approach-of-innovative-solutions

In the aftermath of the Ocean's summit[7], which took place in Brest on February 9–11th, 2022 to promote the protection but also exploitation of oceans, it also appears essential to highlight the importance of oceans and marine ecosystems' protection in the current context of climate change. It is indeed important to shed light on the threats they are facing and to analyze the conservation solutions that can be implemented to reduce these pressures. The intergovernmental conference on high seas (March 7–22nd 2022) was another unsuccessful attempt to find paths for further regulation of international oceans beyond the regulations set within national sea jurisdictions. The UN 2022 Lisbon conference on oceans gathered stakeholders for the protection of life below water (SDG14) and will try to mitigate what now became urgent problems.

As oceans main parts are invisible to non-divers, the impacts of climate change, anthropic pressures, and the climate adaptation of the oceans themselves remain unknown. Meanwhile, the growing influence of the blue economy and 'oceans governance' create additional threats to ocean ecosystems.

Oceans are directly exposed to climate change impacts: Due to the increase in GHG emissions and the dissolution of carbon dioxide in seawaters, the oceans experience a decrease in their pH levels which is called 'ocean acidification.' This acidification combined with the warming of the water, contribute to cause coral bleaching, cyclonic events, and water dilation. According to the IPCC, the current rate of ocean acidification is unprecedented. Warming temperatures, declining pH levels, and declining carbonate ion concentrations represent risks to necessary ecosystem processes such as primary productivity, reef building, and also lead to erosion. These changes also alter the levels of available fish and marine stocks as the fish population shifts to those that can adapt to higher temperatures and lower pH levels and carbonate ion concentrations while the other organisms are greatly reduced or die off completely (Hoegh-Guldberg et al. 2014). The whole food chain cycle is affected leading to the extinction of an increasing number of species. Rapid changes in physical and chemical conditions within ocean sub-regions have already affected the distribution and abundance of marine organisms and ecosystems. In some cases, changes will eliminate entire ecosystems (coral reefs may be among them). These changes will increase risks and vulnerabilities to coastal livelihoods and to human food security (Hoegh-Guldberg et al. 2014).

---

[7] https://www.oneoceansummit.fr/en/

Global warming will continue to result in more frequent extreme events and greater associated risks to ocean ecosystems. Oceans are already exposed to increasing storms and climate related events. The intensification of these events strongly affects the reef ecosystems and their resilience. It takes around ten years for the coral reefs to recover from a cyclonic event. Strong swells cause significant damage to small organisms including seagrass beds, and combined with wind, knock down mangrove trees and other organisms and ecosystems. Additionally, the rise in sea level leads to over-salting of the tidal ecosystems such as mangrove swamps, which kills un-adapted vegetation as well as marine fauna. All these factors are intensified by global warming.

Climate change also has an impact on the diseases faced by the marine and oceans ecosystems. In the Caribbean region, two diseases – the 'black band' and the 'white band' diseases – chronically affect corals and contribute to the decline of coral Acropora palmata populations on the reefs in addition to the coral bleaching incurred by global warming and acidification. This coral is however an important ecosystem for climate adaptation and for marine ecosystems. Global warming and diseases can also contribute to the proliferation of species and to the detriment of others (invasive species). Climate change is therefore the first threat and anthropic pressure on the ocean (Fig. 5).

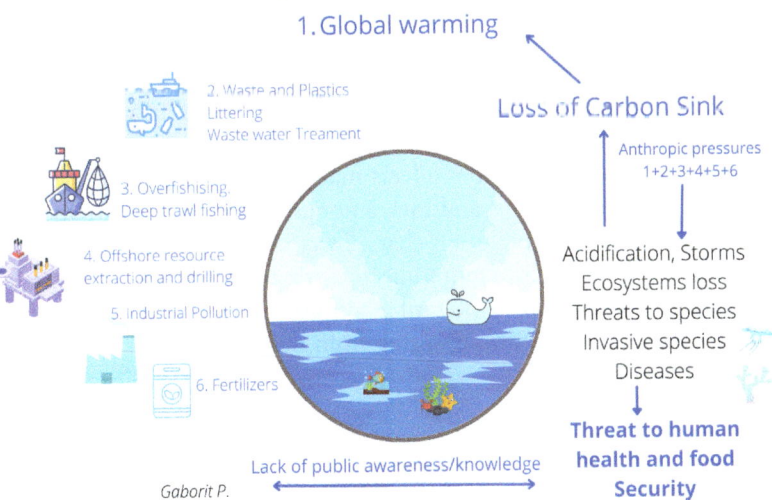

Fig. 5: Anthropic pressures on the oceans

2. *The anthropic pressures (See Fig. 5)*: The oceans are facing invisible but prevalent anthropic pressures. The first of these pressures is climate change, as we have previously detailed. This is amplified by other pressures and threats that are also invisibly acting on oceans. Global waste and water waste largely end up into seas and oceans. This is especially true for plastic. Plastics break down very slowly into the seawaters and are absorbed by the ecosystems and by animals, which explains why it is increasingly discovered in many marine organisms. Recent studies from WWF show that more than 2,141 species have been affected by plastic waste, some of which have been pushed to the brink of extinction[8]. According to a report published by WWF in the first quarter of 2022, 88 % of the marine species in the study were affected by severe contamination of plastic in the ocean. The report said that many animals have ingested these plastics, including animals commonly consumed by humans. This plastic waste also ends up in people (humans) cells and blood streams as we are at the end of the food chain. The littering of plastics called by experts the '6th continent' may only be the hidden side of the iceberg[9].

There are a variety of threats to the composition of marine ecosystems that do not involve plastic. Sewage that implies a variation of water's physicochemical parameters or macro-waste particularly affects beaches and land ecosystems. Overfishing, especially mass industrial fishing techniques (deep sea bottom trailing) are responsible for the depletion of fish and marine ecosystems all over the world, including in European seas. Some elements of blue economy such as the extraction of deep-sea resources can deeply deteriorate marine ecosystems. Industrial pollution, notably discharges directly in the sea and toxic waste, is an externality of many human activities in and near the ocean. Moreover, the intensification of nautical activities affects marine ecosystems with pollution such anchors, boat chains, and fuel, amongst others. These activities are equally causing noise pollution that affects marine animals' abilities to use echolocation and to communicate with other animals. The use of fertilizers in agriculture is polluting soil and water. Agriculture is also contributing to soil erosion resulting from deforestation for agricultural

---

[8] https://www.weforum.org/agenda/2022/02/extinction-threat-ocean-plastic-pollution/

[9] Microplastics found in human blood streams: https://www.sciencedirect.com/science/article/pii/S0160412022001258

crops. The overall lack of public knowledge of coastal areas is part of the challenge.

3. *Nature based solutions as small-scale ecosystems restoration actions*: The possibilities to develop nature-based solutions within marine and ocean environments have been acknowledged by the literature (Seddon et al. 2020). There are several initiatives of restoration actions listed worldwide, including in Europe and for corals, seagrass, and mangroves particularly in Florida and in the Caribbean[10]. Fig. 6 visualizes some examples, but this list is not exhaustive.

Fig. 6: Trying to visualize examples of small scale NBS marine restoration actions-

(a) **Corals**: There are a multitude of methods to restore corals. However, the challenge is how to scale up the solutions. In Guadeloupe, the project LIFE Adapt Island[11] and the Cáyoli program[12] develop a comprehensive action for restoring corals, and mangrove ecosystems, including with nurseries and laboratories, to test innovative solutions. The Fragments of

---

[10] https://www.pilot4dev.com/knowledge/185-policy-brief-n-4-the-preservation-of-marine-ecosystems-a-comparative-approach-of-innovative-solutions
[11] https://www.cayoli.fr/en/life-adaptisland-2/
[12] https://www.cayoli.fr/en/homepage/

Hope, a Belize-based organization, developed several methods of on-site coral cultivation relying on long lasting and hurricane resistant structures tied underwater[13]. Some structures appear more efficient for certain coral species. They also developed several transplantation methods using ropes, masonry nails, and cement discs as well as advice regarding the features of the chosen transplantation sites such as the physicochemical parameters of water, turbidity, and predators.

The program Utila Coral Restoration (UCR) in Honduras offers restoration services but also education/training for scuba divers (PADI) and volunteers. UCR is an organization that restores the Caribbean reefs around Utila (Honduras) to recover the coral populations of the specific species of *Acropora cervicornis* and *Acropora palmata*[14]. They organize trainings on various methodologies of restoration such as ways to maintain corals and monitor the fragments in nurseries and methods of planting the corals back on degraded reefs. This contributes to the diffusion of good practices regarding coral restoration and the preservation of marine ecosystems broadly. These methods could be spread across scuba diver communities.

(b) **Seagrass beds**: at the Florida Atlantic University, marine ecosystem conservation is one of the four priorities in terms of research. The scientists from the university grow seagrass in nurseries and transplant them for local restoration efforts, notably to the Indian River Lagoon. They use the BioRock Technology that involves running a low voltage trickle of electricity through a steel structure which seems to stimulate growth in all types of marine life. However, the data collected thus far is not sufficient to understand the effect of this technology on seagrass in the Indian River Lagoon. Nevertheless, scientists are considering it and already attesting its efficiency in helping the growth of seagrass. This new technique involves the transplantation of grown seagrasses that can affect the donor meadows.

The LIFE Adapt Island project clearly demonstrated that the restoration of diverse marine ecosystems such as corals, but also mangroves and seagrass beds was first depending on the removal of anthropogenic pressures (with eco-anchorages) to avoid a high percentage of loss. New research therefore focuses on in-situ survivorship of in-vitro germinated

---

[13] http://fragmentsofhope.org/wp-content/uploads/2015/12/Acropora-Restoration-Best-Practices-Manual-ABK-v4.pdf

[14] Utila Coral Restoration, http://www.utilacoral.org/about/

seagrasses. A study by Tuya et al. (2017) has shown that Artificial Seagrass Shield (ASS) can be an interesting option to preserve in-vitro germinated seagrass once implanted. The scientists involved in the study planted seagrass surrounded by artificial seagrass made of plastic and concluded that the survivorship of the seagrass increased. This was probably because of the under-estimated influence of herbivorous creatures that were dissuaded by the artificial seagrass shields from consuming the transplanted seagrass beds.

(c) *Habitat and artificial reefs as medium-term solutions to restore several ecosystems*: In order to restore and preserve marine ecosystems, some organizations have decided to work directly on the affected habitats. In Europe, the LIFE Coast Adapt project in Sweden (Skane region) set up rocks made of artificial reefs to mitigate swell, and to strengthen biodiversity in a general effort to make sensitive coastal areas more resilient. The same type of practices have been implemented in Marseille, France, where the municipality implemented the RECIFS PRADO project. In this project 400 reefs of 6 different types have been set up on 200 hectares of seabed. The number of fish species has tripled, and biodiversity has increased by 30 % in ten years (from 2008 – 2018). Another French initiative developed the BioHut Concept that is an artificial habitat made of steel and oyster shell that serves as shelter for marine biodiversity[15]. Particularly, BioHut is allowing fishes to access food and grow safely far from predators.

In Asia initiatives such as the North Bali Reef Conservation focuses on the construction of artificial reef structures on damaged reef to provide a varied habitat for an optimal number of species. 'The deployed structures act as a substrate for coral larvae to attach while also providing a link between natural reef patches, improving connectivity by allowing individuals to move safely from one coral patch to another. [These] structures facilitate coral growth and provide suitable egg laying habitats for reef fish while fish domes also provide protection from predators'[16].

4. *Protection and enhancement of oceans ecosystems as a long-term nature-based solution:* The UN Sustainable Development Goal 14: Life Below Water identifies the most urgent questions and pathways to sustainable oceans. The SDG14 targets include the reduction of marine

---

[15] Ecocean, https://www.ecocean.fr/restauration-ecologique
[16] North Bali Reef Conservation, https://northbalireefconservation.com/

pollution and the sustainable management of coastal ecosystems, addressing overfishing, the sustainable management of ocean resources, and the development of marine technology and research capacity. The long-term proposed solutions are represented in Fig. 7. As detailed in the paragraph about the anthropic pressures, the first priority to protect the oceans would be to limit climate change (1) in Fig. 7. Climate change is having a cascading effect on marine ecosystems, leading to acidification, current modification, and ecosystem losses. The second important solution is an increase of marine protected areas (2) in *Figure 7*. Protected marine areas represent between only 1 %[17] and 7.93 %[18] of all oceans despite ambitions for more areas to be protected. According to the OECD (2017) the protected marine areas represent 4.12 % of the total marine environment. In addition to this, of the 3.41 % global MPA coverage in 2014, only 0.59 % was established as no-take MPAs (Thomas et al., 2014), or areas that prohibit extractive practices such as fishing and mining. Marine ecosystems are theoretically protected under the 1992 Convention on Biological Diversity (Aichi Target) (entering into force in 2014): 'By 2020 [...] 10 % of coastal and marine areas, especially areas of particular importance for biodiversity and ecosystem services, are conserved through effectively and equitably managed, ecologically representative and well-connected systems of protected areas and other effective area-based conservation measures, and integrated into the wider seascape [...]'[19]. This convention has been signed by more than 102 countries. As a means of respecting this target, mostly in the European Union, the Natura 2000 program is the largest network of protected areas in the world. By the end of 2016, 10.8 % of the surface of Europe's seas had been designated as Marine Protected areas, while in 2017, Natura 2000 marine sites covered 515,000 square kilometers, which represent 8.9 % of Europe's seas (European Environment Agency)[20]. Admittedly, European marine protected areas consist mainly of coastal waters and not offshore waters, which makes the preservation operation not yet ecologically representative. Also, international law entails that sea water offshore is international and does not belong to any singular country, which

---

[17] SSI training class marine biology.
[18] According to the IUCN data base, 7,93 % of all marine areas are protected.
[19] https://www.cbd.int/aichi-targets/target/11
[20] https://www.eea.europa.eu/themes/water/europes-seas-and-coasts/assessments/marine-protected-areas

makes preservation and protection more difficult. At the other side of the Atlantic Ocean, the Florida Department of Environmental Protection is also managing 42 aquatic reserves across the state and roughly two-thirds of Florida's coral reef lies within Biscayne National Park and the Florida Keys National Marine Sanctuary, a marine protected area that surrounds the Florida Keys Island chain.

There is a strong movement to increase the size and scope of oceans and marine protected areas beyond national jurisdiction, but also to design them in the most optimal way, so as to cover a wide range of different ecosystems, and to restore the damaged chains in ecosystems.

To reach this goal, the most important measure at the international level will be to regulate the high sea and to protect biodiversity beyond national jurisdiction (3) in Fig. 7. The intergovernmental conference on high seas (7–22 March 2022) was another unsuccessful attempt at the highest level of national governments to regulate the international oceans, beyond national borders to set limits. The aim was indeed to promote the Biodiversity Beyond National Jurisdiction (BBNJ) with a treaty on the high sea, which was unfortunately not successful letting most of the oceans unprotected.

Meanwhile, the conservation of key ecosystems, and several endangered marine species is currently at stake and requires specific measures (4) in Fig. 7.

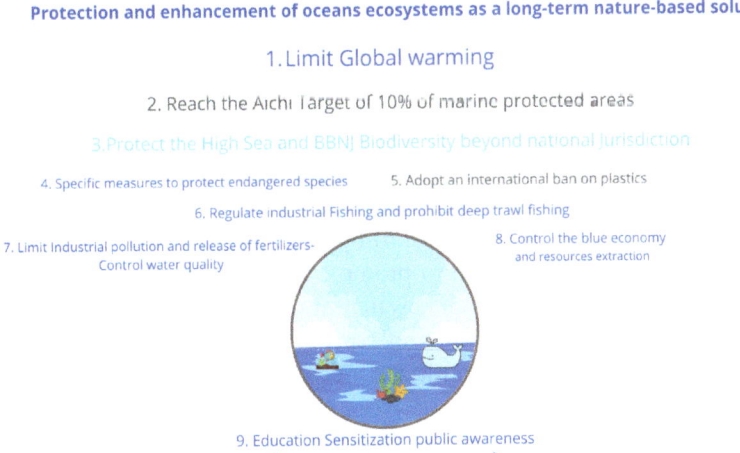

Fig. 7: Long term solutions to protect the oceans

Waste management is an essential tool to preserve marine ecosystems since human activities such as tourism, industries, and agriculture tend to produce considerable amounts of waste that are ejected into the water, including plastics (5) and industrial pollution and fertilizers (7). Waste management relates to a large variety of waste such as: micro waste, macro waste, the specific waste related to fisheries, and water quality. To tackle micro waste some simple regulations can be implemented such as the ban on single use plastics. In February 2022, the United Nations Environment Assembly (UNEA) has taken the first step toward the creation of a landmark treaty to control plastic pollution worldwide, which raises a lot of hope. Specific measures have also been adopted like the ban on plastic bags in countries such as Tanzania (despite the problems of plastic bottles compensating for the lack of general access to drinkable water). The different cleanup operations organized by civil society and the private sector do not have a sufficient impact to tackle the problem.

Belize is often presented as an example or a leader regarding marine environmental regulations. In December 2018, Belize became the first country in the world to put a moratorium on all offshore oil exploration and drilling (8) in Fig. 7. This happened after months of campaigns from civil society organizations, who pressurized the authorities and mobilized local populations in order to obtain this major change in regulation. The moratorium became a ban integrated into the law[21]. This went alongside with other measures to regulate human activities such as environmental taxes redistributed to the conservation and ecotourism market, the protection of parrot fish along the reef by restricting fishing, and a ban on single use plastic since January 2020, amongst others.

A sustainable management of fisheries is an important factor in the protection of marine ecosystems (6) in Fig. 7. In Europe, the Common Fisheries Policy (CFP) is attempting to establish a set of good practices in fisheries, including those related to bycatch. Public, private, and civil society organizations also tend to promote and implement partnerships with the fishers in order to raise awareness on bycatch waste related to fishery. The industrial fisheries, and especially bottom trawl fishing in European seas, are heavily criticized by environmental organizations

---

[21] ROOT Tik (11th of April 2018), How One Country Is Restoring Its Damaged Ocean, National Geographic, https://www.nationalgeographic.com/science/article/belize-restores-coral-reefs-oil-drilling-ban-environment

for their impacts on fishes, ecosystems, dolphins, cetaceans, and turtles. Overfishing is also a major threat worldwide, as the depletion of the oceans by large countries in high seas like in Europe, Japan, and China is having a detrimental effect on local fishers and on food security, for example in Senegal and Somalia.

It is well-known that fisheries damage marine ecosystems, but it is argued that artisanal fishers are actually the ones in the position to implement sustainable fishing. They are thought to have an interest in sustainable fishing as they rely on the reefs for their livelihoods. Moreover, these partnerships provide the possibility to share knowledge on vulnerable fishes through stock assessments, the tracking of patterns of movement, and access to feeding behavior.

Rigorous monitoring is needed for protection and nature-based solutions (7 and 8) in Fig. 7. This monitoring can apply to the oceans' parameters, but also to biodiversity and to endangered wildlife through a robust monitoring of their populations, behaviors, injuries, diseases, or, again, the impact of human activity on the ecosystems. Monitoring environments can create models to observe any changes and/or stimuli. New initiatives like DNA tracing could bring revolutions in the monitoring of biodiversity. Finally, water quality can be controlled by regulations like the Water Framework Directive in the European Union regarding water quality and the discharge of toxic waste[22]. It appears essential to monitor toxic waste and to produce and use the tools that will allow a detailed analysis of water quality. The role of nature-based solutions on water filtering and water quality with sediment traps, for instance, need to be thoroughly and carefully monitored. Research is accumulating new data, methods, techniques, and knowledge every day. It is up to all the stakeholders to build on, spread, and implement this knowledge in order to protect marine ecosystems. This knowledge also needs to be available for possible 'blue economy' investors, which could worsen the situation, such as with the extraction of deep-sea cobalt, depending on the activities' scale and footprint. As already detailed in the previous paragraph, the intergovernmental conference on high seas (7–22 March 2022) was an attempt to regulate the international oceans beyond national borders to set limits with a treaty on the high sea. The aim was indeed to promote BBNJ Biodiversity beyond national jurisdictions but the intergovernmental

---

[22] https://ec.europa.eu/environment/water/water-framework/info/intro_en.htm

aspects have led to failure, letting large swathes swathes of ocean water unprotected. In addition to this, the current large wind offshore energy projects should take into account their impacts on the noise and on the destruction of ecosystems by using adapted technologies such as floating wind farms.

Education and sensitization of a general audience (9)- in Fig. 7- are fundamental to the protection of marine ecosystems. The Guadeloupe Port Caraïbes (GPC) implemented the Cáyoli Junior program dedicated to the sensitization of young people, whereas the North Bali Reef Conservation in Bali, Indonesia, trains volunteers to implement artificial reefs. Governmental programs to promote marine ecosystems to a broad audience have an equally large impact. All these solutions are tested at a small-scale level. The Cáyoli program as well as the LIFE Adapt Island project, among others, are aiming at restoring corals, seagrass beds, and mangroves – the challenge will be to upscale these solutions to larger areas.

It appears clear from the scientific literature and practice that restoration actions will be inefficient if not implemented in synergies with the reduction of pressures such as pollution, global warming, and ocean depletion such as with massive deep trawling fisheries and nets. More protected areas are needed as well as better protection of the oceans as carbon sinks in order to reach a successful climate adaptation and a better coastal resilience.

In the earlier paragraphs, we have moved up from the description of small-scale restoration solutions to the necessity of large-scale ecosystems protection, and to the reduction of anthropic pressures. This is certainly the most efficient nature-based solution against climate change and its impacts. The protection of forests (Part II) and of marine ecosystems and oceans (Part III) are certainly the most impactful and efficient nature-based solutions, while the use of small-scale solutions to cool cities (Part I) and the restoration actions (Part III) can be considered the most useful small-scale solutions. To face the current challenges, however, some large-scale restoration actions have emerged, at very least as projects that are being tested for potential further implementation on larger scales. The most well-known is the Great Green Wall in Sahel, which is a solution proposed to fight against desertification. This Great Green Wall project enters into the scope of nature-based solutions and can underscore some of the challenges to scale up nature-based solutions beyond the protection of particular ecosystems.

IV. Limits and Boundaries: The Example of the Great Green Wall in Sahel

According to the IUCN definition of nature-based solutions, as detailed in the introduction, initiatives to restore ecosystems in areas affected by droughts and desertification entirely fall into the scope of nature-based solutions. Launched in 2007, long before much attention was given to nature-based solutions, the Great Green Wall (GGW) initiative has been set up to restore degraded land across the Sahel Region by planting trees, plants, sustainable land management projects, and to facilitate irrigation. The project involves 20 African countries (including Senegal, Burkina Faso Djibouti, Eritrea, Ethiopia, Mali, Mauritania, Niger, Nigeria, Sudan, and Chad) and many donors (the African Union, the Food and Agriculture Organization of the United Nations (FAO), the Global Environment Facility (GEF), the International Union for conservation of Nature (IUCN-PACO), the World Bank, the Permanent Inter-state Committee for Drought Control in the Sahel (CILSS), the European Union, the Kew Royal Botanical Gardens, the Sahara and Sahel Observatory, and the United Nations amongst others).

1. The context of desertification

Africa's Sahel region faces many consequences of climate change, primarily drought and desertification, which worsen problems with water shortages and food security. According to the COP15 which took place 9–20 May 2022 in Abidjan, 45 % of land in Africa is affected by desertification. Local communities are facing more persistent droughts, food security problems, conflicts often regarding the use of natural resources, armed conflicts, political instability, and migrations. The literature shows that there is a critical lack of data concerning the impacts of climate change in several vulnerable regions of the globe, particularly in the Sahel region (Enenkel et al. 2020). Despite the availability of the information on floods, cyclones, hurricanes, and drought events, they are still often analyzed as isolated events despite the results of the alarming IPCC 2022 report on the impacts of climate change. The impacts of dry spells and weather events such as storms on health, including diseases, malnutrition, and conflicts, is often overlooked because each event is analyzed separately as opposed to a part of a long-term phenomenon. If we approach drought events individually, the challenges arise from its categorization, evaluations in terms of intensity, and the quantification of its impacts on health, death, social livelihoods, conflicts, or even on the sustainability of future ecosystems. The current Global Drought Observatory of the European Union reports, however, the drought events occur in several regions and have interrelated impacts on crops and agriculture[23]. For instance, if we consider the report for January 2022, Somalia declared a national emergency in the wake of three consecutive low rainy seasons with below average rainfall amounts and accumulating drought impacts. More than 3.2 million people in 66 out of the 74 districts in Somalia were affected, of whom 169,000 people have abandoned their homes in search of water, food, and pasture (GDO 2022). According to the Integrated Food Security Phase Classification, about 3.5 million people in Somalia acutely faced high food insecurity in late 2021, due to the shortage of rains, low agricultural production, and high food prices. Over 1.2 million children are likely acutely malnourished[24]. As the Great Green Wall website is declaring *'More than anywhere else on Earth, the Sahel is on the frontline of climate change and millions of locals are already facing its devastating impact. Persistent droughts, lack of food, conflicts over dwindling natural resources, and mass migration to Europe are just some of the many consequences*[25]*'*.

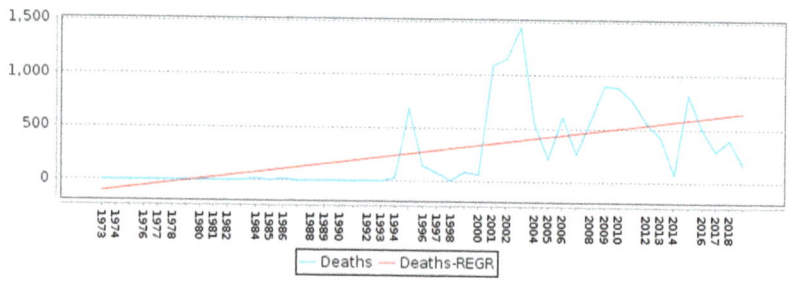

Graph 1: Niger drought related deaths according to the UN Sendai Framework date base. www.desinventar.net

This graph shows the drought related death in a country like Niger over the years, according to the UN Sendai Framework data base.

---

[23] http://edo.jrc.ec.europa.eu/gdo

[24] https://www.iom.int/news/intensifying-drought-threatens-displace-over-1-million-people-somalia

[25] https://www.greatgreenwall.org/about-great-green-wall

2. The Great Green Wall project

In this context, the project of 'Green Wall' appeared as a challenge and a strong ambition in itself. Together, these projects form an 8000 km 'wall', across the width of Africa. The overall aim of the project was to tackle the threat of desertification and the degradation of natural ecosystems, while providing alternative livelihood opportunities for local communities. The GGW initiative supported an integrated approach to sustainable land management, combining agricultural and rural development, food security, biodiversity conservation, and sustainable use. Additionally, climate change mitigation, adaptation, and work with and for the local communities were benefits of the GGW.

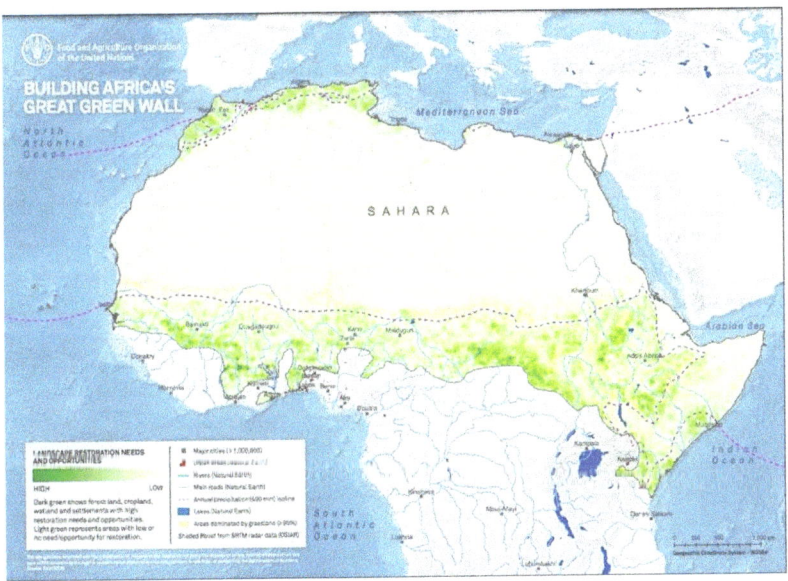

Fig. 8: Great green wall initiative and the Sahel (Source: Food and Agriculture Organization, 2016)

15 years after being launched, the results of the initiative are still unknown. According to the INTPA[26] of the European Commission, the project would have achieved 15 % of the initial objectives. The results

---

[26] International partnership directorate general of the European commission

include 12 million drought resistant trees planted in Senegal alone, 15 million hectares of degraded land restored in Ethiopia, 3 million hectares rehabilitated through local practice in Burkina Faso, 5 million hectares have restored in Nigeria, while the same rehabilitated surface in Niger would bring an additional 500,000 tons of grain per year[27]. According to more pessimistic estimates however, only 4 % of the ambitions have been met[28]. The lack of monitoring and difficulties in accessing funds are issues jeopardizing the project's success. Each country is in charge of providing its own statistics and monitoring internal results. The information is entirely lacking in several countries such as Somalia. According to the NGO Earth.org, the delays and missed timelines would be caused by a number of reasons, including insufficient funds, lack of oversight and lack of allocated technical support. Documentaries in Senegal and in Burkina Faso, amongst others, also show that few of the planted trees would have survived in the absence of stronger irrigation systems (France 24 2021). It is also to be noted that several countries which were initially part of the project experience regular conflicts (Mali, Somalia, Eritrea, and Ethiopia, amongst others). This makes the governance and implementation of such a project even more hazardous. Additionally, according to scientists, resilience in ecological communities requires longer-term perspectives to improve our understanding of ecosystem responses to change (Malhi et al 2020).

---

[27] https://ec.europa.eu/international-partnerships/programmes/growing-great-green-wall-ggw_en
[28] https://www.theguardian.com/environment/2020/sep/07/africa-great-green-wall-just-4-complete-over-halfway-through-schedule

Nature-Based Solutions in Climate Adaptation

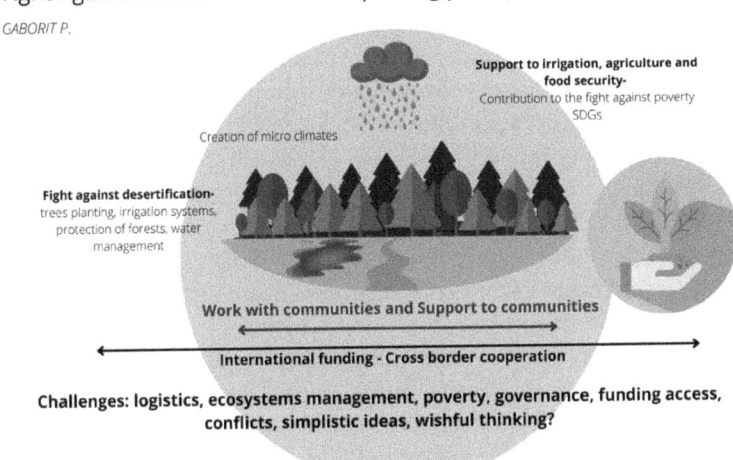

Fig. 9: The fight against desertification: underpinning principles or wishful thinking?

Therefore, the Great Green Wall project has unfortunately not been a success. Fig. 8 shows the underlying principles behind the project, but also the challenges which seem to have been underestimated. Similar locally based projects relying on sustainable land management have shown, on the contrary, that they could be successful. Restoration actions can be failures. In Guadeloupe, the Sirenia LIFE project on the reintroduction of the 'manatee' specie has also failed.

This example shows some concern about the possibilities to scale up nature-based solutions apart from a large and massive conservation of existing ecosystems, particularly forests and marine ecosystems.

Conclusion

This chapter provides a general overview of what nature-based solutions can entail: green infrastructure in cities to fight against heat waves, alternatives for grey solutions like dikes and dams in the protection against floods, and global protection and preservation actions. The current focus on 'nature-based solutions in cities' or 'nature-based solutions for floods prevention' are certainly growing trends showing successes, for example in cooling building temperatures during heat waves or in creating water catchment areas for floods. In terms of climate adaptation for specific climate disaster events, nature-based solutions should be increasingly considered a priority. Nature-based solutions are, however, often implemented as 'small scale' and 'ad hoc' solutions (OECD 2021). The questions of scaling up and replicating NBSs will face tremendous barriers. The project of the Great Green Wall, which should have brought all attention and expertise to its success, has been hampered by conflicts, governance, questions of jurisdictions and boundaries, but also inter institutional arrangements and a lack of technical support (Fig. 8). Perhaps a further analysis on socioeconomic and ecosystem impacts should have been integrated to the project, with facilitated funds transfer. The funding flows have also not been sufficient to cover the needs in real time. 45 % of land in Africa is affected by desertification, and the trend keeps growing. The protection of forests and of marine and land ecosystems is gaining momentum and should be at the heart of nature-based solutions from the local to the international level to be able to achieve substantial gains in terms of climate mitigation but also to preserve ecosystems and to protect food security in the longer run. The development of marine-protected areas (Fig. 7) and the fight against forest fires (Fig. 4) could bring a substantial impact in climate mitigation by enhancing carbon sinks and in climate adaptation with the maintenance of natural barriers. Finally, they underscore the key role of these ecosystems on food provision and on livelihoods. Finally, for the small-scale restoration actions including in greening cities (Fig. 3) and in innovative actions (Fig. 6), more research and analysis of successful examples will be extremely useful in the long run.

References

Barbier E.B., 2017, Marine ecosystem services. *Current Biology*. 2017 Jun 5;27(11):R507–R510. https://doi.org/10.1016/j.cub.2017.03.020. PMID: 28586688

Bertule M., Lloyd G.J., Korsgaard L., 2014, 'Green infrastructure guide for water management: Ecosystem-based management approaches for water-related infrastructure projects' UNEP, WEB-UNEP-DhiGroup-Green-infrastructure-Guide-EN-20140814.pdf (unepdhi.org) Cayoli Program: https://www.cayoli.fr/en/homepage/

Cohen-Shacham, E., Walters, G., Janzen, C., Maginnis, S. (eds.) 2016, 'Nature-based solutions to address global societal challenges'. Gland, Switzerland: IUCN. xiii + 97pp, http://dx.doi.org/10.2305/IUCN.CH.2016.13.en

Enenkel M., Brown M.E., Vogt J.V., McCarty J.L., Reid Bell A., Guha Sapir D., Dorigo W., Vasilaku K., Svoboda M., Bonifacio R., Anderson M., Funk C., Osgood D., Hain C., Vinck P., 2020, 'Why predict climate hazards if we need to understand impacts: Putting humans back into the drought equation', *Climatic Change* 162:1161–1176

European Commission, Directorate-General for Environment 2021, *Land-based wildfire prevention: Principles and experiences on managing landscapes, forests and woodlands for safety and resilience in Europe*. Publications Office. https://data.europa.eu/doi/10.2779/37846

European Environment Agency: 'Marine Protected Areas' https://www.eea.europa.eu/themes/water/europes-seas-and-coasts/assessments/marine-protected-areas

European Commission 2021: Mission Starfish 2030 'Restore our oceans and waters' https://ec.europa.eu/info/publications/mission-starfish-2030-restore-our-ocean-and-waters_en

FAO, 2020 'The state of the world's forests: forests, biodiversity and people', https://www.fao.org/3/ca8985en/CA8985EN.pdf

Florida Atlantic University, Marine Ecosystem Conservation, https://www.fau.edu/hboi/research/marine-ecosystem-conservation/

Florida Department of Environmental Protection, Aquatic Preserve Program, https://fdep.maps.arcgis.com/apps/MapTour/index.html?appid=edf6a73067654ad6a7d3c35c64b6e954#map

France 24, 2021, 'Une fragile muraille verte en Afrique', https://www.youtube.com/watch?v=HXzIUsbAAvQ

Gaborit P., 2014, 'European New Towns', Peter Lang International

Gaborit P. (eds.) 2015, 'European and Asian Sustainable Towns', Peter Lang International

GDO Analytical report, Drought in Somalia, Kenya and Tanzania January 2022 JRC Global Drought Observatory (GDO) of the Copernicus Emergency Management Service (CEMS) – 28/01/2022 https://edo.jrc.ec.europa.eu/documents/news/GDODroughtNews202201_Somalia_Kenya_Tanzania.pdf

Hoegh-Guldberg O., Cai R., Poloczanska E.S., Brewer P.G., Sundby S., Hilmi K., Fabry V.J., Jung S., 2014, The ocean. In: Climate Change 2014: Impacts, Adaptation, and Vulnerability. Part B: Regional Aspects. Contribution of Working Group II to the Fifth Assessment Report of the Intergovernmental Panel on Climate Change [Barros, V.R., Field C.B., Dokken D.J., Mastrandrea M.D., Mach K.J., Bilir T.E., Chatterjee M., Ebi K.L., Estrada Y.O., Genova R.C., Girma B., Kissel E.S., Levy A.N., MacCracken S., Mastrandrea P.R., White, L.L. (eds.)]. Cambridge University Press, Cambridge, United Kingdom and New York, NY, USA, pp. 1655–1731.

Iglesias V., Whitlock C. 2020, If the trees burn, is the forest lost? Past dynamics in temperate forests help inform management strategies. Philosophical. Transactions of the Royal Society B 375: 20190115. http://dx.doi.org/10.1098/rstb.2019.0115 last accessed 22/02/2022

ISOCARP centre for urban excellence, 2020, CRIC 'training on urban sustainable development available' https://www.isocarp-institute.org/knowledge-base/

IUCN (2021), 'Deforestation and forests' degradation', IUCN Policy brief https://www.iucn.org/resources/issues-briefs/deforestation-and-forest-degradation

LIFE Adapt Island program: 'Nature-based Solutions for the adaptation of Caribbean island territories to climate change', https://www.cayoli.fr/en/life-adaptisland-2/

LIFE Coast Adapt Project 2014–2020, Skane Sweden: https://www.lifecoastadaptenglish.se/

Malhi Y., Franklin J., Seddon N., Solan M., Turner M.G., Field C.B., Knowlton N. 2020, Climate change and ecosystems: Threats, opportunities and solutions. Philosophical Transactions of the Royal Society B 375: 20190104. http://dx.doi.org/10.1098/rstb.2019.0104, last accessed 20/02/2022

Masson A. 2015, Synthèse technique pour l'implantation de pépinières à Acropora sp. dans la Réserve Naturelle de Saint-Martin, Master thesis, https://docplayer.fr/35867283-Synthese-technique-pour-l-implantation-de-pepinieres-a-acropora-sp-dans-la-reserve-naturelle-de-saint-martin.html

North Bali Reef Conservation, https://northbalireefconservation.com/

OECD, 2017, 'Marine protected Areas: Policy highlights' https://www.oecd.org/environment/resources/Marine-Protected-Areas-Policy-Highlights.pdf, last accessed 16.03.2022

OECD, 2021, 'Scaling up nature based solutions to tackle water-related climate risks: Insights from Mexico and the United Kingdom', *OECD Publishing*, https://doi.org/10.1787/736638c8-en

Pacheco P., Mo K., Dudley N., Shapiro A., Aguilar-Amuchastegui N., Ling P.Y., Anderson C., Marx A. 2021, 'Deforestation Fronts: Drivers and responses in a changing world'. WWF, Gland, Switzerland

Puntacana Ecological foundation, Fragments of Hope (2015), Best Practices Manual for Caribbean Acropora Restoration, http://fragmentsofhope.org/wp-content/uploads/2015/12/Acropora-Restoration-Best-Practices-Manual-ABK-v4.pdf

Seddon N., Chausson A., Berry P., Girardin C.A.J., Smith A., Turner B. 2020, 'Understanding the value and limits of nature-based solutions to climate change and other global challenges', Philosophical Transactions of the Royal Society B375: 20190120. http://dx.doi.org/10.1098/rstb.2019.0120 [accessed Feb 22 2022]

SSI training 2022, 'Marine biology', https://www.divessi.com/fr

The Nature Conservancy, https://www.nature.org/en-us/ and Coastal Resilience by The Nature Conservancy, https://coastalresilience.org/

Thomas H.L., et al., 2014, 'Evaluating official marine protected area coverage for Aichi Target 11: Appraising the data and methods that define our progress', Aquatic Conservation: Marine and Freshwater Ecosystems, 24 (Suppl. 2), John Wiley & Sons Ltd, New Jersey

Thouvenot Z. 2021, 'Deforestation and forests' degradation: Thinking efficient leverages to tackle this multifactorial issue' Pilot4dev Policy Brief, https://www.pilot4dev.com/images/EN_-_Policy_brief.pdf

Tuya F., Vila F., Bergasa O., Zarrancz M., Espinoa F., Robainac R., 2017, 'Artificial seagrass leaves shield transplanted seagrass seedlings and increase their survivorship', *Aquatic Botany*, 31–34, p. 34, https://doi.org/10.1016/j.aquabot.2016.09.001

Utila Coral Restoration, http://www.utilacoral.org/about/

Wild T., Bulkeley H., Naumann S., Vojinovic Z., Calfapietra C., Whiteoak K. 2020, 'Nature based solutions: State of the art in EU-funded projects', European Commission, Directorate General for Research and Innovation

Chapter 8
The Circular Economy from the Perspective of Local Climate Adaptation and Mitigation

PASCALINE GABORIT PHD

The natural resources are finite, and we are currently consuming more than 3 times the resources which could be renewed naturally[1]. Global consumption of materials such as biomass, fossil fuels, metals and minerals are expected to double in the next forty years, while the annual waste generation is projected to increase by up to 70 % by 2050[2]. Only a minor part of the materials produced globally are recycled. Nevertheless, half of the GHG emissions and more than 90 % of biodiversity loss comes from resource extraction and processing[3]. Drawing on concepts such as 'cradle to cradle', the circular economy is an attempt to propose alternatives to solve this problem by proposing concrete solutions to reduce, repair, recycle, reuse, and recover waste or rethink the current production patterns. These steps and actions are necessary to mitigate climate change emissions linked to an unsustainable production model and other types of pollution, such as air pollution caused by the incineration of waste. It would also support positive transformation toward an economic model adapted to the limited availability of resources and would better integrate 'climate adaptation' as a long-term objective. While looking for solutions, innovative economic actors have proposed to move away from the traditional and linear economic model of producing and using goods and services (take, make, dispose) and to instead develop circular solutions (make, (re)use, recycle). Circularity has a great potential to reduce the use of raw materials and to keep the materials in the loop whenever feasible (Dhawan et al. 2019). The idea is to reduce and to minimize the

---

[1] https://www.un.org/sustainabledevelopment/sustainable-consumption-production
[2] World Bank (2018), *What is a Waste 2.0: A global Snapshot of Solid Waste Management to 2050.*
[3] Communication of the European Commission, The European Green Deal, COM (2019) 640 final.

ecological footprint of every human-made product, decreasing the final impact on the planet. This chapter will investigate how the circular economy concept can be applied in different sectors, such as in plastics and also in the construction sector. It will briefly detail how the concept of circularity is creating opportunities for climate adaptation, particularly for cities, but how it is also leading to false narratives or promises. It will approach how circular economic models could further support waste management and climate adaptation actions. This chapter is proposing a global overview that considers current research and trends.

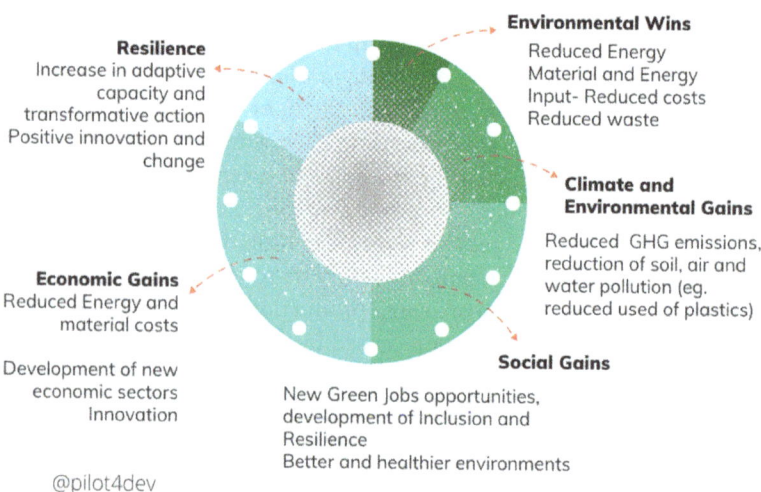

Fig. 1: Circular economy

The objective of this chapter is to present the circular economy as a concept, alongside a few applications and boundaries in order to assess how the circular economy could enhance climate adaptation and mitigation. We will approach the circular economy as a concept (Part I). We will then develop the intractable example of plastics' circularity (Part II) and the potential of the circular economy in the construction sector (Part III), with the possible reuse of the demolition waste. Finally, the chapter will approach the difficulties in engaging with civil society and in raising awareness in a very uncertain context (Part IV). More research would indeed be needed to assess the potential of the circular economy in linkage with waste management and its impacts on climate change, as well as to analyze the concrete opportunities linked to circular economy-based business models in all possible different sectors.

I. The Circular Economy: Stakes and Challenges

The circular economy concept has been gaining momentum as it has been developed to close the loop of product cycle and as a solution for countries, firms, and consumers to reduce harm to the environment. The concept is to take the product from its production to the end of the cycle (waste) and create a loop by promoting the 5 R (reduce, recycle, reuse, refurbish, reinvent). The objectives are several-fold: to move away from a linear economy relying on supply chains, which produce an enormous amount of waste, to increase the interest in developing the recycling sector, and to address a growing concern for the climate and environment. However, the primary objective of the circular economy remains production and profit. The circular economy is also based on the question of waste management. As we will detail in the second part of the chapter, the problem of waste management, for example for plastics, has not been entirely solved in any part of the world. In September 2018, the World Bank announced that global waste production is predicted to rise by 70 % by 2050 unless urgent action is taken. Humankind currently produces two billion tons of waste per year for 7.6 billion people. Population increase may be part of the explanation, but the levels of consumption within a handful of developed nations, and their gross mismanagement of waste, have led to environmental disaster[4].

Example from the European Union: A Nurturing Framework to Develop Circular Economy Patterns

The European Union was not the first to promote the Circular Economy as a concept, as other countries such as Japan, Canada, and China preceded. The first action for a circular economy in the EU, however, began in 2015. The 2015 Circular Economy Action Plan in the

---

[4] https://sensoneo.com/global-waste-index-2019/

European Union was evaluated to show positive outcomes: 'Circularity has opened up new business opportunities, given rise to new business models and developed new markets, domestically and outside the EU. In 2016, circular activities such as repair, reuse or recycling generated almost 147 billion Euros in value-added while standing for around 17.5 billion worth of investments[5]'. On the 11th of March 2020: The European Commission published the communication 'A New Circular Economy Action Plan: For a cleaner and more competitive Europe'[6]. It proposed a realistic action focusing on eco-design, consumers' awareness, digital technologies, electronics and ICT, batteries and vehicles, packaging, plastics, textiles, the built environment, food, water and nutrients, waste management, toxic-free environments, regions and cities, and the market of raw materials. The EU Strategy on Plastics in 2018 is also an important reference document, aiming at tackling the problem of plastics, microplastics, packaging, and global action to better manage plastic waste. Finally, the 2020 DG Research of the European Commission 'Proposed Mission Starfish 2030: Restore our Oceans and Waters' proposed an ambitious program to promote nature-based solutions, enhance the oceans as carbon sinks, reduce the loss of marine biodiversity, and reduce underwater noise with the use of circular systems among others. It also proposed to restore the freshwater bodies and to fight against eutrophication, a source of methane emissions. Indicators and monitoring were equally emphasized. These advancements demonstrate that the European Union has developed a thorough package of tools and regulations regarding the circular economy. These tools allow the further study and evaluation of the viability of the circular economy from a sectoral approach. This package of tools should also be considered against its initial objective, to reduce waste.

A Solution for the Massive Production of Waste

Waste management has become a key problem worldwide as massive amounts of waste are generated daily. According to the World Bank, global waste is expected to grow to 3.40 billion tons by 2050, and to more

---

[5] Report from the Commission to the European Parliament, the Council, the European Economic and Social Committee and the Committee of the Regions on the implementation of the Circular Economy Action Plan Com(2019)190 final p 1.
[6] Com(2020)98 final.

than double population growth over the same period. Overall, there is a positive correlation between waste generation and income level. Daily per capita waste generation in high-income countries is projected to increase by 19 % by 2050, compared to low- and middle-income countries where it is expected to increase by approximately 40 % or more. Waste generation initially decreases at the lowest income levels and then increases at a faster rate for incremental income changes at low-income levels than at high income levels[7]. The total quantity of waste generated in low-income countries is expected to increase by more than three times by 2050. The East Asia and Pacific region is generating most of the world's waste, at 23 %, and the Middle East and North Africa region is producing the least in absolute terms, at 6 %. However, the fastest growing regions are Sub-Saharan Africa, South Asia, and the Middle East with North Africa, where, by 2050, total waste generation is expected to more than triple, double, and double respectively according to the World Bank. In these regions, more than half of waste is currently openly dumped *(on the streets or in rivers)* even though we can assume that the part of organic waste (*thus biodegradable and disappearing quickly*) is more important than elsewhere.

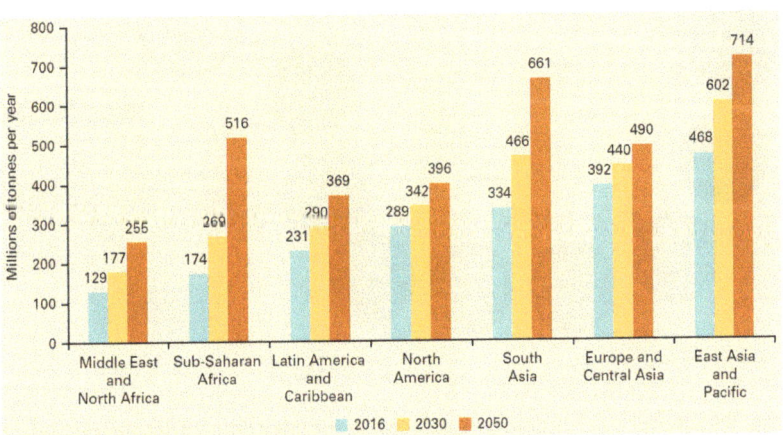

Graph 1: Projected waste generation, by region (millions of tons/year) (Source: World Bank. https://datatopics.worldbank.org/what-a-waste/trends_in_solid_waste_management.html)

---

[7] https://datatopics.worldbank.org/what-a-waste/trends_in_solid_waste_management.html

Waste management is part of climate adaptation. The trajectories of waste growth will have vast implications for the environment, health, and prosperity, thus requiring urgent action. First of all, vast swathes of dumping and littering can create landslides and river water pollution, jeopardizing the access of many communities to clean water. Open dumping of plastics and electric waste is also polluting water and ecosystems, contributing to the loss of biodiversity and jeopardizing many forests, rivers, and oceans (see previous chapter). Finally, waste management and recycling can have an important impact on climate mitigation, by creating problems of access to resources, transformation of resources and related GHG emissions, as well as problems of air pollution, especially caused by waste incineration. Waste incineration is indeed responsible for air pollution in many countries. The situation is critical and is having impacts on respiratory and blood circulation diseases.

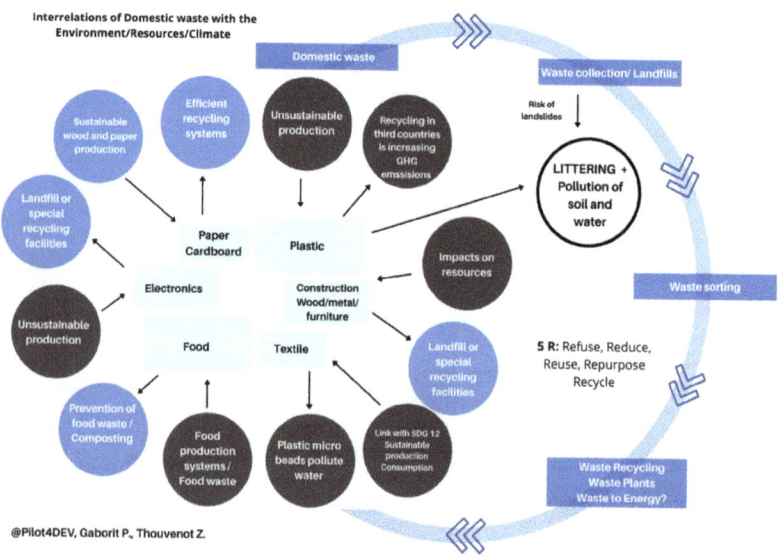

Fig. 2: Interrelation of domestic waste with the environment/resources/climate

In this context of tremendous waste problems, the circular economy has been presented as part of the solution. Here, once again, small scale solutions do exist at the local level both in Europe and in Asia, such as in cities. The potential of the circular economy and its development in Indonesia, for instance, will be developed in the next chapter 'women's role in creating a circular economy'. Cities in countries like Indonesia could indeed be interested in further developing circular economy schemes, as they are developing many recycling facilities for waste recycling such as waste banks. Waste is also recognized as a problem at the national level. The country is facing a problem of population growth, entailing increasing consequences regarding waste production. This creates challenges for which the waste facilities must catch up. According to European Commission data, of the top 20 countries ranked by mass of mismanaged plastic waste in 2010, Indonesia ranks second in waste estimates with approximately 1.29 million of tons of waste per year (De Smet 2019:16). It is also among the 10 producers of waste worldwide. The country benefits from a very long coastline of 54,720 km across the archipelago, which explains why the sheer amount of waste entails a question of littering in the oceans. Solutions are actively being implemented, notably at the city level. Waste banks are indeed common in the country in the form of *Bank Sampah* mechanisms. 'Bank Sampah' actually means a place for collecting and sorting inorganic waste in a way that it can be used, reused, and recycled, which gives waste an economic value (Cyran et al. 2020). Municipalities are creating systems to develop these *Bank Sampah* as well as networks to collect the waste and provide intermediaries between the residents and the waste recycling facilities. Some applications are also available to connect waste sellers and buyers. Successful examples of circular economy implementation exist. However, the question of the circular economy is whether successful practices can be scaled up or used in development models to tackle the issue of waste.

Discussion Points and Skepticism at the Scientific Level

Beyond successful practices, for example in the food short circuit where local suppliers are given priority, or the waste bank systems, it is still difficult to find successful complete circular economy models. Limitations and challenges have emerged: Some examples show that the circular economy is not always applicable without further assessments, such as in the area of biofuels which have proven harmful for the environment and for climate change. Many countries also develop incinerators with or without energy recuperation without taking all externalities into account such as air pollution. Some authors have shown the limitations of the concept of the circular economy. These limitations include that the ideology embraces different conceptions, views and assumptions (Kirchher et al. 2017). The concept would be based on false assumptions (Korhonen et al. 2018), and it also fails to address sustainable development and social impacts (Kirchhher et al. 2017). In other words, it is difficult to move beyond 'trials and errors' and to make sure that 'each Circular Economy Project is considered for its contribution to global net sustainability' or to calculate 'what is left as improvement or positive outcome after an individual project or action, as compared to the situation before the project' (Korhonen et al. 2018: 42).

It is equally difficult to show that the problem and environmental externalities have not been displaced along the supply chain (for instance with some of the biofuels and/or waste incineration). Finally, the circular economy concept should not replace the corporate or producer's individual extended responsibilities, where the waste producers are identified as responsible for negative externalities, GHG and other pollutions induced by their activities.

In the next two parts, we will approach the question of plastics (Part II), as a circular economy based on plastics is recognized as being extremely problematic and is creating intractable problems. Contrarily, the following section (Part III) will address the potential of reusing demolition waste in the construction sector.

II. A Circular Economy of Plastics to Protect Oceans? An Intractable Problem

As we will detail further in Part III, there are concrete examples of circular economy which are already taking off as successful models (e.g. in the construction sector, textile, batteries but also in the reuse of electronic materials…). The case of plastics tends to show on the contrary that the 'circular economy' has been used in this specific case as a 'green washing' concept to hide the difficulties of recycling and the negative impacts (externalities) of waste trade.

In theory, the underpinning idea behind the circular economy (e.g. as rightly developed rightly by important think tanks such as the Mac Arthur foundation) is that it could contribute to the protection of the oceans by reducing the littering of plastics, and of other pollutants such as oils, fertilizers and pesticides, which have a detrimental effect on marine ecosystems.

The first reality is that the oceans are endangered as it was developed in the previous chapter on nature-based solutions. '*Healthy oceans and waters are taken for granted. Yet they are in trouble*' (EC Mission Starfish 2030).

Better ideas to recycle plastics would be needed to protect oceans: According to the European Environment Agency (2021), the annual global plastic production has increased from 2 to 380 million tons since 1950 and is projected to double by 2035 and almost quadruple by 2050. Since 1950, 6.3 billion of plastic waste has been produced, of which 9 % has been recycled and 12 % burnt while the rest has been accumulated in landfills or been littered in the wild (Geyer et al. 2017). In most of the world, plastic waste tends to be burnt, stocked in landfills until saturation, or discarded by roads or rivers. This has negative environmental effects as plastic can take hundreds of years to decompose.

Plastic (including macro, micro, and nano) frequently ends up in oceans and seas worldwide and affects biodiversity and ecosystems. Plastic waste can entangle fish and birds, while micro and nano plastic waste can be ingested by marine animals and absorbed by human beings at the end of the food chain (De Smet et al. 2019). Indeed, human health can also be impacted by plastic pollution. As we have detailed in the previous chapter, oceans are confronted by natural and anthropic pressures, and an increasing number of species are endangered. Their role as climate regulators and as carbon sinks is threatened. Plastic breaks down at a slow pace into the oceans and is increasingly discovered within marine organisms. A recent study from WWF in 2022 shows that more than 2,141 species have been affected by plastic waste, some of which have been pushed to the brink of extinction[8]. According to this report, published in the first quarter of 2022, 88 % of the studied marine species were affected by severe contamination of plastic. The report said that many animals have ingested these plastics, including animals commonly consumed by humans.

Climate change is equally affected by plastic. While plastic production produces GHG emissions, plastic waste management also does. Incineration and plastic waste energy recovery (which may appear as a circular economy model) result in a direct release of GHG emissions and of toxic substances like dioxin, furan, or mercury, which disturb

---

[8] https://www.weforum.org/agenda/2022/02/extinction-threat-ocean-plastic-pollution/

ecosystems and affect human health. As a result, people could be affected by skin and lung diseases due to plastic waste treatment (Vidal, 2019).

Plastic waste, when mismanaged, left uncontrolled, or openly dumped, affects ecosystems and human wellbeing. The scientific insights on the impacts of plastics do not, unfortunately, only cover littering at a large scale but also microplastics and plastics in fibers (De Smet et al. 2019). There are still knowledge gaps, particularly regarding how micro plastics can affect freshwater reservoirs and land systems. The research on the impacts of macro plastics is also limited (De Smet et al. 2019). The impacts of nano plastics still remain unknown. Additionally, the sources of pollution range from the entirety of plastic's life cycle. There are environmental losses throughout the cycle, including during the production phase and the transport. Then the plastic discards are streamed through rivers debris, which eventually end in the oceans (Lechner et al. 2014).

Therefore, there is an opportunity to facilitate the exchange of more knowledge in order to develop solutions for plastic waste. Conversations on what is working and what is not working with regard to decreasing plastic litter, associated emissions, and possible mechanisms linked to the circular economy are needed. Next, the question of waste trade is also hiding the difficulties of recycling plastic components.

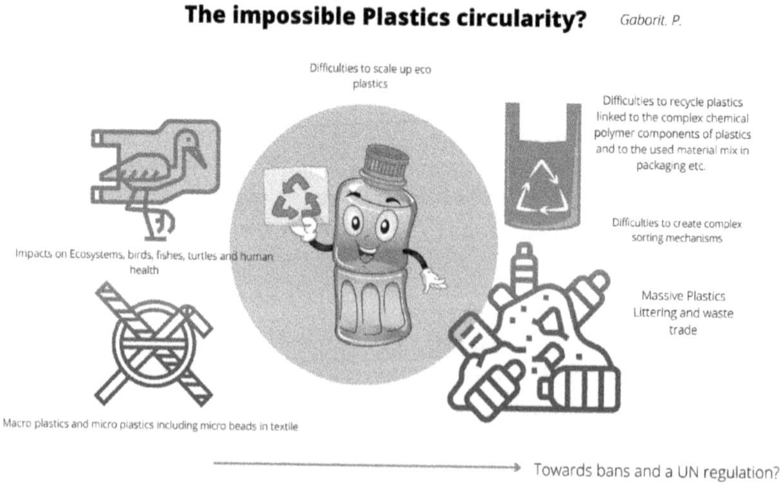

Fig. 3: The difficulties faced by plastics' circularity and recycling

Narratives on the Circular Economy, Hiding a Reality of Waste Trade

Many narratives in Europe have been built around recycling plastics. This is particularly true for many projects funded by the European Union, but this is also true for cities which have developed complex sorting mechanisms for plastics, giving the 'tacit' assumption to residents that these plastics would be recycled. In reality, however, these narratives around sorting plastic and around the circular economy could not prevent another climate and environmental disaster due to plastic trade to Asia. Plastic waste from the Western world has been previously traded to China, which was replaced by Turkey as main destination for export waste from EU countries, and by countries of Southeast Asia such as Indonesia, Malaysia, and the Philippines.

According to the European Environment Agency (2021), exporting plastic waste from the EU to Asia is a means of dealing with insufficient on-site recycling capacities and circular economy schemes in the European Union. Waste import restrictions in China since 2018 have shifted exports to other countries such as Malaysia (Thouvenot 2021). According to Global Alliance for Incinerators Alternatives (GAIA report April 2019),[9] more than 900,000 tons of plastic waste have been imported to Malaysia in 2018, while more than 400,000 tons have gone to Thailand or to Vietnam. This tonnage was mostly originating from the United States, Japan, and Europe, and within Europe was predominantly from Germany, the UK, and Belgium.

Although the European Environment Agency (EEA) explains that the European Union is exporting plastic waste because of limited plastic recycling capacities in the EU in a context of growing production of plastic and plastic waste, the Asian countries also do not always have adequate means of recycling plastic. According to Sahabat Alam Malaysia (SAM), an independent non-profit national organization working on sustainable development and the management of natural resources, Southeast Asia has even less efficient plastic recycling facilities than OECD countries. While the need for an efficient circular economy is acknowledged by EU countries, the exports of plastic waste to Southeast Asia still continue.

In 2019, Malaysia and Indonesia shipped back unregistered or badly registered containers of various types of waste. On the 15th of May 2019,

---

[9] https://www.no-burn.org/

the Philippines president gave an ultimatum to Canada, and sent 69 containers of dumped waste back to Canada. In April-May 2019, at the Basel Convention, Southeastern Asian NGOs, supported by Norway, advocated for the cessation of plastic waste traffic. The proposition has been adopted despite the opposition of several large countries such as the United States. Since then, non-recyclable plastics (polyethylene terephthalate, and polypropylene) cannot be exported anymore without the agreement of the receiving country. The regulation is effective since the 1st of January 2021, but it has not yet stopped the trade. In early 2020 Malaysia sent back 4000 tons of illegal plastic waste to 20 countries including France and also the UK, which actually sent 33,000 tons of waste during the 7 first months of 2020, 81 % more than usual (Thouvenot 2021). The current convention also does not prohibit legal agreements and shipping wherever there is no ban.

Because some types of plastic waste have been added to the United Nations Basel Convention, the option of exporting plastic waste is becoming more difficult but still continues. The European Environment Agency (2021) therefore recommends that 'policymakers, business and other actors build a more robust and circular economy for plastic in Europe'.

Narratives on the Circular Economy Hiding Other Technical Difficulties to Recycle

Advocating for a circular economy based on 'plastics' recycling initially appears to be a good option. However, a more thorough analysis of existing studies demonstrates that the circular economy of plastic is almost impossible to realize. De Smet et al. (2019) references the difficulties and gaps in 'creating a circular economy of plastics'. Among other issues, it states that '*barriers preventing successful implementation of safer alternatives include insufficient understanding of technical performance, incumbent technologies and switching costs and risks.*' The report also mentions '*a lack of knowledge on the chemical composition of plastic articles and their negative impacts on health.*' The authors mention plastic substitutes, like the substitution of EDC BPA by Bisphenols, which have replaced early plastic components without further research on their toxic properties.

New solutions and components do exist to create biodegradable and compostable plastics, which could fit into a circular economy pattern '*through the idea of closing the biological cycle.*' However, again, the combination of these biodegradable plastics with other materials can prevent the 'compostable aspect' of these newly biodegradable plastics from being actualized. According to de Smet et al. (2019), the whole waste management cycle and treatment process of plastics should be specific for each type of plastic and waste to avoid contaminations, such as non-compostable products. Trying to avoid the complexity of sorting mechanisms could lead to false claims. For newly created bioplastics, difficulties of supply already exist and they lack competitive business models when compared to the very low price of regular oil-based plastic production. Finally, what was considered a path for recommendations toward a circular plastic economy actually reveals a life cycle problem of plastic products, entailing usage shifts that would necessitate systemic change. All these problems make the creation of a circular economy based on plastics very unlikely to occur in the near future as the price for the change would be enormous.

As the whole life cycle analysis of plastics demonstrates, the possibilities of recycling plastics are very costly and do not mitigate all the problems and challenges caused by plastics, like the creation of microplastics, molecules and energy leaks in production, and transformation of plastics. In these circumstances, a narrative on circular economy can simply hide larger problems and hamper decisions to implement regulations and bans as a more drastic solution. The IPEN center for a toxic-free environment demonstrates through data analysis that plastic recycling can release toxic materials and be even more detrimental than littering on human health (Beeler 2022). The study also finds out that countries cannot handle such a massive production of plastic waste: 'The studies reveal that countries are unable to handle large volumes of diverse plastics waste streams safely, and the reality is that, without regulations requiring plastic ingredients to be labeled, countries are blindly allowing known toxic chemicals onto their markets in plastic products (…) It calls for public policies to end the recycling of hazardous chemicals in plastics, that poison the circular economy and threaten human health'[10]. This shows that plastic production should absolutely be decreased and lead to more global efforts to prevent plastic waste, instead of increasing massive global innovation, transportation and storage efforts to recycle plastic.

---

[10] https://ipen.org/news/plastic-poisons-circular-economy

In February 2022, the United Nations Environment Assembly (UNEA) took the first step toward the creation of a landmark treaty to control plastic pollution worldwide, which is raising a lot of hope for a viable way forward. As seen with plastics, circular economic projects can mobilize millions of Euros of investment and research and still end up with intractable and unsolvable problems. This shows that a more sustainable production and consumption (SDG12) should come as a priority instead of using the 'circular economy' as a buzzword to hide polluting practices. There is, however, a strong potential for the circular economy to be used in other sectors such as the construction sector. In the construction sector the scale of construction and demolition could decarbonize part of the sector and could also create a successful economic model with the creation of waste banks for construction and demolition materials.

III. The Reuse of Demolition Waste: An Opportunity to Enhance Climate Mitigation and Adaptation

The reuse of construction materials has proven environmental and economic efficiency (89 % of construction and demolition waste in the EU is reused[11]) which offers a good example of how the circular economy could address problems of climate adaptation and enhance the environment. In many countries, cities and public authorities are directly responsible for large housing programs and public works. This entails construction, demolition, infrastructure, facilities, and also the dragging of soils. Such demand for infrastructure relies heavily on raw materials like sand for concrete and mortar, soil, stone for aggregates, and limestone for cement. The key challenge is to make materials available in a manner that considers the exhaustible nature of these resources and addresses ecological impacts associated with their extraction and processing without omitting the impacts for ecosystems and for an inclusive economy. The recycling of construction materials and of the earth, as well as the restoration of ecosystems can contribute to reducing the environmental impact of construction and its impact on resources. Studies have shown the extraordinary potential of recycling construction materials

---

[11] Eurostat Newsrelease, 'Record recycling rates and use of recycled materials in the EU', 439/2019, 4 March.

in countries with a high urbanization rate like India. Organizations like ACR+[12] have also published guidelines for public local authorities on the recycling and transformation of demolition waste (Marengo 2019).

Buildings and construction have a central impact on SDG 12, which focuses on sustainable production and consumption, but also SDG11 on sustainable cities, SDG13 on climate change, SDG8 on inclusive and sustainable growth, and SD9 on resilient infrastructure.

Construction and Demolition Waste for Growing Urbanization

In many Asian countries urbanization is booming, increasing the needs for urban dwellings and housing. In India, more than 30 % of the population lives in urban areas and it is projected that over 40 % of the population will be living in urban areas by 2030, forecasting a very high growth in urbanization[13]. The pace of urbanization in the entire Asian continent is astounding half of the world's total urban population is now in Asia (Moench et al. 2011). With urban growth averaging 4.4 % annually, Indonesia experiences the highest rate of urbanization in Asia, even higher than in India and China. It is predicted that 68 % of the population will live in cities in the next ten years. Urban population densities exceed 15,000 habitants/km² in cities like Jakarta and more than 30,000 habitants/km² in some inner-city areas[14]. Cities are heavily impacted by climate change, although they could be seeds for solutions. Countries like Vietnam (36 % of urban population) and the Philippines (45 %) have large urban area concentrations, with fast-growing urban centers. Urban areas host the majority of the vulnerable populations as well as vital infrastructure and social infrastructure. Local governments have increased pressure to develop services, infrastructure, employment, and housing in heavily urban areas (UN-Habitat).

Construction works are mushrooming in many Indian cities. It is estimated that in India almost 70 % of buildings supposed to exist by 2030 are yet to be built[15]. There are huge opportunities to reduce the use of

---

[12] The European association of cities and regions for a sustainable use of resources.
[13] According to the think tank Mac Kinsey 2016.
[14] www.resilient-cities.com/about
[15] According to the think tank Mac Kinsey 2016.

polluting materials, to reuse the earth material induced by the digging and dragging of the works, and to create waste banks for smaller construction companies. Some regions are already experiencing serious environmental degradation due to the extraction of materials, leading to more scarcity of resources (Roychowdhury et al., CSE 2020). Additionally, the possibilities of retrofitting existing buildings and addressing the informal housing sector are needed. Although few efficient technologies remain in use, there is a great opportunity to invest in sustainable, green and low carbon 3 R schemes through the reuse of materials. Likewise, there is a great potential for water saving in the construction sector. Planning and infrastructure development need to be integrated in circularity to address underdeveloped basic services, water, sanitation, waste management, and resilience. Finally, the GHG emissions created by the transportation of the construction materials can also be reduced by GIS technologies and medium-sized enterprises.

The Life Cycle Approach for Building and Construction Sector verifies typical environmental concerns and other resource efficiency-related issues with the life cycle stages associated with the value chain of the construction sector in respect to inclusiveness. This sectoral strategy primarily focuses on the end-of-life stage of the sector.

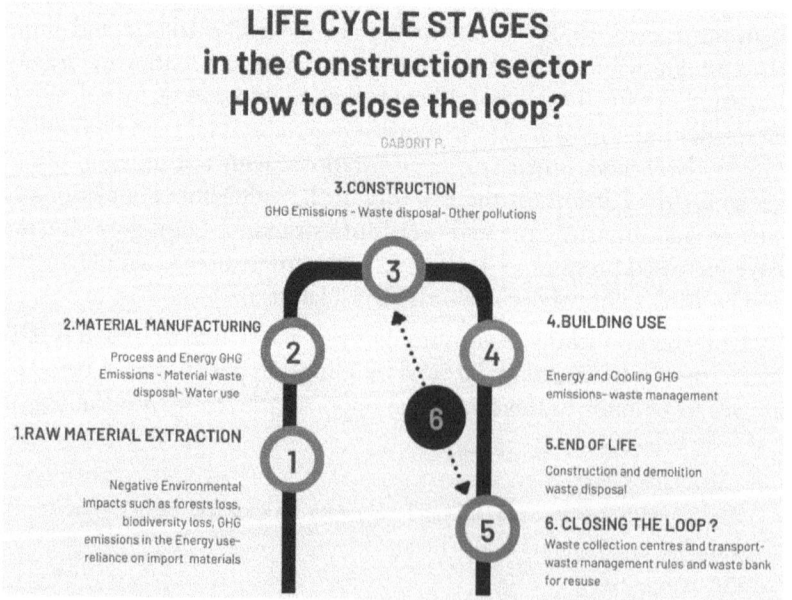

Fig. 4: Life Cycle stages in the construction sector: how to close the loop?

Fig. 4 shows the different steps of construction and demolition. It highlights the impacts on the environment, on the resource extractions, and on the GHG emissions. The challenge is **to close the loop** and to create waste collection centers and waste banks for the different construction materials.

The key challenge in creating circularity will be to make building materials more available and extracted in a manner that considers the exhaustible nature of these resources, while addressing the ecological impacts associated with their extraction and processing. Additionally, the reuse of demolition materials can support positive impacts for an inclusive economy, taking into account informal settlements. The informal housing sector still represents an important share of many fast-growing cities in Southeast Asia and South Asia. For example, it still represents between 17 % and 36 % in a city like Ahmedabad. India's construction industry is one of the world's largest consumers of construction materials for both natural materials like stone, sand, clay, lime and manufactured and synthetic materials. The country manages to recover and recycle only about 1 % of its construction and demolition (C&D) waste (Roychowdhury et al. CSE India 2020). The rest remains in landfills or littered in the landscape. Demolition of a building in an average city may lead to waste generation of 425 kg/m², which means that tons of material could be sorted and reused (Roychowdhury et al. CSE India 2020). The national annual consumption of sand in countries like India and in Indonesia is estimated to be a million tons, while a million square meters of soil are used only for the manufacturing of clay fired bricks. These figures are not even including the soil materials induced by public works like dragging. There are solutions to reuse the soils including in European cities (Diab 2016). But soil is not the only material which could be recycled. 2 billion tons of stones are also used for making aggregates annually in India. Furthermore, millions of tons of limestones are consumed in cement plants. Additionally, water conservation techniques, with specific application to buildings, could be further developed to close the loop (Fig. 4). There is a large focus on groundwater recharging and creating rainwater harvesting structures without focusing on water efficiency and usage within buildings. Water scarcity remains an issue in many large growing cities (SDG 6). Water, energy and other resources could be saved by a sustainable reuse of demolition materials. A substantial share of the new demand in construction materials can indeed be met using the demolition waste of the buildings' existing stock. For the resource-efficient construction sector, companies need to manufacture sustainable construction products from recycled materials. Dependence on virgin materials needs to be gradually reduced to the profit of an enhanced reuse of the demolition waste. This will reduce the extraction of new resources, further deforestation and better protect water resources.

The potential for the reuse of construction and demolition waste is tremendous. Contrarily to plastics, there are no important externalities such as the release of toxic components, with the exception of paint and asbestos which can be mitigated through taking the necessary precautionary steps. The scale of buildings' construction and demolition could also make them useful for the creation of successful business models. This can positively influence the maintenance of fresh water and groundwater reserves and reduce the drainage of soils as a protection against floods. Therefore, the schemes adopted for a circular economy in the construction sector would directly contribute not only to climate mitigation, but also to climate adaptation with more protection of the ecosystems, and better protection against heat waves and floods. The positive impacts on climate mitigation are also tremendous in the reduction of GHG emissions linked to extraction, deforestation, transportation, and energy.

As detailed in the two previous parts of this chapter (Part II and Part III), the circular economy is a complex concept, which can hide realities of the pollution in the plastics sector but can also create tremendous opportunities in the sector of construction and demolition waste. The question of civil engagement and raising awareness in civil society and in the private sector is, however, extremely complex in an uncertain scientific environment.

IV. Awareness Raising: A Complex Approach in an Uncertain Scientific Environment

Better public information and a rise of awareness seem necessary to further sensitize private customers and industries to decrease their environmental impacts. This implies the sorting of waste, the reduction of materials uses, and, in the case of plastics, reduction of marine litter and mitigation of the further endangerment of marine species and ecosystems. We have also detailed in Part III how the circular economy could enhance climate adaptation and mitigation in the reuse of the construction and demolition waste through circular models. The reuse of materials could indeed enhance adaptive climate planning, reduce resource extraction, and bring positive developments for the stabilization of soils, the fight against deforestation, and climate change overall. However, the market undertaking of such a shift in production and consumption is not automatic. Fig. 4 illustrates how 'closing the loop' could be challenging.

How can transformative change be achieved to create a better circular economy on one hand and climate adaptation on the other hand? How could the different private actors coordinate their efforts toward better adaptive models? Parts of existing research are keen on producing 'behavioral change' patterns which would explain human decisions with a collection of algorithms. There also seem to be expectations to have the data collected by human scientists or sociologists. In our research and book, we approach the question from a 'human science' angle, by giving particular importance to the question of logics, communication, dialogue, education, and multi-stakeholders' involvement. Nevertheless, when faced with complex concepts such as 'the circular economy,' what we lack are simple messages. Indeed, the problems are so complex and reticulated that they often contradict one another. To take the case study of beach clean-ups and plastic littering, how can people be sensitized to plastic littering if scientists rightly tell them that beach and oceans cleanup is inefficient in proportion to the problem? How can people promote environmentally friendly behaviors when their access to drinkable water is reduced to bottled water wrapped into plastic? Do citizens know how their plastic sorting waste is recycled (or not) as we have detailed in Part II of this chapter? There are sometimes false messages in the narratives promulgated by international organizations and civil society to simplify the problems. Uncertainty and lack of knowledge lead to simplistic messages, and these simplistic messages sometimes cause uninformed decision-making. This is what happened when the interest in biofuels had been overestimated which led to the development of large plantations causing deforestation.

Simple but concrete messages can be key in promoting the circular economy. Composting, for instance, is a good example of how to communicate on a circular economy of organic waste. Composting enables organic waste to be transformed into organic fertilizers for soils. The project bio circular cities in which the cities' and regions' network ACR+[16] is taking part will investigate with several pilot projects

---

[16] The Association for a sustainable use of Resources.

the circular economy potential of unexploited bio-based waste streams by exploring the development of economically and environmentally efficient models for organic waste like food and kitchen waste, garden waste, agricultural waste from agro-based industrial sector, wood waste, and forestry residues[17]. This composting is especially interesting in countries like Indonesia where organic waste remains important (see the next chapter). Organizations like 'The Circular Collective' and others also propose a series of successful concrete examples[18].

The circular economy can be useful when concrete proposals and local solutions are identified. The risk is, however, as was the case for the circular economy of plastics, that simplistic messages are used to make a scientific reality less complex, and to postpone more necessary solutions like a drastic reduction of the use and future extraction of fossil fuels (and of plastic production) In these alternative models, though, the available knowledge needs to guarantee that the identified solutions decrease the carbon footprint, the environmental footprint, and the toxicity to human health by decreasing air pollution. Another identified problem mentioned by research (De Smet et al. 2019) is the cost for every shift in the production, transport, and in recycling, and the absolute necessity to integrate a whole life cycle approach of products in the circular economy model, as well as to develop and nurture a producer's extended responsibility.

Conclusion

'The circular economy' is a broad term which has recently gained momentum. This chapter presents the concept, its boundaries, and emerging opportunities to enhance circular economy streams, such as the reuse of demolition waste, not only to reduce GHG emissions but also to protect resources, soils, ecosystems, and to stir the resilience of territories to climate change. Our analysis entails that the circular economy can only exist through very practical applications and initiatives. The reuse of soil and construction materials are excellent examples of how the 'circular economy' patterns can contribute to enhancing the resilience potentials of ecosystems like marine ecosystems, soils, and forests. This can also

---

[17] https://biocircularcities.eu/
[18] www.thecircularcollective.com

positively influence the maintenance of fresh water and groundwater reserves and reduce the drainage of soils as a protection against floods. The challenges, however, remain for the creation of circular economy patterns for highly polluting materials such as plastics for which the current impacts and alternatives do not yet provide enough outcomes for a possible shift towards better and more responsible production and consumption (eg. progressively shifting away from fossil fuels). In this framework, clear and accurate information is needed to avoid simplistic messages and to create informed decision making. Investigation of waste management practices like waste trade is equally necessary to target unsustainable practices that are detrimental to global efforts for lowering carbon production and circular approaches being applied to more economic sectors.

References

Beeler B. 2022, *How plastics poison the circular economy*, IPEN, https://ipen.org/news/plastic-poisons-circular-economy

Cyran K., Viviyanti V., Hayati K. 'Waste Management in Pekanbaru Indonesia', Policy paper, proposal for a future sustainable city

Dhawan P., Beckmann J.P. 2019, *Circular Economy Guidebook for Cities*, CSCP, www.thecircularcollective.com

De Smet M., Linder M. 2019 'A circular Economy for plastics: insights from research and innovation to inform policy and funding decisions'. European Commission, DG Research https://op.europa.eu/en/publication-detail/-/publication/33251cf9-3b0b-11e9-8d04-01aa75ed71a1/language-en/format-PDF/source-87705298, last accessed 2022

Diab Y. 2016, *La terre dans tous ses états*, Paris, Presse des Ponts, https://www.unitheque.com/la-terre-dans-tous-ses-etats/presses-ecole-nationale-des-ponts-chaussees/Livre/284525

DG Research of the European Commission, 2020, 'Proposed Mission Starfish 2030 Restore our Oceans and Waters'.

EC Commission Report from the Commission to the European Parliament, the Council, the European Economic and Social Committee and the Committee of the Regions on the implementation of the Circular Economy Action Plan COM (2019)190 final

EC Commission 2020, 'A New Circular Economy Action Plan: For a cleaner and more competitive Europe'. Com(2020)98 final

European Environment Agency, 2019, modified in 2021, 'The plastic waste trade in the circular economy', Policy briefing, https://www.eea.europa.eu/publications/the-plastic-waste-trade-in

Eurostat Newsrelease, 'Record recycling rates and use of recycled materials in the EU', 439/2019, 4 March, GIZ training manual http://re.urbanindustrial.in/live/hrdpmp/hrdpmaster/igep/content/e64918/e64922/e67075/e67087/GIZ_CD_eTrainingManual.pdf

Geyer R., Jambreck J., Lavender Law K. 2017, 'Production, use and fate of all plastics ever made', *Sciences Advances*, Vol. 3, Issue 7, quoted by *Le Monde Diplomatique*

Kircherr J., Reike D., Hekkert M. 2017, 'Conceptualizing the circular economy: An analysis of 114 definitions' in Resources, *Conservation and Recycling*, 127 (2017), 221–232

Kolk, A. 2016, 'The social responsibility of international business: From ethics and environment to CSR and sustainable development', *Journal of World Business*, 51(1), 23–34, https://doi.org/10.1016/j.jwb.2015.08.010

Korhonen J., Honkasalo A., Seppälä J. 2018, 'Circular economy: The concept and its limitations', *Ecological Economics*, 143 (2018), 37–46

Lechner A., Keckeis H., Lumesberger-Loisl F., Zens B., Krusch R., Tritthart M.,..., Schuldermann E. 2014, 'The Danube so coulorful: A potpourri of plastic litter, outnumbers fish larvae in Europe's second largest river', *Environmental Pollution*, 188: 117.

Marengo P. 2019, 'Sustainable Construction guidelines for public authorities' ACR+ publications, https://www.acrplus.org/images/technical-reports/2019_ACR_Sustainable_construction_guidelines_for_public_authorities.pdf

Moench M., Tyler S., Lage J. 2011, 'Catalyzing urban governance: Applying resilience concepts to planning practice in the ACCCRN Program 2009-2011', ACCRN publication

Roychowdhury A., Somvanshi A., Verna A. 2020, 'Another brick in the wall', CSE India (center for Science and Environment), https://www.cseindia.org/another-brick-off-the-wall-10325

Thouvenot Z., 2021, 'Plastic waste Trade: Shifts, Impacts and Solutions', Policy Brief www.pilot4dev.com/knowledge

Vidal A., May 2021, 'Des exportations occidentales qui ruinent les efforts des écologistes locaux, déferlement de déchets plastiques en Asie du Sud-Est', Le Monde Diplomatique

Wuppertal institute: 'Factsheet construction and demolition', https://www.urban-pathways.org/uploads/4/8/9/5/48950199/factsheet_up_construction_demolition-small_3.pdf

Chapter 9
Indonesian Women in a Circular Economy: Waste Management Programs

BULAN PRABAWANI, PhD, WIWANDARI HANDAYANI, PhD,
DESSY ARIYANTI, PhD, DIANA NUR AFIFAH, PhD

Women are significant actors in boosting a country's economy through various industrialization and business activities as they represent half of the working force. Prior to Covid 19, approximately 48.75 million women worked in various sectors in Indonesia (Pusparisa, 2020). Women have also played a significant role in the education sector, with a greater number attending school than men since the 1970s. The participation rate of women in higher education (universities) is higher than men, although they face more challenges such as gender biases, cultural and social pressure, work-life balance, and health issues (World Bank, 2000).

In 2019, 48.9 % of women in rural areas worked in the agriculture, forestry, and fishery sectors, while those living in urban areas were employed in the processing industry (Indonesian Central Bureau of Statistics, 2019). However, women still experience inequality in the formal and informal sectors with numerous cases of violence at the workplace, income disparity, and poor health and working conditions (Tacoli, 2012). They are still considered second-class citizens, especially in developing countries, although according to FP Analytics (2020), women have higher incomes than men in male-dominated sectors, such as electricity, gas procurement, construction, real estate, transportation, and warehousing sectors. These data indicate that women play a central role in the country's economy and deserve greater opportunities.

The circular economy was initiated in 1966 as defined by Kenneth E Boulding in a book chapter titled 'The Economics of the Coming Spaceship Earth' (Salvador et al., 2021) but was only brought into the conversation regarding development in Indonesia around 2015. This initiative was compelled by the phenomena of exploitation, environmental damage, and increasingly massive waste generation. It is imperative to develop the circular economy in Indonesia because, apart from its ability to mitigate and potentially overcome a variety of environmental problems, it also has the potential to generate a GDP equivalent to 45 billion US Dollars by 2030. This increased revenue would thereby create over 4.4 million new employment opportunities, 75 % of which would be for women in 5 sectors: food and beverages, textiles, wholesale and retail trade in plastic, construction, and electronic packaging. The Indonesian Ministry of National Planning and Development assumes that women have great potential as employees and important actors in the circular economy and more disadvantages in a linear economy system because they are more physically, economically, and socially vulnerable than men (Ministry of National Planning and Development, 2021).

This book chapter describes the challenges as well as the opportunities of Indonesian women in the circular economy, alongside their initial efforts to participate in the circular economy. The first sub-chapter analyzes the importance of involving women by changing business and work patterns. This discusses the position of women in terms of social, economic, and political perspectives. The second sub-chapter presents strategic accelerating steps toward changing the role of women in the circular economy through elements of education, supporting agencies, insight, creativity, and systematic initiatives. The third sub-chapter presents several lessons on the circular economy and the products initiated by Indonesian women with potential for further development. The last sub-chapter presents the importance of envisioning women through long-term thinking, IT support, and business re-design. This discussion is important to ensure that women would have better facilitation and access to be involved in the success of the circular economy in Indonesia.

I. The Changing Business

A linear economy is a one-way economic pattern or a take-make-and-dispose economic process which uses resources and energy extensively and produces large amounts of waste. This economic pattern also encourages competition for natural resources and cheap labor while ignoring environmentally friendly practices (Sariatli, 2017). In the context of the Kluckhohn Strodtbeck Framework (Hills, 2002) humans in a linear economy are masters of nature, in which nature is an object to explore and not to nurture. In this economy, part of the material is lost in the production process, is not used, and has high inventory costs. This is not sustainable. Thus, an alternative economic pattern is needed: a circular economy in which economic actors use as few resources as possible for as long as possible, increase their value optimally, and regenerate products and resources at each stage of use (Korhonen et al., 2018). This new concept offers 5R-based economic activities, namely Reduce, Reuse, Recycle, Refurbish, and Renew. These use fewer resources to extend the product life cycle and reduce the impact on the natural and social environment (Neves et al., 2022). In Indonesia, the circular economy promotes the establishment of new industries from spare-parts products to improve the supply chains of the waste process. Humans maintain harmony with nature by developing and maintaining social and natural integration in this situation. This economic pattern fundamentally suppresses resource inefficiency to achieve energy, material, and water savings. Additionally, the circular economy ensures less pollution and waste, specifically in textile, construction, retail, and electronic industries, which largely employ women, particularly in the green sector (Ministry of National Planning and Development Indonesia, 2021). Moreover, the Neves et al. (2022) study shows that young people are more concerned with the circular economy than older people, and that the higher a person's income is, the less they are likely to receive circular economy products. As of 2017, 74 % of Indonesians are distributed in the productive age of the 262.41 million total population (Mubarok et al., 2019). Because Indonesia would enjoy a demographic bonus as the age of society is dominated by the productive age (or young people) at its peak in 2029–2030 (Mubarok et al., 2019), the country has a high potential for the development of a circular economy if the government and industry have a commitment to initiating processes. It takes knowledge, skills, and sensitivity from various parties to the circular economy to ensure that positive and environmentally friendly economic growth can be achieved.

Along with the high population, Indonesia faces a waste problem mainly stemming from household and traditional markets while the majority of waste is still processed linearly and has no added value. Therefore, smart and sustainable waste management is essential (Fatimah et al., 2020). A circular economy in Indonesia increasingly becomes interesting for the development field considering that the main source of waste is the household and that women are mainly the ones who manage a household's domestic affairs. Therefore, it is imperative to handle the waste chain by involving women.

Women's Socioeconomic Positions in Indonesia

Indonesian women must take on three roles that are carried out at the same time: the domestic function of managing household work, the production function as workers who make a living for their families (Arifin et al., 2019), and the community management function through their active social roles in their communities (Zuhdi, 2018). Nevertheless, according to OECD (2020), Indonesian women are disadvantaged domestic employees with low wages. Many women are exposed to chemicals because they often work in the manufacturing sector, especially with garments, textiles, cigarettes, and herbal medicine. Those in the business sector, another part of the linear economy system, are exposed to more exploitative behavior because the material, capital, or even human resources used are exploited to an optimal scale, consumed in the short term, and, finally, disposed. Indonesian women experience discrimination in the household, workplace, and social environment. A number of policies have been formulated to protect the rights of women, especially workers, but in practice women do not have authority of their rights. In an imperfect market competition, the industry has greater power to regulate revenue and control. For example, women have menstrual leave, but they tend not to choose to use their rights because they are afraid and because they want to avoid poor performance appraisals (Hanif & Savitri, 2018). Efforts to eliminate discrimination and non-equality have shown a positive trend (Krisnalita, 2018), but there is still a disparity in the income of men and women (Ekaningtyas, 2020). In the informal sector, women are not granted adequate social protection: they are vulnerable to exploitation, violence, discrimination, and even trafficking. Women receive salaries that are not in accordance with government provisions, have erratic and long working hours, and do not receive the paid leave to which they are entitled in full (Zuhdi, 2018). Wages of the informal sector are not even reported as income, signifying the lack of vale society places on women's labor (Krisnalita, 2018).

In order to meet the demands of the triple roles, Indonesian women could be more empowered as entrepreneurs in micro, small, and medium industries. Entrepreneurship is assumed to be one of the quickest ways to produce female leaders, enabling them to determine their own policies, salaries, and work-life balance without depending on others. They also have the opportunity to employ other women. The earliest trace of businesswomen dates back to 1870 BC in Northern Iraq in the trade sector (Hardach, 2021). Current business trends show that more women occupy CEO positions in large companies, such as IBM and General Motors, than was the case a few years ago (Larcker & Tayan, 2013), although the number is still less than 5 % (PwC, 2016). This is a positive signal for business because gender diversity promotes creativity and innovation through a diverse thinking perspective, which eventually drives growth (PwC, 2016). Besides, 43 % of entrepreneurs in Indonesia are women, and almost 50 % of small businesses are controlled by this gender (International Finance Corporation, 2016). These figures demonstrate the power of women in business. However, Indonesia is still ranked 85th in gender equality out of 135 countries which is assessed to be only better than Malaysia and Myanmar in Asia (The World Bank, 2016). Indonesian women entrepreneurs need greater financial inclusion, increased digital access, and access to expanded sources of suppliers. Digitization is an opportunity for women's businesses because it develops market access and facilitates the supply chain.

Furthermore, Indonesian women have high social capital skills that are used to communicate and build relationships to improve the family's economic wellbeing through various social gatherings. (Arifin et al., 2019; Puspitasari, 2012). However, women have limited resources in the family (Puspitasari, 2012), are not sufficiently able to respect themselves (Arifin et al., 2019), and the majority of Indonesian women are entrepreneurs due to needs and not opportunities (Puspitasari, 2012). Therefore, women need access to more business facilities.

The government has made efforts to facilitate women's empowerment through programs including PKH (family hope program), Revitalization of Posyandu through PKK and Health Cards, fuel compensation programs such as Raskin (rice for the poor), SLT (direct cash subsidies), BOS (school operational assistance), and PDM DKE (Community Empowerment in Overcoming Crisis Impacts). Nevertheless, these programs are not specific for empowering women while women have weak access to information, economic resources, and job opportunities (Zuhdi, 2018). Therefore, women entrepreneurs need access to adequate facilities and support so that they have optimal work opportunities in a circular economy to select jobs that are more suitable, safe, comfortable, and equal.

Women's Leadership as a Political Agenda

Women's participation in the political sphere is part of the 2030 agenda for Sustainable Development Goals because, despite their increase in political positions, women as heads of state and government agencies are still relatively small compared to men: Women are heads of state in only 26 of the total countries of the world. Additionally, the number of women in national governments only grows by 0.52 % per year, and this even occurs in a limited area where they take care of children, old age, social, and employment affairs, with a limited role in the field of natural resources. The number of women as members of parliament is not more than 25 % in various countries. Only a few countries in the world have a balance between men and women, such as the United Emirates, Cuba, and a number of countries in Europe, Latin America, and Africa due to gender quotas (UN Women, 2021). According to preliminary studies, women's leadership brings fresh air, diverse thinking, and new ideas to governance, as demonstrated by the implementation of new development programs following a significant number of women in governance, which have a variety of goals and orientations. These development programs include projects and programs for drinking water, childcare (Chattopadhyay & Duflo, 2004), violence elimination, gender equality, and electoral reform in women-led countries (Ballington, 2008).

In Indonesia, the political transition after the 1998 reform period promoted gender equality in which women were required to have greater roles in political representation. Gender mainstreaming is one of the mandatory agendas in development as stated in some long-term and short-term development plan documents, on a national and local scale, according to the Presidential Instruction No. 9 of 2000 (International Finance Corporation, 2016; Pengarusutamaan Gender Dalam Pembangunan Nasional, 2000). This instruction requires a gender perspective as an integral part of the functional activities of government institutions, from planning, drafting, implementing, monitoring to evaluating development program policies. Therefore, to complement this instruction, the government provides technical assistance that reports directly to the President through the Ministry of Women's Empowerment and Child Protection. Law no. 10 of 2008 confirms the 30 % quota for women's representation in party management which in 2017 only reached 19.8 % and was lower than the Philippines, Laos, Vietnam, Cambodia, and Singapore. Through the SDGs Goal, Indonesia aims to achieve the equal participation of women in politics by 2030 (50 %).

Women's participation in politics is essentially significant as it eases their pathway to playing a significant role in the circular economy. Quotas for women in politics need to be accompanied by policies in the economic and business sectors such as technology, telecommunications, infrastructure, oil, gas, and health and pharmaceutical services. Policies are needed to protect and accommodate the interests of women, such as the right to an equal standard of wages and the protection from violence. This is important considering that the current population structure of Indonesia is dominated by young voters from 17 to 34 years of age. This group has higher access to information than others, therefore, they have more developed socioeconomic and business concerns, including those related to climate change, education, and life in urban areas. Indonesia is the country projected to have the strongest economy in 2050, alongside China, India, and the US. There is an opportunity and a challenge for policymakers to empower more women to have equitable roles and opportunities, especially in the era of the circular economy.

II. Strategic Steps in Enhancing Women's Role in Circular Economy: A Perspective from Indonesia

Role of Women in the Indonesian Economy

Women are an important economic asset in Indonesia due to their ability to contribute significantly to the various investment plans. There is a progressive trend of women's involvement in Indonesian development. According to the Ministry for Human Development and Cultural Affairs (2019), Indonesia's Gender Development Index (GDI)[1] increased from 90.82 in 2016 to 90.99 in 2018. A GDI score close to 100 indicates that the development gap between men and women is getting smaller. Meanwhile, the country's Gender Empowerment Index (GEI)[2] shows that women's active role in economic and political life also rose from 71.39 in 2016 to 71.74 in 2017. Another underlying fact is that despite the low average length of schooling, which ranges from 6.96 to 9.31 in rural and urban areas, women's average level of education is higher than men's

---

[1] GDI is an indicator that describes the ratio of achievements between HDI women and HDI men.
[2] GEI focuses on participation, by measuring gender inequality in the economy, political participation, and decision-making.

(Dini et al., 2020). Contrarily, however, the female labor force participation rate of 55 % is lower than that of the male. According to the 2018 National Labor Force Survey (Indonesian Central Bureau of Statistics, 2019), the proportion of men in the formal work sector is almost double that of women. Unfortunately, in the last 10 years, the trend of this proportion has remained stagnate, and even working women are still very vulnerable to being exposed to economic shocks. In 2019, CBS reported that 26 % of women are domestic sector employees with medium to low skills whose proportion reaches 89 % or around 43.8 million people.

Generally, along with the ups and downs of the development practices, Indonesian women are more likely to participate in various economic activities. Although most of them are poorly educated and engaged mostly in informal sectors (MoWECP, 2020; ILO, 2018), their contribution needs to be considered a strategic asset in promoting circular economy-related initiatives. The following are some highlighted characteristics that might be important in proposing strategies capable of enhancing their role in the circular economy:

(a) Women have greater consumption impact over smaller consumer products such as food, clothing, and household utensils at the household level while men tend to have decision-making power over more expensive purchases.

(b) Women are more receptive to circular actions, such as the repair and reuse of certain home utensils and are capable of extending the use of an item.

Based on these two examples, Witek & Kuźniar (2021) and Foster (2016) demonstrate that women are likely to contribute positively and promote better consumption behavior in circularity culture. On the contrary, there are also some limitations:

(a) Women are likely to place greater weight on ecological products. Nevertheless, women tend to be more price-sensitive while ecological products are generally offered at higher prices compared to others.

(b) Digitization is also an obstacle as many forms of circularity require more digital technology. However, digital literacy is more prominent among men than women.

From a producer's perspective, women's roles are quite specific compared to men's:

(a) Women are often engaged with and dominantly represented in the production chains for cheap goods, such as the textile and cigarette industries.
(b) Women are more active in and in charge of waste management. The majority of these roles are in the form of informal work with higher risk.
(c) Women are also heavily employed by underpaid or unpaid jobs compared to men.

Concerning consumer and producer perspectives, various proposed strategies are needed in at least three areas of intervention, including education, supporting agencies, and innovative funding/initiatives.

Education and Campaign as a Strategic Pillar to Promote Circular Economy

Policies targeting education campaigns for women increase their power as conscious consumers that demand a circular economy shift for sustainability and health. There is also a need for innovative initiation and practical solutions related to the role of women in the circular economy, such as the triple solution concepts for gender, economic, and environmental problems in development activities. According to the Ministry of National Development Planning (2021), the implementation of a circular economy in five industrial sectors, namely food and beverage, textile, construction, retail, and electronics, can create 4.4 million green jobs in Indonesia between 2021 and 2030. There is increased potential for a circular economy as indicated by more than 200,000 new employees in waste processing, recycling, and cleaning services who have been employed between 2012 and 2018 (Ministry of National Planning and Development Indonesia, 2021). Women have the potential to play a role in the circular economy with green jobs in the food and beverage industry and in the waste processing sector. The Indonesian Ministry of Environment and Forestry (2021b) stated that Waste Banks in Indonesia are a manifestation of the Minister of Environment and Forestry Regulation number 14 of 2021, with 11,603 Waste Bank units spread across 365 districts and cities in Indonesia and a total of 382,778 customers comprising 40,841 women and 22,230 men.

One of the important steps that the government, practitioners, and environmentalists can take, alongside the community, is to promote the improvement of women's capabilities and capacities in the circular economy through education and campaign through formal and non-formal education strategies. Initially, education was a fundamental approach to raising awareness, knowledge development, and creating innovative thinking. Entrepreneurship is still an important subject in the circular economy. Current entrepreneurship learning materials or crafts have become mandatory or optional content in various schools at all levels of education, from elementary to university. In addition to fostering innovation, the importance of skills education and entrepreneurship for women is based on the issue of competition for job opportunities between genders. Current economic activities need to accommodate the massive changes in market innovation that are quick and dynamic with many types of work change in accordance with the state of the industry, one of which is the presence of many startup companies and MSMEs. Women are expected to gain additional motivation, skills, creativity, and willingness to become entrepreneurs through education, which helps them enter into and take on bigger roles and challenges in the circular economy. Non-formal education strategies can also be conducted for those unable to access formal education due to age restrictions. These stratgies include socialization, coaching, and seminars. Entrepreneurship training and facilitation can effectively improve women's skills in the field of business and foster an entrepreneurial spirit.

Role of Governance and Women's organizations

Strategies to increase the role of women in development can be carried out in both top-down and bottom-up approaches. The top-down strategies are derived by the government through its policies and community development plan, and the bottom-up initiatives stem from the community. As a policy, Indonesia has emphasized the role of women and the importance of empowering them. As a reference document for provinces to districts, the National Medium-Term Development Plan (RPJMN) IV 2020–2024 places women's and children's rights as the background for preparing future development plans. The protection of women is an important factor in ensuring their involvement in every development sector. RPJMN provides a clear direction of the development plans at the regional and local levels, which include and accommodate gender mainstreaming.

In addition to the national scope, the role of women in small spheres such as family and the environment continues to be improved through empowerment efforts. This is conducted in Family Welfare Empowerment programs through a women's organization known as 'PKK'. PKK promotes female empowerment by facilitating programs that encourage its members to create new products in the food and and beverage industry as well as handicraft sectors. According to the Ministry of Home Affairs, family empowerment in the form of women's organization at the neighborhood or community level is carried out using a learning model to foster, guide, and empower them as Agents of Knowledge. Institutionally, PKK is an organization with structured institutions consisting of Family Character Development, Family Economic Education and Improvement, Strengthening Food Security, Family and Environmental Health. The PKK has also indirectly promoted the role of women in circular economic activities. Currently, the functions and objectives of PKK have been expanded to adapt to the dynamic of social and economic changes, which led to the formation of women's organizations that are intended to explore and develop women's potential, foster entrepreneurship, and encourage women to contribute more to the environment. PKK organizations help their members to create new products in the food and beverage sector as well as handicrafts. Waste Banks, MSMEs, and home industries are outputs that provide direct benefits for PKK members, especially women, amid the pandemic (Geisendorf & Pietrulla, 2018; Hanis & Marzaman, 2020; Riana et al., 2014). The PKK also has a strategic role to become the government partners with the private sector and other community organizations to contribute to counseling, facilitators, planners, implementers, and movers.

Innovative Funding (Micro-Finance) and Creative Economy

Women are able to enhance their role through the microfinance support scheme for home, as well as small and medium industries related to circular economic activities (Grdic et al., 2020). According to Gazzola et al. (2020) and Lopes et al. (2021), the part of the circular economy most often entered into by women is in the food and beverage industry, and this is supported by the expansion of the use of applications and e-commerce that bring small industrial producers closer to consumers (Sánchez-Torres et al., 2021; Tolstoy et al., 2021). During the pandemic, the circular economy was one of the priorities for driving the National Economy initiated by the Ministry of Tourism and Creative Economy. It consists of 17 sub-sectors of the creative economy, such as game developers, architects, interior designers, musicians, and those in the industries of fine arts, product design, fashion, culinary arts, film and video animation, photography, visual communication design, television and radio, craft, advertising, performing arts, publishing, and applications. Based on current data, the majority of women in the creative economy are in the culinary field (Bonis-Profumo et al., 2021). Therefore, to increase income and competitiveness, it is necessary to raise the capability and capacity of women in utilizing technology to expand the reach of the product market (Symons & Hurley, 2018).

In Indonesia, MSMEs and women are interdependent. Based on the data of micro, small, medium, and large enterprises in Indonesia from 2014 to 2018, approximately 99.99 % of the 64 million business units in Indonesia are MSMEs, with 60 % managed by women (Kemenko PMK in 3 Broadcasts Press Number: B-024/SETMEN/HM.02.04/02/2021). The three sectors of the MSMEs that women dominate are fashion, culinary, and craft, and it is also expected that these MSMEs can further be developed through changing the value chain to be more circular.

The government needs to continuously strive to support and strengthen MSMEs by establishing a commitment to 'Strengthen Economic Resilience for Quality Growth' as one of the seven development agendas in the National Medium Term Development Plan (RPJMN) 2020–2024. In an effort to increase the role of women in microfinance, they need to be supported with easy access to entrepreneurship information, product branding, banking facilities, e-commerce, and other soft skills training. Additionally, successful and technologically literate women microfinance entrepreneurs are expected to strengthen networks as well as act as innovators in developing products and equitable access to business capital for small-scale businesses by joining cooperatives. The role of women in the Indonesian economy, specifically microfinance, needs to be increased through synergy and collaboration between ministries, women's observer institutions, the community, and other relevant stakeholders, to accelerate the post-Covid-19 Economic Recovery.

III. The Lessons Learned of Circular Economy Products/ Activities

Irrespective of the uncertain tendency of women to be advantaged or disadvantaged in the circular economy in Indonesia, there are numerous circular economy-based activities that will be most naturally initiated by women, particularly in informal sectors. According to a report published by the Ministry of National Planning and Development Indonesia in 2021, five sectors have large potential for circular economy implementation: food and beverage, textiles, construction, wholesale and retail trade, as well as electrical and electronic equipment (Indonesian Ministry of National Development Planning et al., 2021). From these sectors, numerous programs and activities have been initiated by women, which are ongoing within the community and have large potentials to be developed as part of the implementation of the circular economy.

Waste Bank for Promoting Sustainable Living

A Waste Bank system is one of the potential circular economy implementation methods applied for wholesale and retail traders. It is a solid waste management strategy that adopts the simple banking system of replacing money through solid waste recycling (Indonesian Ministry of Environment and Forestry, 2021a; Waste4Change, 2021). When customers deposit solid waste, they are rewarded with money, and this process is considered a means of saving money. Some of the Waste Banks also allow customers to borrow money from the bank and return it in the form of solid waste worth the borrowed amount (Waste4Change, 2021). The Indonesian Government Regulation established the concept of Waste Bank through the Ministerial Regulation of Environment no 13 2012 on The Implementation Guidelines for Reduce, Reuse and Recycle on Waste Banks, Acts no 18 2008 on waste management, and Regulation of Environment no 14 2021 on the Waste Management Bank (Indonesian Ministry of Environment and Forestry, 2021a). Initiation and awareness on municipal waste management, specifically in implementing recycling household waste in the form of Waste Bank, starts at the community level. Women, with their social capital in communities, have the power to carry out campaign activities. Many Waste Banks in Indonesia were started by women in their neighborhoods and evolved as economic instruments and income sources for users. Asteria and Herdiansyah (2020) examined women's vital roles in Waste Bank management and provided their significant impact in the community. According to the Indonesian Waste Bank Management Information System (2021a) managed by the Indonesian Ministry of Environment and Forestry, out of the total customer registered, 64.97 % are women. Furthermore, there are 11,612 units of Waste

Banks spread across the country that manages almost 1,500 tons of solid waste per month, which consist of paper, organic materials, plastics, plastic bottles, duplexes, boxes, and metals, with 2.1 billion IDR turnovers monthly.

A substantial example is the Gedawang Asri Waste Bank located in Semarang, Central Java, Indonesia (Gedawang Asri, 2020) with their activities as shown in Fig. 1. The Waste Bank activities were initiated by five women in the neighborhood in 2014. And two years after, after a continuous campaign, the numbers of users increased from 50 to 860 in 2021. This has become an alternative source of income for women (Budihardjo et al., 2022). Apart from economic support, Gedawang Asri Waste Bank also continuously conducts community development and engagement to increase awareness of waste management by reducing, reusing, and recycling for sustainable living (Sinaga et al., 2021).

Fig. 1: Activities of Gedawang Asri Waste Bank

Integrated Black Soldier Flies (BSF)-Compost for Sustainable Farming

Food waste in Indonesia has become another critical problem and potential sector for circular economy implementation at the same time. Based on the report by the Ministry of National Planning and Development (CNN Indonesia, 2021), the amount of food waste over a decade is 23–48 billion tons per year, which equals 1,702.9 Mt CO_2 equivalent of greenhouse gasses emissions. Several approaches have been used to manage the food waste generated, such as the integrated Black Soldier Flies (BSF)-Compost for sustainable farming. This is appealing because it is feasible from a technical and economic point of view. The women's organizations first initiated many established units of integrated BSF-Compost sustainable farming in Indonesia within their communities. As previously stated, the women's organization 'PKK' is an established organization that maintains its role at the community level, with a substancial involvement in the successful development of the circular economy program BSF-Compost sustainable farming. Integrated Black Soldier Flies (BSF)-Compost for sustainable farming are associated with the use of solid food waste, such as vegetable residues, catering waste, municipal organic by-products, and biomass as a substrate to produce insects (*Hermetia illucens L.*) that are used as a source of high protein content for animal feed, such as for fish and poultry (Shumo et al., 2019). The compost is also used as organic fertilizer for BSF production. One of the best practice examples of the integrated BSF-Compost sustainable farming is units run by Kelompok Wanita Tani (KWT) Mekar Ayu, located in Lubuk Terentang Village, Tanjung Jabung Barat, Jambi Province, as shown in Fig. 2. Company CSR, NGOs, and the local government partnered to introduce the sustainable farming concepts and build the facilities. Approximately 25 women were taught how to produce BSF or maggots from the food collected from the waste management facility nearby and use them as compost for their organic farms. From this integration, they are able to reduce the consumption of fertilizers up to 40 %, with a significant increase of income from vegetables and fruits such as kale, cucumber, long beans, chilies, peanuts, spinach, and papaya – up to 13 million IDR or 879 US Dollars in the range of 4 months and hundreds of kilograms of catfish. Therefore, by using this type of circular economy implementation, women in the Kelompok Wanita Tani (KWT) Mekar Ayu are able to improve the quality and quantity of their foods and to have additional income for the family.

Fig. 2: Activities of Kelompok Wanita Tani (KWT) Mekar Ayu

Plastic and Textile Waste Upcycling as small scale good practices

The objective of a circular economy is to create economic growth by sustaining the use of raw materials and products as long as possible, thereby minimizing social and environmental damage. The concept is known as the 5R, comprising of reduce, reuse, recycle, refurbish, and renew, which can be easily implemented in the informal sectors to create a major impact on the development of a potential unit of economy circular activity in Indonesia (Indonesian Ministry of National Development Planning et al., 2021). Textile and plastic waste upcycling are methods conducted by informal communities or organizations, which, in Indonesia, mostly comprise of women, as part of awareness-raising for environmental issues caused by the fast-growing food and beverage, textile, and wholesale and retail trade sectors of the economy, although it will not be sufficient to solve the waste management problem. Small scale good practices are observed in many regions in Indonesia, with some successfully transformed from informal activity to small and medium-sized enterprises (SMEs) with high-quality products. These include a brand of bags called Kreskros in collaboration with local women community in several regions, such as Ambarawa, Jakarta, and Bogor, used to recycle the plastic waste into high-value products (Kreskros, 2021). Others are in the form of community development for women's empowerment in many places in Indonesia (BBC Indonesia, 2021; Hakim, 2021). Another example is in Semarang, Central Java, where a group of women concerned about the amount of plastic in their neighborhood established Nayyara to help recycle plastic. Fig. 3 shows the various activities in these units, including sorting plastic according to type and then converting them into various handicrafts, such as bags, wallets, sandals, and other accessories.

Fig. 3: Activities in Nayyara Creation, Upcycling Plastic Waste

The role of women in creating, establishing, and maintaining circular economy-based activities is very important. Sectors predicted to have massive potentials for circular economy implementation are closely aligned with women" daily activities, as the majority of women are in charge of the management of the products used in their households. Additionally, the roles of women within the community and their representation and involvement in an organization such as PKK (Family Welfare Empowerment) drive their further involvement in creating successful circular economy-based activities. Therefore, applying this concept of a circular economy for household waste management triggers many activities and programs for the community's sustainable living in Indonesia, as shown in Fig. 4.

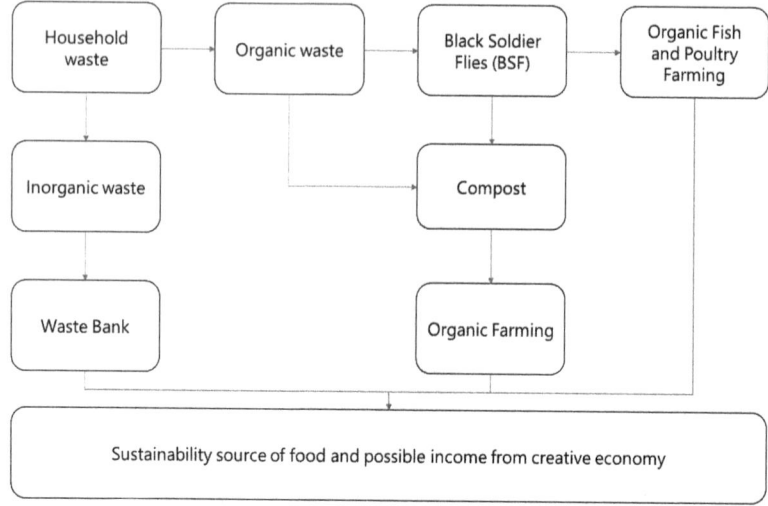

Fig. 4: Schematic Diagram of Waste Management System Based on a Circular Economy Concept

IV. The Importance of Envisioning for Women

Women's empowerment is a critical point in a circular economy because they are good at educating their families and communities. The practice of circular economy is not only about waste management, but also about resource efficiency, although the experience remains small scale. The experience can determine ways in which people can use products for prolonged periods. Women's empowerment can be a key agent of change to help promote economic growth without harming the environment to improve resource efficiency and promote sustainable lifestyles (OECD, 2021).

The role of women in Micro, Small, and Medium Enterprises (MSMEs) for the nation's progress is a necessity. For Indonesia, MSMEs are the main pillar of the economy, with more than 65 million units that contribute 61 % of the national economy. At the same time, 64 % of Indonesian MSME actors are women, empowering them with quite high resilience, especially amid the global pandemic. According to President Joko Widodo in a speech at the side event of the G20 Summit, Indonesia is poised to strengthen the role of MSMEs and women-owned businesses (Cabinet Secretariat of the Republic of Indonesia, 2021).

Increasing the Financial Inclusion of MSMEs and Women for Circular Economy

Financial inclusion is Indonesia's priority, and its index has reached 81 %, with a 90 % target by 2024. Therefore, to achieve this, friendly financing and access to finance for MSMEs in Indonesia need to be strengthened. Indonesia has allocated 17.8 billion US Dollars as loans for MSMEs (KUR), and more than 2.4 million women entrepreneurs have received this assistance. Additionally, the country has also launched US$1.1 billion for the Productive Micro Business Program, and 63.5 % of this was received by women entrepreneurs (Cabinet Secretariat of the Republic of Indonesia, 2021). A special capital scheme called the *Mekaar* (*Membina Ekonomi Keluarga Sejahtera* or 'Build a Prosperous Family Economy') program has also been developed for micro and ultra-micro women entrepreneurs. In English, 'Mekar' means to grow and blossom, therefore, it also symbolizes the spirit of growth and development for women's economic roles. There are currently more than 10.4 beneficiaries with a total financing of US $1.48 billion. Non-performing loans are very low: only 0.1 % of the total. This proves the ability of women entrepreneurs to manage funds and be entrepreneurially competent in business development (Permodalan Nasional Madani, 2019). With this program, women will have sustainable businesses with long-term thinking manifested in clear and structured visions and missions for growth.

Digitalization to Support the Economic Transformation of MSMEs and Women

Technological developments have changed the way people communicate and trade. For example, a number of e-commerce websites that are useful as places for buying and selling goods online by entrepreneurs have been developed. Additionally, this opportunity is not missed by MSMEs, where small and medium business owners utilize this marketplace. Generally, the obstacles faced by SMEs are marketing problems, specifically for MSMEs in rural areas. Therefore, e-commerce can reach the wider market efficiently, at a cheaper rate.

In order to improve the country's economy, the government has also prepared various programs that can be used for both old and new entrepreneurs. One of the government programs is aimed to provide strategic directions and guidelines in accelerating the implementation of e-commerce. Indonesian Central Bank (BI) noted that approximately 87.5 % of MSMEs were affected by the Covid-19 pandemic (Saputra, 2021). Those that survive tend to have switched to e-commerce, which was one of the drivers of the Indonesian economy during the pandemic, with a yearly GDP of US $24.8 billion. During the pandemic, 8.4 million Indonesian MSMEs entered the digital ecosystem, including 54 % of females (Cabinet Secretariat of the Republic of Indonesia, 2021). The advantage of selling in an online marketplace is that the market is wider and provides opportunity to attract more customers: Everyone connected to the internet can see promotions without being limited by location. Therefore, the products offered can be viewed not only by potential domestic consumers but also by international buyers. This is in contrast to opening an offline business, as in an online business the costs incurred are quite minimal because there are no expenses for renting a storefront amongst other expenses. This allows entrepreneurs to be more focused on developing their products. Promotions can be performed at home in online marketing while conducting household activities. Currently, there are many e-commerce services that entrepreneurs can select from, such as https://www.tokopedia.com/, and https://www.bukalapak.com/ , amongst others. However, most of these sites are used by large entrepreneurs and a few local MSMEs. Some of the marketplaces specifically for MSME entrepreneurs include https://ukmmarket.co.id/ and https://usahadesa.com/. In Indonesia, there are also companies that focus on developing these enterprises by collaborating with the government, such as GOUKM.

MSMEs in Indonesia face at least seven challenges in switching to digital services, namely lack of digital innovation, inadequate financial reports making it difficult to get loans from banks, low productivity, poor marketing strategies, licensing, product quality, and the manager's reluctant attitude. Therefore, the government needs to promote more MSMEs to switch to digital platforms by focusing on three steps: improving regulations, promoting the development of supporting infrastructure, and providing training. Presently, the Ministry of Communication and Information Technology (Kominfo) is building more Base Transceiver Station (BTS) towers in frontier, remote, and underdeveloped areas (3T). Furthermore, Kominfo is cooperating with some providers, such as Telkomsel and XL Axiata, to build BTS at 7,904 points in nine areas (Setyowati, 2021). Some training for MSMEs, specifically women entrepreneurs, have been carried out by the government, universities, and companies through their CSR.

As a good practice, Universitas Diponegoro has en Entrepreneurship Clinic and Business Incubator (KKIB) established in 2010 by the economics and business faculties, which has continued to grow directly under the coordination of the Vice Chancellor for Communication and Business. The institution has also improved the incubator capabilities by including mentoring services, human resources, infrastructure, improved startup quality, and capability. This is in addition to the provision of special assistants provided for MSMEs that are trying to advance and fix their business problems.

Conclusion

The role of women in social life and involvement with household tasks, such as selecting what to consume by the family and managing food and non-food waste, is likely to benefit the community from a circular economy and better waste management, although the question of waste will remain as the waste production is increasing quickly. These activities will reduce environmental damage, waste generation, and pollution and will also increase their income in an environmentally friendly environment. Therefore, a circular economy generates economic opportunities for women and increases employment and green entrepreneurship (OECD, 2021).

As stated in Part 3 of this chapter, some of the small-scale activities carried out by women in several locations in Indonesia, including Waste Banks, Integrated Black Soldier Flies (BSF)-Compost for promoting sustainable living, for farming, and the creative economy support the circular economy. These activities can be more thoroughly developed and scaled up in the coming years, assuming they are implemented to address environmental problems in their areas and benefit the wider community. Therefore, universities and corporate organizations need to provide and allow for increased social responsibility, financing, and market expansion. With IT support, the businesses can also be re-designed conveniently, while waste banks will be able provide the recycled amounts of materials.

References

Arifin, A. N., Marwanti, T. M., & Haryani, A. 2019, 'Keterampilan Sosial di kalangan Perempuan Rawan Sosial Ekonomi: Satu Kajian di Kota Bandung, Indonesia', *Malaysian Journal of Social Sciences and Humanities (MJSSH)*, Vol. 4, Issue 3, pp. 128–139. https://doi.org/10.47405/mjssh.v4i3.219

Asteria, D., & Herdiansyah, H. 2022, 'The role of women in managing waste banks and supporting waste management in local communities', *Community Development Journal*, Vol. 57, Issue 1, pp. 74–92. https://doi.org/10.1093/cdj/bsaa025

Ballington, J. 2008, Equality in Politics: A Survey of Women and Men in Parliaments, (No.54), Interparliamentary union.

BBC Indonesia, 2021, From Waste to Millions of Rupiah Products: How the Lombok Island Community Overcomes Plastic Waste, *BBC Indonesia*, https://www.bbc.com/indonesia/majalah-55946198

Bonis-Profumo, G., Stacey, N., & Brimblecombe, J. 2021, 'Measuring women's empowerment in agriculture, food production, and child and maternal dietary diversity in Timor-Leste', *Food Policy*, Vol. 102, pp. 1–13. https://doi.org/10.1016/j.foodpol.2021.102102

Budihardjo, M. A., Ardiansyah, S. Y., & Ramadan, B. S. 2022, 'Community-driven material recovery facility (CdMRF) for sustainable economic incentives of waste management: Evidence from Semarang City, Indonesia', *Habitat International*, Vol. 119, pp. 1–10. https://doi.org/10.1016/J.HABITATINT.2021.102488

Cabinet Secretariat of the Republic of Indonesia, 2021, President Jokowi: G20 must strengthen the Role of MSMEs and Women through real action.

Chattopadhyay, R., & Duflo, E. 2004, 'Women as policy makers: Evidence from a randomized policy experiment in India', *Econometrica*, Vol. 72, Issue 5, pp. 1409–1443. https://doi.org/10.1111/j.1468-0262.2004.00539.x

CNN Indonesia, 2021, Indonesian Ministry of National Planning and Development: RI Food Waste Reaches 48 Million Tons Per Year, *CNN Indonesia*, https://www.cnnindonesia.com/

Dini, I. M., Fajriyah, Mahdiah, Y., Fahmadia, E., & Lukitasari, I. 2020, *Gender-Based Human Development*.

Ekaningtyas, R. M. 2020, 'Persaingan Dan Diskriminasi Upah Gender Di Industri Manufaktur Indonesia', *Ilmiah Ekonomi Dan Bisnis*, Vol. 17, Issue 2, pp. 168–175. https://doi.org/10.31849/jieb.v17i2.4085

Fatimah, Y. A., Govindan, K., Murniningsih, R., & Setiawan, A. 2020, 'Industry 4.0 based sustainable circular economy approach for smart waste management system to achieve sustainable development goals: A case study of Indonesia', *Journal of Cleaner Production*, Vol. 269. https://doi.org/10.1016/j.jclepro.2020.122263

FP Analytics, 2020, Women as levers of change: Unleashing the power of women to transform male-dominated industries.

Future Learn, 2021, How women are shaping the future of business.

Gazzola, P., Pavione, E., Pezzetti, R., & Grechi, D. 2020, 'Trends in the fashion industry. The perception of sustainability and circular economy: A gender/generation quantitative approach', *Sustainability (Switzerland)*, Vol. 12, Issue 7, pp. 1–19. https://doi.org/10.3390/su12072809

Gedawang A., 2020, Gedawang Asri Garbage Bank, https://bsgedawangasri.blogspot.com/

Geisendorf, S., & Pietrulla, F. 2018, 'The circular economy and circular economic concepts—a literature analysis and redefinition', *Thunderbird International Business Review*, Vol. 60, Issue 5, pp. 771–782. https://doi.org/10.1002/tie.21924

Grdic, Z. S., Nizic, M. K., & Rudan, E. 2020, 'Circular economy concept in the context of economic development in EU countries', *Sustainability (Switzerland)*, Vol. 12, Issue 7, https://doi.org/10.3390/su12073060

Hakim, L. 2021, This Woman Recycles Textile Waste Into High Economic Value Products, *IDX Channel*, https://www.idxchannel.com/

Hanif, A., & Savitri, L. A. 2018, Di Balik Sepatu Berkelas: Cuti Haid dan Otoritas Tubuh Perempuan Buruh Pabrik Sepatu di Tangerang, Universitas Gadjah Mada.

Hanis, N. W., & Marzaman, A. 2020, 'Peran Pemberdayaan Kesejahteraan Keluarga dalam Pemberdayaan Perempuan di Kecamatan Telaga', *Publik (Jurnal Ilmu Administrasi)*, Vol. 8, Issue 2, pp. 123–135. https://doi.org/10.31314/pjia.8.2.123-135.2019

Hardach, S. 2021, The secret letters of history's first-known businesswomen, *BBC*, https://www.bbc.com/worklife/article/20210111-the-secret-letters-of-historys-first-businesswomen

Indonesian Central Bureau of Statistics, 2019, Indonesian Woman Profile.

Indonesian Ministry of Environment and Forestry, 2021a, Waste Bank Management Information System (Simba.id), https://simba.id/

Indonesian Ministry of Environment and Forestry, 2021b, Waste Management at the Waste Bank.

Indonesian Ministry of National Development Planning, Embassy of Denmark & UNDP, 2021, Summary for Policymakers – the Economic, Social and Environmental Benefits of a Circular Economy.

International Finance Corporation, 2016, UKM yang dimiliki Wanita di Indonesia: Kesempatan Emas untuk Institusi Keuangan Lokal.

Kemenpppa, 2019, Gender-based Human Development, In Pembangunan Manusia Berbasis Gender.

Korhonen, J., Honkasalo, A., & Seppälä, J. 2018, 'Circular Economy: The Concept and its Limitations', *Ecological Economics*, Vol. 143, pp. 37–46. https://doi.org/10.1016/j.ecolecon.2017.06.041

Kreskros, 2021, Kreskros, https://kreskros.com/

Krisnalita, L. Y. 2018, 'Perempuan, HAM dan Permasalahannya di Indonesia', *Binamulia Hukum*, Vol. 7, Issue 1, pp. 71–81. https://doi.org/10.37893/jbh.v7i1.15

Larcker, D. F., & Tayan, B. 2013, 'Pioneering Women on Boards: Pathways of the First Female Directors', *Stanford Closer Look Series, Sept*(Topics, Issues, and Controversies in Corporate Governance and Leadership), pp. 1–9. https://papers.ssrn.com/sol3/papers.cfm?abstract_id=2325026

Lopes, R., Santos, R., Videira, N., & Antunes, P. 2021, 'Co-creating a vision and roadmap for circular economy in the food and beverages packaging sector', *Circular Economy and Sustainability*, Vol. 1, Issue 2, pp. 873–893. https://doi.org/10.1007/s43615-021-00042-z

Ministry of National Planning and Development Indonesia, 2021, The Economic, Social and Environmental Benefits of a Circular Economy in Indonesia.

Mubarok, B., Hanita, M., & Rohman, S. 2019, 'Demographic and youth bonus; creative economic development based on local culture', *ICSGS*. https://doi.org/10.4108/eai.24-10-2018.2289656

Neves, A. et al, 2022, 'Drivers and barriers in the transition from linear economy to a circular economy', *Journal of Cleaner Production*, Vol. 341. https://doi.org/10.1016/j.jclepro.2022.130865

OECD, 2020, Mainstreaming Gender and Empowering Women for Environmental Sustainability. 2020 Global Forum on Environment, 5–6.

OECD, 2021, Gender and the Environment: Building Evidence and Policies to Achieve the SDGs, In *Gender and the Environment*, OECD Publishing. https://doi.org/10.1787/3d32ca39-en

Pengarusutamaan Gender dalam Pembangunan Nasional, 2004 1 (2000).

Permodalan Nasional Madani, 2019, *PNM Mekaar & PNM Mekaar Syariah*, https://www.pnm.co.id/

Pusparisa, Y. 2020, Women Dominate Service Business Workforce, Katadata. Co.Id, https://databoks.katadata.co.id/datapublish/2020/10/14/peremp uan-mendominasi-tenaga-kerja-usaha-jasa

Puspitasari, D. C. 2012, 'Modal Sosial Perempuan dalam Peran Penguatan Ekonomi Keluarga', *Jurnal Pemikiran Sosiologi*, Vol. 1, Issue 1, pp. 69–80. https://doi.org/10.22146/jps.v1i2.23445

PwC, 2016, The PwC Diversity Journey, *September* (Issue September).

Riana, N. R., Sjamsuddin, S., & Hayat, A. 2014, 'Studi tentang Program Pendidikan dan Keterampilan', *Jurnal Administrasi Publik (JAP)*, Vol. 2, Issue 5, pp. 851–856.

Salvador, R. et al, 2021, 'Circular economy strategies on business modeling: Identifying the greatest influences', *Journal of Cleaner Production*, Vol. 299. https://doi.org/10.1016/j.jclepro.2021.126918

Sánchez-Torres, J. A., Berrío, S. P. R., & Rendón, P. A. O. 2021, 'The adoption of E-commerce in SMEs: The colombian case', *Journal of Telecommunications and the Digital Economy*, Vol. 9, Issue 3, pp. 110–135. http://doi.org/10.18080/jtde.v9n3.403

Saputra, D. 2021, Survei BI : 87,5 Persen UMKM Indonesia Terdampak Pandemi Covid-19.

Sariatli, F. 2017, 'Linear economy versus circular economy: A comparative and analyzer study for optimization of economy for sustainability', *Visegrad Journal on Bioeconomy and Sustainable Development*, Vol. 6, Issue 1, pp. 31–34. https://doi.org/10.1515/vjbsd-2017-0005

Setyowati, D. 2021, Hampir 16 Juta UMKM Rambah E-Commerce, tapi Hadapi 7 Tantangan.

Shumo, M. et al, 2019, 'The nutritive value of black soldier fly larvae reared on common organic waste streams in Kenya', *Scientific Reports*, Vol. 9, Issue 1, pp. 1–13. https://doi.org/10.1038/s41598-019-46603-z

Sinaga, M. L. A. et al, 2021, 'Study on waste bank capacity building plan and development strategies in Semarang City', *IOP Conference Series: Earth and Environmental Science*, Vol. 896, Issue 1. https://doi.org/10.1088/1755-1315/896/1/012082

Symons, J., & Hurley, U. 2018, 'Strategies for connecting low income communities to the creative economy through play: Two case studies in Northern England', *Creative Industries Journal*, Vol. 11, Issue 2, pp. 121–136. https://doi.org/10.1080/17510694.2018.1453770

Tacoli, C. 2012, 'Urbanization, gender and urban poverty: paid work and unpaid care work in the city', *Urbanization and Emerging Population Issues* (Issue March).

The World Bank, 2016, *Women Entrepreneurs in Indonesia: A Pathway to Increasing Shared Prosperity.*

Tolstoy, D. et al, 2021, 'The development of international e-commerce in retail SMEs: An effectuation perspective', *Journal of World Business*, Vol. 56, Issue 3. https://doi.org/10.1016/j.jwb.2020.101165

UN Women, 2021, Facts and figures: Women's leadership and political participation, https://www.unwomen.org

Universitas Diponegoro Entrepreneurship Clinic and Business Incubator, 2021, Overview: The History of the Incubator, https://kkib.undip.ac.id/

Waste4Change, 2021, Waste Bank to Support Indonesia Clean-from-Waste 2025, https://waste4change.com/

World Bank, 2000, *Resume: Development of the Gender Perspective.*

Zuhdi, S. 2018, 'Membincang Peran Ganda Perempuan Dalam Masyarakat Industri', *Jurnal Jurisprudence*, Vol. 8, Issue 2, pp. 81–86. https://doi.org/10.23917/jurisprudence.v8i2.7327

Part IV

FUNDING CLIMATE ADAPTATION

Chapter 10

Sustainable Finance: The Hindered Potential of ESG Investing in Funding Climate Mitigation, Climate Adaptation and Resilience

ZOÉ THOUVENOT

Climate adaptation is defined by the United Nations Framework Convention on Climate Change (UNFCCC) as the entirety of the 'adjustments in ecological, social, or economic systems in response to actual or expected climatic stimuli and their effects or impacts. It refers to changes in processes, practices, and structures to moderate potential damages or to benefit from opportunities associated with climate change'[1]. In practice, these adaptations take place through various initiatives – the implementation of early warning systems such as technologies, nature-based solutions and management and organizational solutions, amongst others, the adjustment of planting patterns, the development of crops and species that are resistant to water stress, the adaptation of infrastructures and urban designs, or, again, the improvement of energy-related supply chains and infrastructures to make them climate-resilient. The IPCC defines resilience as 'the capacity of social, economic and environmental systems to cope with a hazardous event or trend or disturbance, responding or reorganizing in ways that maintain their essential function, identity and structure, while also maintaining the capacity for adaptation, learning and transformation'[2]. A critical need to fund climate

---

[1] https://unfccc.int/topics/adaptation-and-resilience/the-big-picture/what-do-adaptation-to-climate-change-and-climate-resilience-mean

[2] IPCC, 2018: Annex I: Glossary [Matthews, J.B.R. (Ed.)]. In: Global Warming of 1.5°C. An IPCC Special Report on the impacts of global warming of 1.5°C above pre-industrial levels and related global greenhouse gas emission pathways, in the context of strengthening the global response to the threat of climate change, sustainable development, and efforts to eradicate poverty [Masson-Delmotte, V., Zhai P., Pörtner H.-O., Roberts D., Skea J., Shukla P.R., Pirani A., Moufouma-Okia W., Péan C., Pidcock R., Connors S., Matthews J.B.R., Chen Y., Zhou X., Gomis M.I., Lonnoy E., Maycock T., Tignor M., and Waterfield T. (Eds.)].

adaptation and resilience is identified (IPCC, 2022, World Bank, 2021). Financial flows devoted to climate adaptation are increasing at a slower pace than adaptation costs in a context of increasing and accelerating climate change, which tends to widen the gap between costs and financial flows in climate adaptation (UNEP, 2020). Climate adaptation financial flows were equivalent to 30 billion USD in 2017/2018 and increased to 46 billion USD in 2019/2020, which is 'far below the scale necessary to respond to existing and future climate change'[3] (CPI, 2021).

> *"An increase of at least 590% in annual climate finance is required to meet internationally agreed climate objectives by 2030 and to avoid the most dangerous impacts of climate change"*[4]

If the financial flows dedicated to climate adaptation 'increased by 35 percent between 2015–2016 and 2017–2018 (…), they still fall short of what is needed to avoid severe economic and human impacts'. Indeed, climate related events tend to have strong economic impacts, not only for the public sector, but also for the private sector that might find their infrastructures or supply chains impacted for instance. Wellbeing and livelihoods can also be strongly affected due to population displacements, primary necessity products shortages, and human losses, amongst other issues. Moreover, the Climate Policy Initiative (CPI) (2021) states that 'flows have slowed down in the last few years. (…), [and that] the increase in annual climate finance flows between 2017/2018 and 2019/2020 was only 10 % compared to previous periods, when it grew more than 24 %'. Flows dedicated to climate adaptation are expected to decline because of the COVID-19 pandemic (UNEP, 2021, CPI, 2021). It is expected that adaptation costs in developing countries alone will be around 70 billion USD per year for 2020–2030, and up to 140–300 billion USD by 2030, if we consider the upper range of the calculation (UNEP, 2021, UNEP 2020). It is also estimated that these adaptation costs could reach from 280 to 500 billion USD in 2050 (UNEP, 2020).

The public sector provides almost all of the financial flows dedicated to climate adaptation (CPI, 2021). Multilateral development banks are currently the most prominent sources of public climate adaptation funding (World Bank, 2021). Nevertheless, climate adaptation represents only 14 % of total public finance (CPI, 2021). The private sector has gained

---

[3] Climate Policy Initiative (2021). Global Landscape of Climate Finance 2021, p. 3.
[4] Ibid, p. 2.

a lot of attention as a possibility to fulfill adaptation finance needs, but it has been very limited, especially in developing countries. Indeed, if the private sector financial flows for climate change have significantly grown since 2015, the proportion dedicated to adaptation remains small – in 2018, these private sector flows represented 0.05 % of total climate finance and 1 % of adaptation finance (IPCC, 2022). Private finance has the greatest potential for expansion (World Bank, 2021). To do so, the regulatory environment is evolving towards the voluntary or mandatory disclosure of climate-related risk – an initiative led by the European Union but spreading globally (World Bank, 2021). As examples, the UN Task Force on Climate-Related Financial Disclosures (TCFD) is developing voluntary climate-related financial risk disclosures for companies, banks, and investors. Additionally, the Sustainable Banking Network, facilitated by the International Finance Corporation (IFC), is gathering financial sector regulatory agencies and banking associations from emerging markets to improve their performance on environmental, social, and governance (ESG) criteria and their climate risk management, and to increase capital flows for climate action (World Bank, 2021). There is a growing interest in sustainable finance, including adaptation finance, but this needs to be scaled up and supported by funds in climate mitigation.

According to the common definition of UNEP, NASA, and the European Environment Agency, climate change mitigation refers to the efforts to reduce and/or prevent greenhouse gases (GHG) emissions. Climate mitigation finance refers to every financial engagement taken to reduce GHG emissions – it can relate to the use of more energy-efficient new technologies and/or equipment, the shift to renewable energies, changes in management practices (energy efficiency), non-energy GHG reductions, waste management, transport management or other cross-cutting issues like the support to local, national, regional and international policies. Climate mitigation finance is identified as to lack financing, including to support climate adaptation. Climate adaptation finance is targeted as the 'poorest' sector and is called in brief 'adaptation finance' (World Bank, 2021).

Sustainable finance has gained a growing attention from investors and policy makers as well as from civil society stakeholders as a way 'to deliver financial returns, align with societal values, and contribute to sustainability and climate-related objectives' and is increasingly associated with environmental, social and governance (ESG) investing. Environmental, Social and Governance (ESG) criteria are an important tool of sustainable finance. They are non-financial factors integrated in the financial decision-making process (Ziolo et al, 2019). At a practical level, they are scores attributed to companies to assess how they impact these criteria. They are not precisely fixed and listed, but they relate to the 17 United Nations Sustainable Development Goals (UN SDGs) launched in 2015 (Ziolo, 2021) and they are managed by the UN Principles for Responsible Investment (UN PRI) signed by 1750 professionals from over 50 countries (Friede, Busch, Bassen, 2015, Rusu, 2020). ESG scores progressively emerged at a global level (Kuzmina, 2017) in a society that 'has reached a point where caring for the environment, having concerns for labour, human rights, and corruption is not just good to have, but takes the form of necessity'.

This chapter is a review of literature seeking to understand the rise and function of ESG investing, its links to climate adaptation needs, its use in the context of climate mitigation, climate adaptation, and resilience and under what conditions, if any, it can be beneficial in tackling climate change and participating in the emergence of resilient societies.

The plan is detailed as follows: historical background (I), the question of profitability (II), the issues of regulation and monitoring (III), and climate adaptation investments (IV). We argue in this chapter that ESG criteria would need a clear blueprint to better contribute to answering the needs of climate adaptation.

I. Historical Background

In terms of terminology, it is worth understanding Kuzmina and Lindemane's (2017) main distinction between:

- Corporate Social Investment (CSI)/Responsible Investment (RI): referring to investment funds that combine financial objectives and commitment to social and environmental issues.
- Socially Responsible Investing (SRI)/Ethical Investing (EI): a very popular term describing investment funds relying on social criteria, avoiding 'scandalous industries such as tobacco or alcohol.'[5]
- ESG Investing/Sustainable Investing (SI): very popular as of 2022, describing investments that match the three criteria quoted above.

---

[5] KUZMINA Jekaterina, LINDEMANE Marija (2017), ESG investing: new challenges and new opportunities, Journal of Business Management, No. 14, p. 87.

The modern form of what is termed Socially Responsible Investing (SRI) appeared during the post-World War II area with the empowerment of civil society and the rise of the civil rights movement, the anti-Vietnam War movement and increased consumers rights activism, and environmentalism (Puaschunder, 2019). Already in the 1960s and 1970s corporate social and environmental performances were rated by the Council on Economic Priorities, an international non-profit and non-partisan economic think tank, and governments started enacting 'shareholder rights to address corporate activities that caused social injury.' Environmental disasters like Chernobyl or oil spills enhanced the movement towards socially and environmentally responsible finance, leading to the idea to 'set[ting] standards for corporate social engagement and environmentally conscientious conduct', which started in the 1990s (Puaschunder, 2019). These standards also consider failures 'to take into consideration the social and the environmental costs of a project along with its economic benefits', like in the famous case of the huge damages caused by hurricane Sandy in 2011 in the United States while the American Society of Engineers had warned since 2009 of the necessity to install surge barriers, especially in New York City (Fatemi, Fooladi, 2013). These pressures were coupled with globalization, the 2008 recession, and, more broadly, the societal trend towards a greener, more human, socially equal and fair world (Rusu, 2020), which created the opportunity for SRI and ESG to emerge. This seems to have paved the way for the emergence of a new paradigm, sustainable finance (Fatemi, Fooladi, 2013). The term sustainable finance was first proposed by Howard Bowen (1953) that introduces the idea of a moral obligation in the business and financial world to give back to society (Rusu, 2020) and has culminated in the evolution described earlier.

ESG criteria are part of a new paradigm of sustainable finance (Le Monde, 2022), what Janicke and Jacob (2009) call the Third Industrial Revolution: a long-term path, new policies, especially at the economic level, and a new productivity model that enhances innovation, technologies, labor conditions, and human capital while decreasing the total consumption of resources. Creating Shared Value (CSV) by Porter and Kramer (2011) is related to ESG through encouraging policies and practices that allow competitiveness but that also take into consideration the global environment in which economic growth happens: 'shared value creation focuses on identifying and expanding the connections between societal and economic progress.' Therefore, societal issues can be approached from a value perspective, meaning that 'it is not philanthropy but self-interested behaviour to create economic value by creating societal and environmental value.'

The leading idea of ESG criteria is that of integrating non-financial factors into financial processes. Finance could then become a driver of societal change and sustainability (Scholten, 2006, Weber, 2014).

II. Profitability/Return on Investment (ROI)?

Building on Porter and Kramer's idea of creating shared value, a first large body of literature on ESG criteria focuses on profitability and return on investment. Professionals and academics realize the impacts of climate change (Financial Times, 2021) and ESG criteria are the result of a demand for more sustainable habits (Rusu, 2020, Ziolo et al, 2021, Financial Times, 2021). This appears even truer for the youth, who tend to be educated and informed about the environment and care about their consumption patterns, which has been amplified by the Covid-19 pandemic (Financial Times, 2021). Some observers explain how strong political and social movements push companies to label their products as sustainable and/or produce ESG report (Guien, 2021). These labels and reports appear as powerful marketing tools, or 'immaterial capital [and] intangible assets'[6]. Societal demand seems to pave the way for changes in the financial sector. Indeed, demand is the second most important criteria when asset managers decide whether to consider ESG criteria in their decision making (Amel-Zadeh, 2018).

But the question remains: How are investment performances – in terms of profit/returns – affected by these criteria? Overall, in the scientific literature, there is no statistically significant difference in returns on investment between ethical funds and conventional ones (Bauer, 2005), or socially responsible funds and conventional ones (Bello, 2005, Revelli, Viviani, 2015). The issue is hotly debated: It is not clear in the literature if sustainable investments represent a benefit to financial performance or not, and there are still questions of how costs and reputational benefits work together (Revelli, Viviani, 2015).

---

[6] 'Capital immatériel, actifs incorporels' translated from GUIEN Jeanne (2021), Le consumérisme à travers ses objects, Editions Divergences, Paris, p. 213.

Specifically regarding ESG, several methods have been employed to assess the returns on investment of ESG investments and the results are debated, even if the majority of the scientific literature on the topic recognizes no significant difference between ESG investing and conventional investing, if ESG investing does not entail better returns. Surveys on 419 asset managers seem to reveal that the primary reason why they choose to follow ESG criteria is performance: 63.1 % of the 82.1 % that consider ESG information when making investment decisions do so for the performance, while only 4 % of the 17.9 % of investors that do not consider ESG information when making investment decisions do not consider it because of the performance (Amel-Zadeh, 2018). The aggregated study based on around 2200 quantitative empirical studies from the 1970s to 2015 by Friede, Busch and Bassen (2015) reveals that there is mostly no difference between ESG investments and classic ones (Corporate Financial Performance, CFP). An experiment led by Rusu (2020) on how investors incorporate financial and sustainability information in their decision-making has shown that 'being a sustainable organization is not significantly influencing investors' decision'[7], which admittedly supports 'Friedman's theory of maximizing wealth for the company's owners.'[8] A study by the French EDHEC Business School quoted by the journalist Aurélie Fardeau in *Le Monde* estimates that climatic data represent only 12 % of the determinants of portfolios weightings. In other words, 88 % of the construction of these indexes rely on classical financial determinants. It appears even truer in ESG investing where climatic criteria represent only 7 % of the determinants of portfolios weightings (Fardeau, *Le Monde*, 2021[9]). These results make sense given that only 10 % of professionals received a training about ESG and how to integrate them in investment processes (CFA Institute, 2019).

One of the major challenges in the private sector of climate-related finance is to demonstrate a profitability, financial returns as avoided damages (or costs) of climate change, increasing productivity and innovation, and the social and environmental benefits (IPCC, 2022, World Bank,

---

[7] RUSU Dana Ioanna (2020), The impact of environmental, social and governance factors on investors behaviour – an experimental study in the realm of sustainable investment, Journal of Public Administration, Finance and Law, Issue 17, p. 317.
[8] Ibid.
[9] FARDEAU Aurélie, 18.10.2021, Placements en faveur du climat: gare au « greenwashing », Le Monde.

2021). This is even more challenging in developing countries where both perceived and real risks appear higher to investors (IPCC, 2022). The Global Commission on Adaptation (2019) quoted by the UNEP (2020) estimates that a 1.8 trillion USD investment in adaptation measures – such as early warning systems, climate resilient infrastructure, mangroves protection, resilient water resources and improved dryland agriculture – would bring a return of 7.1 trillion USD of avoided costs as well as non-monetary social and environmental benefits (non-monetary) (UNEP, 2020). The World Bank (2021) estimates that 'every USD 1 invested in resilient infrastructure in low- and middle-income countries yields 4 USD in net benefits.'[10]

However, beyond the avoided costs relating to climate change, the World Bank (2021) states that climate adaptation finance also brings opportunities for the private sector. Notably 'the development of new products and services that fill market gaps; new expanded markets for existing products and services; cost savings across the whole value chain; collaborations through supply chains; and reputation and brand value, among others.'

Nevertheless, the context of the pandemic appears to have slightly changed the orientation of considerations for ESG in investing. A study by Global Data has shown that the Covid-19 pandemic is at the center of financial decision-making – 69 % against 57 % for ESG criteria. It is a debated and controversial issue taking place in a changing financial world. The question of profitability in the study of ESG criteria is important in the academic field. In order to introduce changes in financial behaviors, the first rational step is to assess how safe or even beneficial new types of investment could be. The financial viability of new investment strategies is especially important as the primary motives of investors appear to be financial rather than ethical (Amel-Zadeh, 2018). Therefore, in the same vein in this literature, some models appeared to allow investors to calculate the potential performances, costs, and benefits stemming from ESG investments (Pederson, 2020). This leads to a sub-category of this body of literature regarding the issues of tools, measurements and standardization that limit the possibilities of assessment of ESG investing performances.

---

[10] World Bank (2021), Enabling private investment in climate adaptation and resilience, p. 7.

III. Regulation and Monitoring Hindering ESG Investing's Potential in Climate Mitigation

Companies report ESG-related data following national or international regulations while they report their environmental, social, and governance-related objectives to provide a long-term view of their actions. Using this data, rating agencies provide an assessment of ESG scores for each company submitting an application. But ESG criteria, as well as reporting and assessment methods, are heterogenous which means that their credibility can sometimes be questioned. A recent study by Gyönyörová, Stachoň and Stašek (2021) assesses the validity of ESG scores provided by international rating agencies and points out that this validity depends greatly on the industry type and the country of domicile. The analysis emphasizes the heterogeneity of local and national specificities and legal systems which define various frameworks relating to sustainability data disclosure and ESG scores. Also, there are important differences of regulations depending on the companies' sector of activity (Fardeau, *Le Monde*, 2021[11]). This 'lack of cross-company comparability'[12] is identified as a missing tool in ESG analytics (Amel-Zadeh, 2018, Ziolo et al, 2019, OECD, 2021). Moreover, rating agencies appear to lack transparency: they do not publicly deliver information about the process leading to ESG scores and corporate sustainability performance. There are also issues of commensurability since ESG rating agencies could measure the same concept using different methods. Another issue is the 'trade-offs among criteria'[13]: some methodologies can compensate for low scores in some categories by having higher scores in others. Finally, these rating agencies do not consider investors and stakeholders preferences, which means that the scores are lacking in influence and usefulness (Escrig-Olmedo et al., 2019).

---

[11] FARDEAU Aurélie, 05.08.2020, 'Que se cache derrière l'étiquette « investissement socialement responsable', Le Monde.
[12] AMEL-ZADEH Amir, SERAFEIM George (2018), Why and how investors use ESG information: evidence from a global survey, Harvard Business School Accounting and Management Unit, Working Paper Financial Analysts Journal, 74:3, p. 92.
[13] ESCRIG-OLMEDO Elena, FERNANDEZ-IZQUIERDO Maria Angeles, FERRERO Idoya, RIVERA-LIRIO Juana Maria, MUNOZ-TORRES Maria Jesus (2019), Rating the Raters: evaluating how ESG Rating Agencies integrate sustainability principles, Sustainability, 11, 915, p. 7.

Investments are different between regions, notably between Europe and the USA (Friede, Busch, Bassen, 2015). For example, in the oil and gas industries, an article from the French newspaper *Le Monde* that could be translated as 'Energy: Exxon does not want to change profession' explains that the sustainability path chosen by ExxonMobil (USA) is quite different from the one chosen by European oil and gas companies. If ExxonMobil is going to invest 20 billion USD in low carbon production, these investments will be more oriented towards the development of technologies to capture CO2, synthetic fuel or hydrogen while European companies like Total Energies, Shell, and BP tend to invest in offshore wind farm or solar energy[14]. That being written, these same European companies are not exempt from increasing problematic fossil fuel projects (such as the EA COP East African Crude Oil Pipeline project in Uganda in which Total Energies is involved[15], project questioned for its impacts on climate and on local communities, or their previous disruptive investment in Cabo Delgado Mozambique). Indeed if Total Energies is having a presence in Mozambique[16] their project in Cabo Delgado has been taken as a target for terrorists groups in the region, questioning the social sustainability of the project. Harmonization of the reporting and the rating methods appears to be greatly needed to have a better approach of real performances.

If there is no way to compare ESG criteria and ESG scores, confident and easy investment in climate resilience appears even more difficult. Also, this lack of comparability on methods and tools makes any type of monitoring and/or evaluation complicated, thus greenwashing is enhanced. No matter the financial flows directed to climate resilience, if there is no way to systematically monitor their real impact through clear and standardized procedures, they have less, if any, real substance.

Moreover, Eric Albert, a French journalist for *Le Monde*, explains that polluting companies are the ones that use ESG criteria the most. In the European Union, polluting companies (mostly from the oil and gas and mining industries) display 43 % of all green bonds, against 7 % for companies considered 'clean'. It is not fundamentally negative: it means that the market can push companies to reduce the environmental and social

---

[14] ESCANDE Phillippe, Energie: 'Exxon ne veut pas changer de métier', Le Monde, 2nd of February 2022, https://www.lemonde.fr/economie/article/2022/02/02/energie-exxon-ne-veut-pas-changer-de-metier_6111979_3234.html

[15] StopEACOP | East African Crude Oil Pipeline https://www.stopeacop.net/

[16] Welcome in TotalEnergies Mozambique |

issues they have created. But it also means that investors choosing a 'green investment' have high chances of investing in Total, Shell or BP – the money will go to a green project (a wind farm, for instance) but financed by a major fossil fuel actor. In *'L'illusion de la finance verte'* (2021), Julien Lefournier and Alain Grandjean explain the logic of Sustainability Linked Bonds (SLB) – an investor does not invest for a specific green project, he relies on the reporting of the company regarding its overall environmental and social progresses. However, the objectives are freely defined, and the sanctions in cases of non-compliance are minimal. For example, in February 2021 TotalEnergies committed to display only SLB and declared that these supposedly green products will finance all their activities … including activities of fossil fuel exploitations. Desiree Fixler, former head of sustainable development at DWS (Deutsche Bank), explains that her former employer presented itself as much 'greener' than its investment policies entailed, overestimating ESG investments (Albert, Chocron, *Le Monde*, 2021). Greenwashing is a real risk as sustainable finance is increasingly popular. Ignazio Visco, Governor of the Central Bank of Italy, says that 'sustainable finance has grown so much that one can wonder if it is really sustainable'[17] – there is a need for a harmonization between regional actors and strong established norms and regulations (Albert, Chocron, *Le Monde*, 2021).

"The climate transition presents a historic investment opportunity"
(Larry Fink, BlackRock's[18] CEO, "Larry Fink's 2021 letter to CEOs"[19])

If Larry Fink sees climate transition as an opportunity, Tariq Fancy – former BlackRock ESG manager for 2 years – explains that ESG criteria are applied to investments as an administrative formality. If investors were sensitive to the cause, their role was to focus on profits and risks, 'not to mention that their remuneration was indexed to their return on investment, not to their ecological virtue…'[20]. Tariq Fancy explains that

---

[17] Literally translated from French, 'La finance durable a tellement grossi qu'il faut se demander si c'est vraiment durable', ALBERT Eric, CHOCRON Véronique, 21.10.2021, Le mirage de la 'finance verte', *Le Monde*.
[18] BlackRock is the biggest assets management agency in the World (USD 10 000 billion).
[19] https://www.blackrock.com/corporate/investor-relations/2021-larry-fink-ceo-letter
[20] Literally translated from French, 'sans compter que leur rémunération était indexée sur leur retour sur investissement, pas sur leur vertu écologique…', ALBERT Eric, CHOCRON Véronique, 21.10.2021, Le mirage de la 'finance verte', Le Monde.

in the 2010s ESG investing increased alongside GHG emissions. He raises the issue of the real effect of these financial tools. Indeed, according to a study led by Boffo (2020) quoted in the 2021 OECD report 'ESG investing and Climate Transition', sometimes, high E pillar scores are actually correlated with high CO2 emissions which 'suggests that firms' plans to reduce emissions play a significant (and positive) role in determining their E pillar scores, rather than their current level of emissions.'[21]

Recent evolutions have shown progress towards a harmonization of the methodologies of reporting and assessment of ESG investing. There is a trend toward more precision and a multiplication of criteria regarding the environment. In their study, Escrig-Olmedo et al. (2019) show that in 2008, ESG criteria focused on emissions and climate change while, 10 years after, in 2018, the focus was also on water use and management, protection of biodiversity, and waste management/reduction. However, they still do not fully integrate the sustainability principles that were established by Escrig-Olmedo et al. (2019) and rely on existing literature on sustainability. These principles are listed as follows:

1. A normative principle, entailing that strong norms and values lead the way to sustainable development,
2. The equity principle between generations and species, whereas current and future generations of species can enjoy an equitable quality of the planet,
3. An integration principle that is opposed to the idea of 'balancing' or 'trading off',
4. The idea that sustainability is a changing process that will have to adapt.

Rating agencies should rely more on these principles to enhance sustainable development (Escrig-Olmedo et al., 2019). Moreover, the new European taxonomy is a great innovation in sustainable finance in the harmonization and regulations of ESG investing. This taxonomy has nonetheless raised debates because the Commission accepted the inclusion of nuclear and gas as 'sustainable' solutions. Gyönyörová, Stachoň

---

[21] OECD (2021), ESG Investing and Climate Transition: Market Practices, Issues and Policy, Considerations, Paris, p. 12.

and Stašek (2021) tend to conclude that a 'naïve use'[22] of ESG scores provides a 'misleading clue'[23].

More than harmonization, strict monitoring and strong guidelines are needed. Tariq Fancy and Julien Lefournier argue that the private sector has a major role in the energy transition, but states need to create the right normative conditions and '(…) then investments will follow the profitability within the new rules of the game.'[24] In France, the *Inspection générale des finances* (IGF) admits the confusing nature of the Socially Responsible Investment label that relies on requirements built on companies' reporting and unequal rating agencies which 'cannot guarantee an effective earmarking of funding towards activities falling within a sustainable economic model'[25]. This is leading to a loss of credibility.

The financial world is changing – Stéphane Lapiquonne, BlackRock Europe director on sustainability announces a radical transformation of the economy (Albert, *Le Monde*, 2022). But he insists: 'there is money available, but it has to be directed. Government decisions are essential for this'[26].

IV. Adaptation Finance: The Lack of Blueprint

According to the Adaptation Gap Report (UNEP, 2021), there is a growing involvement of the private sector in developing and delivering climate adaptation through new tools and incentives, but these efforts are not enough to face the expected climate crisis. A first issue relates to public incentives to encourage private investments in adaptation (UNEP, 2021, IPCC, 2022). In the context of the pandemic, the stimulus

---

[22] GYONYOROVA Lucie, STACHON Martin, STASEK Daniel (2021), ESG ratings: relevant information or misleading clue? Evidence from the S&P Global 1200, Journal of Sustainable Finance & Investment, p. 1.
[23] Ibid.
[24] Literally translated from French, '(…) ensuite, les investissements iront là où se trouvera la rentabilité à l'intérieur des nouvelles règles du jeu.', ALBERT Eric, CHOCRON Véronique, 21.10.2021, Le mirage de la 'finance verte', *Le Monde*.
[25] Literally translated from French, 'ne sauraient garantir un fléchage effectif des financements vers des activités relevant d'un modèle économique durable', ALBERT Eric, CHOCRON Véronique, 21.10.2021, Le mirage de la 'finance verte', *Le Monde*.
[26] Literally translated from French, 'l'argent est là, disponible, mais il faut le flécher. Les décisions des pouvoirs publics sont pour cela essentielles', ALBERT Eric, 02.02.2022, Le changement climatique va secouer les marchés financiers, *Le Monde*.

packages provided by public institutions in the aftermath of the COVID-19 pandemic appeared as a potential 'window of opportunity to invest in a green, resilient and inclusive economic recovery'. But of the 66 countries studied by the UNEP (2021), only 17 flagged physical climate risks, adaptation, and/or resilience in their investments' priorities.

Moreover, some countries even invested more in activities that will increase climate change vulnerability. The Vivid Economics Greenness of Stimulus Index, quoted by the UNEP (2021) shows that, 'as of the 1st of February 2021, only 141 billion of 667 billion USD of tracked green stimulus had been directed towards 'nature and biodiversity' compared to 262 billion USD of stimulus directly associated with pollution or activities expected to negatively impact biodiversity'[27]. They add that 'the packages announced by 15 of the G20 nations will have a net negative environmental impact and even in the National Resilience and Recovery Plans in Europe, there is more spending that will damage nature than enhance it'[28]. This context can appear as an indirect incentive not to act, or at least as a general indifference to the environmental emergency. Public and private investments orientations need to be complementary. Direct public incentives are essential to enable private investment in climate adaptation and resilience. Indeed, public institutions are well placed to provide incentives notably by modifying regulatory frameworks and/or leveraging support from multilateral development banks and donors, amongst others (World Bank, 2021).

CASE 1 The Climate Resilient and Inclusive Cities (CRIC) project

The Climate Resilient and Inclusive Cities (CRIC) project is funded by the European Union for 5 years in Indonesia. It is led by UCLG ASPAC Asia in a partnership with ACR+ the cities association for the use of sustainable resources, ECOLISE the association of grassroot movements for the environment, the University of Gustave Eiffel (UGE), and Pilot4DEV. The overall objective of the project is to propose a long-lasting cooperation between cities and research centers in Europe and in Indonesia (and beyond), and to contribute substantially to sustainable integrated urban development, good governance, and climate adaptation and mitigation

---

[27] Ibid, p. 57.
[28] Ibid.

> through cooperation and tools such as sustainable local action plans, early warning tools and experts' panels.
>
> As the project started in January 2020, the first phase of the project dedicated to the study of 10 Indonesian pilot cities has led the partners to the conclusion that there is a stake of private funding in climate mitigation and adaptation in these cities and that such investments would need to be supported by strong public regulations and incentives.
>
> The full reports on these pilot cities are available on www.resilient-cities.com/knowledge

Public institutions are also in a good position to provide information on adaptation finance gaps and climate risks. The UNEP (2021) and the World Bank (2021) identify a major issue of data and information availability. Indeed, country-level climate risk and vulnerability data are lacking, which is stamping out investment decision-making (World Bank, 2021). Moreover, there is a lack of clarity on where gaps need to be filled. Governments' investments gaps are unclear, if public institutions could provide clear information on where investments are needed, adaptation goals could be more easily reached (World Bank, 2021). This is coupled with a lack of data on private flows (IPCC, 2022) and uncertainties remaining in both public and private decision-making regarding climate adaptation (UNEP, 2021). Finally, tools and methods are missing, 'investors currently have limited analytical capacity to price climate risks and to integrate the 'value' of adaptation outcomes and averted climate impacts into project assessments or return calculations'[29]. As stated before, the issue of profitability – in terms of profit/returns as well as social and environmental benefits – remains a large issue for these types of investments. Even if 'an IMF working paper estimated that for every dollar spent on ecosystem conservation (a nature-based solution), almost seven more were generated in the economy over five years'[30], there is an issue of 'perceived or actual returns on investment'[31] as well as 'perceived or

---

[29] World Bank (2021), Enabling private investment in climate adaptation and resilience, p. 30.
[30] United Nations Environment Programme (2021). Adaptation Gap Report 2021: The gathering storm – Adapting to climate change in a post-pandemic world. Nairobi, p. 56.
[31] World Bank (2021), Enabling private investment in climate adaptation and resilience, p. 27.

actual environmental and social benefits.'[32] Private investments tend to gravitate toward opportunities where profit is higher and risks are lower as investors also appear to struggle to capture the profitability and environmental/social benefits generated by climate adaptation investing. To leverage private investment, the World Bank (2021) and UNEP (2021) tend to encourage additional work on showing evidence of the benefits of climate adaptation and its effectiveness. Indeed, an additional concern and barrier to private sector climate adaptation investing is the perception of real outcomes of these investments. First, reporting on adaptation implementation is only at its nascent stages – data is too scarce to efficiently evaluate the state of implementation of adaptation actions worldwide (UNEP, 2021). Tools are needed to give a sense of the reality of the implementation – many countries, including (and probably mainly) developing countries, have few, if any, regulations and metrics to ensure that investments are meeting adaptation needs (World Bank, 2021). Additionally, it is unclear if current adaptation approaches have long-term beneficial effects. Adaptation implementation needs to be monitored and evaluated to ensure its efficiency and to give a fuller perception of its real impact. If not, there is a risk of 'maladaptation' (UNEP, 2021) – meaning the implementation of actions that, in the long run, may increase risks relating to climate change (for instance, actions that increase GHG emissions).

CASE 2 (World Bank, 2021) The Energy Development Corporation (EDC), the Philippines

In its 2021 report 'Enabling private investments in climate adaptation and resilience', the World Bank displays an interesting case study on the Energy Development Corporation (EDC) – the biggest energy company in the Philippines. Indeed, the Philippines are vulnerable to climate change, especially cyclones, floods, droughts and landslides. The EDC has been affected by the Urduja tropical storm in 2017 that brought one meter of rain in three days in the region and reduced the capacity of one the company's power plants by 50 %. This event made the EDC realizes that its infrastructures were not resilient enough and started integrating climate risk into its decision making by changing its modeling and risk analysis to include climate-related natural events (which confirmed that

---

[32] Ibid.

Sustainable Finance 313

infrastructures needed to be more resilient). Therefore, EDC invested USD 6.2 million in 2018 in climate adaptation measures 'to improve the company's resilience, minimize risk exposure and ensure a continuous energy supply to consumers and the local community.' The 2017 event made the ECD realize that resilience would allow for the preservation of the company's activities, the protection of future revenues, and the minimization of costs.

What is interesting is that the 'EDC Philippines had a supportive enabling environment' (World Bank, 2021). Indeed, the company worked with municipal agencies to better understand climate risk factors and best practices – which also led to the creation of the Disaster Prevention and Recovery Unit providing training around EDC's host communities to boost community resiliency. Moreover, they had the financial and political support of the IFC (multilateral development institution).

In this case, the cooperation between public agencies and the private sector tackled the issues related to climate risk data availability, climate risk integration in decision making and profitability.

CASE 3 The Indonesia Impact Fund (IIF) in Indonesia

The IIF is a private-led investment fund investing in early-stage start-ups participating to the achievement of SDGs (poverty, healthcare, education, inclusivity, gender-equality, sustainable cities, affordable housing, climate-smart technologies).

This private fund is benefiting from a partnership with the United Nations Development Programme (UNDP) that is easing the exchanges of data and information usually lacking. This public-private partnership also provides an Impact Measurement and Management (IMM) ensuring an efficient contribution to SDGs.

The IIF appears to provide an example of the type or partnerships that can be thought to respond to the current identified issues in adaptation finance.

The UNEP (2021) concludes that 'more ambitious adaptation will be critical going forward'[33] and indeed, this type of initiatives need to

---

[33] United Nations Environment Programme (2021). Adaptation Gap Report 2021: The gathering storm – Adapting to climate change in a post-pandemic world. Nairobi, p. 69.

be scaled up. If the climate adaptation funds in cooperation with public agencies seem to have adopted the right type of mechanism to implement efficient climate adaptation actions, this adaptation needs to become more systematic, and these actions need to be scaled up. Indeed, the scientific literature reports mostly minor modifications/measures in dealing with extreme weather events, which does not reflect a real shift in practices or values (IPCC, 2022). There are some examples – such as village relocations or creation of new multi-stakeholder resource governance systems – but these are rare and adaptations measures that are broad in scope tend to be slow, suggesting that achieving high transformation in depth, scope, speed, and limits may be particularly challenging or even involve trade-offs (IPCC, 2022).

Finally, it is fundamental to avoid the risk of maladaptation – initiatives that would increase climate risk or create new risks – as evidence of maladaptation is increasing (UNEP, 2021, IPCC, 2022). Coupled with adaptation initiatives, a framework to assess the adequacy of these initiatives appears essential. At the time of publication, no studies have assessed the adequacy and effectiveness of adaptation initiatives at a global scale, and thus there is no existing framework (IPCC, 2022).

Conclusion

Sustainable investing is gaining influence and visibility in the financial, political and public spheres – especially through the popular ESG investing. As being sustainable is more and more at high stakes for companies to respond to public policies as well as to civil society demand, the essential question of profitability of ESG investments is raised. There is currently no unanimous answer as to whether and how ESG investing affects profitability at the moment. Nevertheless, these debates are fundamental to incentivize investors into making decisions on ESG criteria, although more work is needed to come to a consensus. Globally, procedures in the formulation of ESG criteria, a transparent monitoring of ESG investing, and guidelines/incentives for the orientation of investments are lacking.

ESG investing might have the potential to direct financial flows towards climate mitigation, climate adaptation and resilience. However, as of 2022, it appears that there still needs to be a clear blueprint on ESG criteria as well as strong incentives from public agencies in order for ESG investing to be successful in achieving their goals. These strong incentives and structures entail indications of climate risks and financial gaps, an enabling environment that supports best practices, and the sanctioning of greenwashing and polluting activities. Moreover, a globally coherent environment, where public and private flows jointly fund climate mitigation, climate adaptation and resilience, is urgently needed.

References

Amel Zadeh, A. & Serafeim, G. 2018, 'Why and how investors use ESG information: Evidence from a global survey', *Financial Analysts Journal*, Working paper 17 - 079, pp. 87–103

Bauer, R., Koedjik, K. & Otten, R. 2005, 'International evidence on ethical mutual fund performance and investment style', *Journal of Banking and Finance*, Vol. 29, Issue 7, pp. 1751–1767. https://doi.org/10.1016/j.jbankfin.2004.06.035

Bello, Z. 2005, 'Socially responsible investing and portfolio diversification', *Journal of Financial Research*, Vol. 28, no. 1, pp. 41–57. https://doi.org/10.1111/j.1475-6803.2005.00113.x

Escrig-Olmedo, E., Fernandez-Izquierdo, M.A., Ferrero, I., Rivera-Lirio, J.M. & Munoz-Torres M.J. 2019, 'Rating the raters: Evaluating how ESH rating agencies integrate sustainability principles', *Sustainability*, Vol. 11, no. 3, pp. 2–16. https://doi.org/10.3390/su11030915

Fatemi, A. & Fooladi, I. 2013, 'Sustainable finance: A new paradigm', *Global Finance Journal*, Vol. 24, Issue 2, pp. 101–113. https://doi.org/10.1016/j.gfj.2013.07.006

Friede, G., Bush, T. & Bassen, A. 2015, 'ESG and financial performance: Aggregated evidence from more than 2000 empirical studies', *Journal of Sustainable Finance & Investment*, Vol. 5, Issue 4, pp. 210–233. https://doi.org/10.1080/20430795.2015.1118917

Guien, J. 2021, 'Le consumérisme à travers ses objets ', Editions Divergences, Paris, 222 pages.

Gyonyorova, L., Stachon, M. & Stasek, D. 2021, 'ESG ratings: Relevant information or misleading clue? Evidence from the S&P Global 1200', *Journal of Sustainable Finance & Investment*, pp. 1–35. https://doi.org/10.1080/20430795.2021.1922062

Kuzmina, J. & Lindemane, M. 2017, 'ESG investing: New challenges and new opportunities', *Journal of Business Management*, no. 14, pp. 85–98

Pederson, L.H., Fitzgibbons, S. & Pomorski, L. 2020, 'Responsible investing: The ESG-efficient frontier', *Journal of Financial Economics*, Vol. 142, Issue 2, pp. 0–37. https://doi.org/10.1016/j.jfineco.2020.11.001

Porter, M. & Kramer, M. 2011, 'Creating shared value: How to reinvent capitalism and unleash a wave of innovation and growth', part IV, article 16, Book: Managing Sustainable Business by Gilbert G. Lessen & N. Craig Smith (Editors), pp. 323–346

Puaschunder, J. 2019, 'The history of ethical, environmental, social and governance-oriented investments as a key to sustainable prosperity in the finance world', *Public Integrity*, Vol. 1, pp. 1–21. http://dx.doi.org/10.2139/ssrn.2957367

Rusu, D.I. 2020, 'The impact of environmental, social and governance factors on investors behaviour – an experimental study in the realm of sustainable investment', *Journal of Public Administration, Finance and Law*, Vol 17, pp. 301–319

Scholtens, B. 2006, 'Finance as a driver of corporate social responsibility', *Journal of Business Ethics*, Vol. 68, pp. 19–33. https://doi.org/10.1007/s10551-006-9037-1

Weber, O. 2014, 'The financial sector's impact on sustainable development', *Journal of Sustainable Finance and Investment*, Vol. 4, Issue 1, pp. 1–8. https://doi.org/10.1080/20430795.2014.887345

Ziolo M., Filipiak, B.Z., Bak, I. & Cheba, K. 2019, 'How to design more sustainable financial systems: The roles of environmental, social and governance factors in the decision-making process', *Sustainability*, Vol. 11, Issue 20, pp. 1–34. https://doi.org/10.3390/su11205604

Ziolo M., Bak, I. & Cheba, K. 2021, 'The role of sustainable finance in achieving sustainable development goals: Does it work?', *Technological and Economic Development of Economy*, Vol. 27, no. 1, pp. 45–70. https://doi.org/10.3846/tede.2020.13863

Other Resources

Albert, E. 26th of October 2021, 'Un investisseur qui achète un produit vert a de fortes changes d'aider Total, Shell ou BP à se financer', *Le Monde*, https://www.lemonde.fr/idees/article/2021/10/26/un-investisseur-qui-achete-un-produit-vert-a-de-fortes-chances-d-aider-total-shell-ou-bp-a-se-financer_6099885_3232.html

Albert, E. & Chocron V. 21st of October 2021, 'Le mirage de la « finance verte »', *Le Monde*, https://www.lemonde.fr/economie/article/2021/10/21/le-mirage-de-la-finance-verte_6099347_3234.html

CFA Institute, 2020, Sustainable Finance Disclosure Regulation (SFDR), https://www.pwc.ch/en/publications/2020/sustainable-finance-disclosure-regulation.pdf

Climate Policy Initiative, December 2021, 'Global Landscape of Climate Finance 2021', https://www.climatepolicyinitiative.org/wp-content/uploads/2021/10/Full-report-Global-Landscape-of-Climate-Finance-2021.pdf

Escande P. 2nd of February 2022, 'Energie: «Exxon ne veut pas changer de métier »', *Le Monde*, https://www.lemonde.fr/economie/article/2022/02/02/energie-exxon-ne-veut-pas-changer-de-metier_6111979_3234.html

Fardeau A. 5th of August 2020, 'Que se cache derrière l'étiquette « investissement socialement responsable »', *Le Monde*, https://www.lemonde.fr/argent/article/2020/08/05/que-se-cache-derriere-l-etiquette-investissement-socialement-responsable_6048163_1657007.html

Fink L. 2021, 'Letter to CEOs', https://www.blackrock.com/corporate/investor-relations/2021-larry-fink-ceo-letter

Gomez PY. 30th of November 2021, 'Entrée dans l'ère « de la performance durable »', *Le Monde*, https://www.lemonde.fr/emploi/article/2021/11/30/l-evaluation-de-la-performance-durable_6104123_1698637.html

Indonesia Impact Fund, https://indonesiaimpactfund.com/

IPCC, 2018, 'Annex I: Glossary. In: Global Warming of 1.5°C. An IPCC Special Report on the impacts of global warming of 1.5°C above pre-industrial levels and related global greenhouse gas emission pathways, in the context of strengthening the global response to the threat of climate change, sustainable development, and efforts to eradicate poverty', https://www.ipcc.ch/site/assets/uploads/sites/2/2019/06/SR15_AnnexI_Glossary.pdf

IPCC, 2022, 'Climate Change 2022, Impacts, Adaptation and Vulnerability', https://www.ipcc.ch/report/ar6/wg2/downloads/report/IPCC_AR6_WGII_FinalDraft_FullReport.pdf

Janicke, M & Jacob, K 2009, 'A third industrial revolution? Solutions to the crisis of resource-intensive growth', FFU Report, Environmental Policy Research Centre, Free University of Berlin. https://ssrn.com/abstract=2023121

Mooney A., 10th of March 2021, 'Greenwashing in finance: Europe's push to police ESG investing', Financial Times, url: https://www.ft.com/content/74888921-368d-42e1-91cd-c3c8ce64a05e?desktop=true&segmentId=7c8f09b9-9b61-4fbb-9430- 9208a9e233c8

OECD, 2021, 'ESG Investing and Climate Transition: Market Practices, Issues and Policy Considerations', Paris, https://www.oecd.org/finance/ESG-investing-and-climate-transition-Market-practices-issues-and-policy-considerations.pdf

United Nations Environment Programme, 2020, 'Adaptation Gap Report 2020', Nairobi, https://www.unep.org/resources/adaptation-gap-report-2020

United Nations Environment Programme, 2021, 'Adaptation Gap Report 2021: the gathering storm – Adapting to climate change in a post-pandemic world ', Nairobi, https://www.unep.org/resources/adaptation-gap-report-2021

World Bank, 2021, 'Enabling private investment in climate adaptation and resilience', https://openknowledge.worldbank.org/bitstream/handle/10986/35203/Enabling-Private-Investment-in-Climate-Adaptation-and-Resilience-Current-Status-Barriers-to-Investment-and-Blueprint-for-Action.pdf?sequence=5

Chapter 11

South Asia Region: Expanding Economy with Resilience and Adaptation Financing Challenges

KAMLESH KUMAR PATHAK

Adaptation can be defined as the process of adjustment to actual or expected climate change impacts and their effects. In human systems, climate adaptation seeks to moderate or to avoid harm or to exploit beneficial opportunities. In some natural systems, human intervention may facilitate the adjustment to expected climate and its effects[1].

Adaptation involves activities and approaches on responding to climate change by assessing its impacts, vulnerability, and risks to human and natural systems and accordingly planning the implementation of contingency arrangements when impacts occur by addressing losses and monitoring and evaluating adaptation efforts[2].

Specific arrangements have been developed under the UNFCCC within the Cancùn Adaptation Framework to formulate and to implement national adaptation plans (NAPs) as a means of identifying medium- and long-term adaptation needs and as a way of developing and implementing strategies and programs to address those needs. It is a continuous, progressive, and iterative process which follows a country-driven, gender-sensitive, participatory and fully transparent approach. Each country needs to set up NDCs or national determined contributions. Several structures and groups have been developed to support climate adaptation, such as the lead developed countries expert groups and the adaptation committee, amongst others.

---

[1] Reflecting progress in science, this glossary entry differs in breadth and focus from the entry used in the Fourth Assessment Report and other IPCC reports. {WGII, III}.

[2] https://unfccc.int/topics/adaptation-and-resilience/workstreams/national-adaptation-plans#:~:text=UNFCCC%20Nav&text=It%20enables%20Parties%20to%20formulate,programmes%20to%20address%20those%20needs.

This chapter will look into the countries' programming on climate adaptation with a particular focus on selected countries in South Asia. Specific emphasis will be given to the funding streams for infrastructure and urban resilience.

The chapter will apply a blended approach to estimate capital needs by developing countries through secondary research, consultation, validation, and reporting, with an objective to evaluate the climate finance needed by the different countries in the adaptation sector. The chapter will look into medium term targets and financing needs by 2030. The chapter will be developed with the objective to look into technical barriers and support for climate finance needed in the Urban and Infrastructure sector, in order to meet the NDCs goals.

This version of the synthesis report on nationally determined contributions synthesizes information from the 164 latest available nationally determined contributions communicated by the 191 Parties to the Paris Agreement and recorded in the interim registry of nationally determined contributions as of the 30th of July 2021.

I. How Adaptation is Perceived and Reported by Parties to the Paris Climate Agreement[3]

Most of the parties to the Paris Climate Agreement have prominently included the adaptation component in their NDCs (National Determined Contributions), as an attempt to communicate about the vulnerabilities and national circumstances linked to climate change in order to mobilize financial resources and technical capacity. They have communicated on needs to encourage adaptation related research, National Action Plans, sectoral actions, contingency measures and visible synergies between related global frameworks, including monitoring and evaluation.

Impacts, Risk and Vulnerability

The National Determined Contributions submitted by all the parties of the Paris Agreement (PA) have prominently captured the key adaption components and characteristics with particular emphasis on temperature,

---

[3] Nationally determined contributions under the Paris Agreement Synthesis report by the Secretariat, P 30,154.

precipitation and sea level rise as a major cause for extreme weather (flooding, long droughts, heatwaves, frequent wildfires, damage to ocean ecosystems and permafrost) events. The increased frequency of extreme weather events due to climate change led to weakening adaptation capacity of nature and posed serious risks to food security, biodiversity in ecosystems, health and urban infrastructure.

Enhancing Adaptation-Related Research for Policymaking

Information related to climate change like tracking meteorological patterns, forecasting impacts, and assessing risks is crucial for planning climate change adaptation in order to enhance adaptation -relevant data, information and monitoring systems, for evidence-based climate information. Countries in the NDCs have described their intention, efforts, priorities and requirements for enhanced adaptation activities in the form of adaptation centric climate data, climate risk modeling, mapping, regional scenarios and international cooperation. Countries have set priorities and agendas for research and action in land, ocean, and coastal areas, to implement solutions for floods or multi-hazards, hydrometeorological extremes and oceanic disturbances, and have at the same time raised the need for international cooperation to join the ecosystem for scientific, research and financing.

National Actions on Climate Adaptation

The Nationally Determined Contributions under the Paris Agreement Synthesis report noted that many parties have formulated their National Adaptation Plan (NAP) and their status with various stages in the adaptation planning process. This includes its documentation, status of preparation, and a description of how NAP could be an instrument in meeting NDC goals aligned with adaptation actions.

Some Parties have described the scope of their NAP, including risk and vulnerability, climate adaptation in development planning, climate data and adaptation information systems, technical capacity and the costs of adaptation, amongst others.

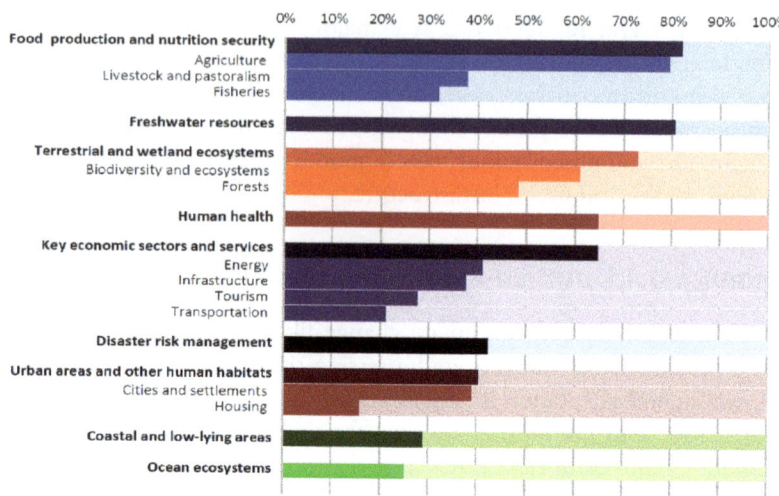

Fig.: 1: Distributed adaption components of adaption priorities and sectors based on the NDCs (Source: Nationally determined contributions under the Paris Agreement Synthesis report)

Fig. 1 is a synthesis of specific measures and quantified targets in adaptation priority areas. It provides an overview of priority areas and sectors specified which were included by the national governments in the nationally determined contributions. According to this table, food production and food security are identified as top priority in adaptation needs. This entails a range of specific measure across agriculture, livestock and fisheries.

Some of the important considerations relevant to the urban and infrastructure sectors could be highlighted as follows:

Global populations experiencing water scarcity and affected by major river floods are projected to increase, and climate change is expected to reduce raw water quality and pose risks to drinking water quality (https://www.ipcc.ch/report/ar5/wg2/). Availability, efficiency and quality of water is identified as a major component for adaptation, as well as the access to freshwater resource infrastructure.

Many countries from Asian Regions have low-lying coastal areas that face higher exposure to climate risks and disasters. These inherent risks require urgent attention, financial support, and solutions.

Countries also identified actions against the loss of land as a main adaptation objective, with efforts including assessing and monitoring impacts on coastal areas and national plans for coastal protection and management. These include nature-based solutions for coastal restoration and protection, and standards, regulations and guidelines for construction and flood protection.

In 2021 the United Nations Framework Convention on Climate Change (UNFCCC) found that the emissions reductions that were estimated based on targets communicated through countries' new or updated Nationally Determined Contributions (NDCs) 'fall far short of what is required' to limit global warming to 1.5°C or even 2.0°C above pre-industrial levels (UNFCCC 2021a). These findings underscore the urgency of developing – and subsequently implementing – adequate and effective adaptation plans to reduce vulnerability and build resilience in order to withstand the current and future impacts of climate change (Adaptation Gap Report 2021: The Gathering Storm).

All of the Parties to the Paris Agreement (UNFCCC 2016) have committed to engage in adaptation planning processes and in the implementation of actions. As part of the Global Stocktake under the UNFCCC process[4], parties will review the adequacy and effectiveness of adaptation and progress towards the global goal on adaptation (articles 7.14 and 14).

The Adaptation Gap Report 2021 (Fig. 2) assessed the global status of adaptation planning by examining the number of adaptation plans and strategies produced by 196 Parties to the UNFCCC and the extent to which these plans and strategies are effective and adequate (UNEP 2021).

Globally, 79 % of countries have addressed adaptation at the national level through a plan, strategy, policy, or law. This is an increase over the analysis from 2020, when 72 % of countries had a national adaptation instrument in place. A further 9 % of countries are in the process of developing their first national instrument[5].

---

[4] The global stocktake of the Paris Agreement (GST) is a process for taking stock of the implementation of the Paris Agreement with the aim to assess the world's collective progress towards achieving the purpose of the agreement and its long-term goals (Article 14). Decision 19/CMA.

[5] This includes national plans, strategies, policies or laws explicitly and primarily focused on adaptation or focused on climate change more broadly, with a significant adaptation component. National adaptation programmes of action were not included in the tally due to their unique role as a tool for LDCs to identify and act on urgent priority adaptation activities, rather than as an instrument to facilitate an overarching or holistic adaptation response.

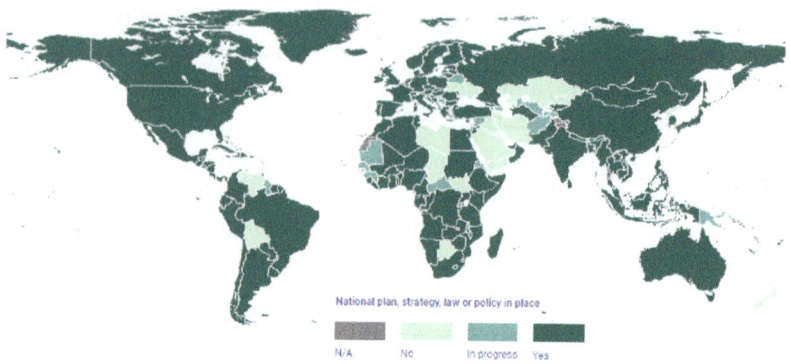

Figure 2. UNEP Gap report 2021 Effect of new or updated nationally determined contributions on 2030 greenhouse gas emissions relative to previous nationally determined contributions- initial source climate watch 2021

However as the UNEP emissions gap report shows (Figure 2), not all countries have reported new or reported NDCs creating some concerns on the process.

Under UNFCCC, formalization of NAP process through implementable action, strategies and investment is considered as a cornerstone for developing countries (UNFCCC 2020). We can conclude from the analysis, that developing countries have progressed in the NAP process since 2010, with significant progress demonstrated through commitment at scale in terms of technical assistance, institutional strengthening, and finance from 2015 onward.

As reported in September 2021, 125 of the 154 developing countries had undertaken activities related to the implementation of NAPs. Some countries had developed and submitted sectoral and thematic strategies and other relevant outputs. 22 countries had enacted or were working on enacting their Monitoring and Evaluation frameworks or systems for the NAPs[6]

The pace of national level action on adaptation planning across the world has shown remarkable acceleration since 2000. More than 50 % of countries have developed additional supplementary adaptation

---

[6] P. 19. Adaptation Gap Report 2021: The Gathering Storm.

instruments, or have replaced previous ones, to complement and to give more teeth to initial adaptation plans, policies, strategies or regulations. The growth in adaptation planning throughout the world could be attributed to the dire warning of IPCC, the scientific community, civil society and UNFCCC. The results are however insufficient as reminded by the IPCC in 2022, especially that the evaluation of NDCs is also depending on governments. The warning of the global leaders highlighted the plight of their countries in countering catastrophic episodes of climate change with severe consequences to survivability, economic viability, and human existence.

Looking ahead, the presence of clearly defined national adaptation goals and quantitative and qualitative adaptation targets could be an important way of gauging where adaptation planning has now become outcome-oriented and is measurable. Indeed, new and updated NDCs suggest that countries are already moving in this direction by including more quantitative and time-bound targets as part of their adaptation contributions.

Based on the various studies, the COVID-19 pandemic and the current tensed geopolitical situation and conflicts have seriously weakened the progress on multilevel governance and climate adaptation planning. Nevertheless, countries along with UN Institutions, multi-lateral development banks, and private sector initiatives, work consistently to enhance the adaptive capacity.

The 2021 report takes stock of positive progress both in terms of the number of nationals plans, as well as their adequacy and effectiveness. Significant progress has been achieved in 2021 against the criteria and indicators of adequate and effective planning. There are serious questions however whether we will collectively achieve the goals.

II. Regional Impact of Climate Change and Adaptation Priorities in South Asia

Climate Facts of South Asia and NDC of South Asian Countries

South Asian regions are facing alarming challenges of climate risks in the form of super climate disasters with high levels of vulnerability. With an urgent need to adapt and to respond to climate risks, the region was affected by disastrous weather events between 1999 and 2018, according to the Global Climate Risk Index[7].

The World Bank has assessed and reported the downgrading of living standards and well-being of populations in major South Asian countries such as India, Bangladesh, Sri Lanka and Pakistan. With a disproportionate impact on the poor and vulnerable populations, South Asia is going through a 'New Climate Normal' with highly intense heat waves, longer cyclones and droughts, and frequent major urban floods. This situation has challenged the capacity of government, authorities, the private sector, and global coalitions to test and address the crisis by providing a safer and more resilient society[8].

Another study by the Potsdam Institute for Climate Impact Research and Climate Analytics has predicted that the sea level rise in the region is likely to make parts of highly populated coastal areas uninhabitable in the future and has estimated 40 million climate migrants in South Asia by 2050[9].

Another study from the IPCC has found that 64 % of the world population is living in areas with a high exposure to devastating climate change impacts. Many of these vulnerable areas are located in South Asia: Afghanistan, Bangladesh, India, Nepal and Pakistan with inherent risks of hydrometeorological phenomenon (which increases already existing seismic risks). It has been visible with the super cyclone Amphan in Bangladesh & India in May 2020. Nevertheless, the issue of governance, the economic situation, regional instabilities and COVID-19 has significantly weakened the capacity to deal with climate crisis.

---

[7] https://www.germanwatch.org/en/17307

[8] Eckstein, David, Vera Künzel, and Laura Schäfer. 2021. 'Global Climate Risk Index 2021. Who suffers Most from Extreme Weather Events? Weather-related Loss Events in 2019 and 2000–2019.' Briefing Paper, Germanwatch, Bonn. https://germanwatch.org/en/19777.https://openknowledge.worldbank.org/handle/10986/29461

[9] Potsdam Institute for Climate Impact Research and Climate Analytics. 2013. Turn Down the Heat: Climate Extremes, Regional Impacts, and the Case for Resilience. Washington, DC: World Bank. https://openknowledge.worldbank.org/handle/10986/14000

Climate Finance Needs in South Asia for Addressing Adaptation Challenges

Financial resources will be the most desirable catalysts for climate resilient actions with clear goals and targeted approaches. It will be necessary to mobilize a diversity of financial instruments and climate funds and to scale-up the private sector's contributions. Various reports suggested that an estimated 40 billion USD will be required annually to address the climate change risks through adaptive actions. Another estimate from the Asian Development Banks has projected a 2 billion USD per year for adaptation needs in the south Asia region[10].

Another report published by *Ahmed, M. and S. Suphachalasai, Assessing the Costs of Climate Change and Adaptation in South Asia (2014)* has shown adaptation costs in different emission scenarios, indicating high investment needs in any pathway. The analysis concluded that, the region requires funding of the magnitude of 1.3 % of total GDP on average per annum between 2010 and 2050 under the BAU1 scenario (see Tab. 1). To avoid climate change impact under the BAU2 scenario toward 2100, the study concluded that adaptation funding of around $73 billion per annum on the average would be required between 2014 and 2050. However, adapting to a lower temperature under the BAU3 scenario (i.e., 2.5°C temperature rise and 0.3-meter sea level rise) only would require an adaptation funding of roughly $40 billion (0.48 % of GDP) on average per year. If the adaptation funding under the C–C1 scenario, with 0.24 %–0.82 % of GDP per annum (entailing a mean outcome of annual costs equating to 0.48 % of GDP) was actualized, the region would only require $31 billion on average per year[11].

---

[10] ADB & DfID, 2014, Assessing the Costs of Climate Change and Adaptation in South Asia 2020, CPI, A Snapshot of Global Adaptation Investment and Tracking Methods; RMI, 2019, Climate Finance Access Network: Program Document.

[11] https://www.adb.org/sites/default/files/ publication/42811/assessing-costs-climate-change-and adaptation-south-asia.pdf, 81

Policy Scenario	Adaptation Target	$ billion		GDP (%)	
		Annual Average Cost	Range	Annual Average Cost	Range
BAU₁	2100 worst case (6.9°C, 1.1 m SLR)	110.9	51.2–198.0	1.32	0.64–2.29
BAU₂	2100 (4.5°C, 0.70 m SLR)	72.6	33.1–127.8	0.86	0.42–1.46
BAU₃	2050 (2.5°C, 0.30 m SLR)	40.2	18.3–71.5	0.48	0.23–0.81
C–C₁	2100 (2.5°C, 0.55 m SLR)	40.6	18.8–71.4	0.48	0.24–0.82
C–C₂	2050 (1.9°C, 0.30 m SLR)	31.0	14.2–54.5	0.36	0.18–0.62

Tab. 1: Annual average adaptation cost, 2010–2050, South Asia (Source: Assessing the Costs of Climate Change and Adaptation in South Asia)

Economic Costs of Climate Adaptation in South Asian Region

The capital needs of South Asian countries to finance action in adaptation will demand innovative approaches to climate and green financing with active engagement of financing and investment ecosystems.

The NDC submitted by countries globally and by South Asian countries in this case have failed to integrate the private sector budgets in their NDCs. Thus, more emphasis will be desirable by countries, regulators, and NDC coordinating committees in countries, to articulate the needs of the private sector and to direct the investments into the climate adaptation pipeline (see the stakes and challenges in the previous chapter).

The impact of climate change is predicted to be highly disastrous in South Asian countries, which is noted by members of the coalition of South Asian Association for Regional Cooperation (SAARC) countries. The countries in business-as-usual scenarios are expected to lose nearly 2 % of GDP by 2050, rising to a loss of nearly 9 % by 2100.

Tab. 2: Adaptation Finance Need Assessment of South Asian Countries (Source: author)

Qualitative Assessment on Nationally Determined Contribution Tracking Studies and Status									
Country	Income Level	NDC Reviewed	Most Recent NDC Submitted	Most Recent Published 1st or 2nd (Updated)	NDC Implementation Period	Conditional	Unconditional	Both	Total Adaptation Cost
Bangladesh	Lower Middle Income	1st and 2nd	2021	2nd	2020–2030			Both	Yes
Bhutan	Lower Middle Income	1st and 2nd	2021	2nd	2020–2050			Both	Shall be provided in NAP
India	Lower Middle Income	1st	2016	1st	2021–2030			Both	Yes
Nepal	Lower Middle Income	1st and 2nd	2020	2nd	2021–2030			Both	Provided in the NAP
Pakistan	Lower Middle Income	1st	2016	1st	2016–2030			None	ADB Report – South Asian Region
Sri Lanka	Lower Middle Income	1st and 2nd	2021	2nd	2021–2030			Both	Yes

If no action is taken to adapt to and mitigate global climate change, the average total economic losses are projected to be 9.4 % for Bangladesh, 6.6 % for Bhutan, 8.7 % for India, 12.6 % for the Maldives, 9.9 % for Nepal, and 6.5 % for Sri Lanka. The Maldives could encounter as high as 38.1 % GDP loss equivalent (5 % chance) by 2100. While, by 2050, annual GDP losses are projected under the BAU scenario for Bangladesh (2.0 %), Bhutan (1.4 %), India (1.8 %), Nepal (2.2 %), and Sri Lanka (1.2 %)[12].

However, an annual average of only US$ 2 billion in adaptation investment was made in the region across 2015–2016, as tracked by the Climate Policy Initiative for the Global Climate Adaptation[13].

Governance and Financing Challenges in Ocean Ecosystems: Need a More Ambitious Agenda

The Special Report of IPCC on the Ocean and Cryosphere in a Changing Climate titled as 'Changing Ocean, Marine Ecosystem and Dependent Communities' clearly cited anthropogenic carbon emission as major agent for ocean warming, oxygen loss, deviation in nutrient cycling and loss of primary production ecosystem. Moreover, the loss of natural cycle in ocean has resulted extraordinary risks to marine biodiversity in general, and to fisheries and blue carbon vegetation in particular.

There is a growing consensus building up, to improve ocean governance and to contribute to the protection of the oceans. We could picture this as a time, where humans need to pay back the oceans and the earth for the services they took for many thousand years, and thus the finance for sustainable governance and sustainable use of ocean and earth resources is an emergency.

Several studies estimate that not less than USD 300 billion will be annually required for the conservation of oceans and marine ecosystems, but the exact expenditure tagging in this context is still missing. This clearly indicates the need for a 'Sustainable Ocean Economy Financing Needs Assessment'. In a report, Sumaila et al. found out that only 0.002 % of global GDP is invested in the conservation and sustainable

---

[12] ADB & DfID, 2014, Assessing the Costs of Climate Change and Adaptation in South Asia.

[13] 2020, CPI, A Snapshot of Global Adaptation Investment and Tracking Methods.

use of biodiversity generally. This study also indicates that investment in sustainable ocean economy is completely insufficient and only amounts to 13 billion USD from collective sources of philanthropy Official Development Assistance (ODA) over the past 10 years.

It is also imperative to highlight the fact that Marine plastic pollution is an enormous and growing problem. About 150 million tons of million tons plastic is already littered in the ocean, and this amount increases every year from about 8 million to 12 million with a large share coming from Asian countries. Eight of the 10 rivers that transport 88 %–95 % of the global load of plastics into the sea are in Asia: the Yangtze, Yellow, Hai, Pearl, Amur, Mekong, Indus, and Ganges rivers[14].

Global efforts to address marine plastic pollution have been gaining momentum, and the Asian region has seen increasing calls for action in recent years, including the Statement on Combating Marine Plastic Debris by the East Asian Summit Leaders in 2018 and the Bangkok Declaration on Combating Marine Debris in 2019.

All this is expected to remain a distant possibility, unless the plan for sustainable ocean economy didn't get integrated in the Nationally Determined Contributions' and supported through various forms of financing instruments. It is expected that countries with large marine and marine protected areas' will review their NDCs for a fair integration of a sustainable blue economy, with priorities in protecting the oceans, marine ecosystems as well as to keep them sustainable.

Conclusion and Recommendations for Countries in the South Asian Region

The study by the Asian Development Bank in 2018 concluded that South Asian countries will need to invest around 9 % of their total GDPs on infrastructure development during the period of 2016–2030 to adapt to climate change[15].

Countries in South Asia have some of the largest at-scale interventions in all aspects of adaptation, including disaster-risk reduction,

---

[14] C. Schmidt, T. Krauth, and S. Wagner. 2017. Export of Plastic Debris by Rivers into the Sea. *Environmental Science & Technology*, 51(21), pp. 12246–12253.
[15] Infrastructure Financing in South Asia ADB South Asia Working Paper Series, September 2018

nature-based solutions community and locally led action. Knowledge sharing, including of these solutions with other countries in the region and beyond, would be an important step in strengthening adaptation responses. A particularly strong topic for knowledge sharing is the region's globally recognized civil-society involvement for locally led adaptation actions such as in Bangladesh.

To prevent climate change's worst effects and to adapt to its impacts, a clear assessment of the investment level is required for the region of South Asia. It would be desirable for countries in South Asia to create policies, incentives, and strategies for channeling investment in the climate adaptation sector, with enough provisions for technical assistance and investment funding by incorporating elements of international assistance, domestic budgetary allocation and private sector capital. Nevertheless, south-south, north-south and triangular cooperation shall open new avenues and opportunities for joint action and cooperation.

It would be desirable that cities and regions create climate finance tracking and monitoring procedures to understand the need of public and private sector approaches, and the allocation procedures for adaptation finance. A more reliable and accurate scoping of the availability of climate finance for investment in the urban resilience and infrastructure sector is needed.

References

Ahmed, Mahfuz, and Suphachol Suphachalasai. *Assessing the Costs of Climate Change and Adaptation in South Asia. www.adb.org*, Asian Development Bank, 30 June 2014, www.adb.org/publications/assessing-costs-climate-change-and-adaptation-south-asia.

Asian Development Bank (ADB), 2014, Assessing the Costs of Climate Change and Adaptation in South Asia, Mandaluyong City, Philippines, https://www.adb.org/sites/default/files/publication/42811/assessing-costs-climate-change-and-adaptation-south-asia.pdf

Climate Policy Initiative (CPI), 2020, A Snapshot of Global Adaptation Investment and Tracking Methods, https://www.climatepolicyinitiative.org/wp-content/uploads/2020/04/A-Snapshot-of-Global-Adaptation-Investment-and-Tracking-Methods-April-2020.pdf

Germanwatch, 2020, Global Climate Risk Index, Briefing paper, Bonn, https://www.germanwatch.org/sites/default/files/20-2-01e%20Global%20Climate%20Risk%20Index%202020_14.pdf

Germanwatch, 2021, Global Climate Risk Index 2021: Who suffers most from extreme weather events? Weather-related loss events in 2019 and 2000–2019, Briefing paper, Bonn, https://www.germanwatch.org/en/19777.https://openknowledge.worldbank.org/handle/10986/29461

Global Center on Adaptation, 2020, State and Trends in Adaptation Report 2020, url: gca.org/reports/state-and-trends-in-adaptation-report-2020/.

Government of Bangladesh, 26th of August 2021, Bangladesh First NDC.

Government of Bhutan, 24th of June 2021, Bhutan First and Second NDC Report.

Government of India, 2nd of October 2016, India First NDC Submitted.

Government of Nepal, 8th of December 2022, Nepal First and Second NDC Report.

Government of Pakistan, 10th of November 2016, Pakistan First NDC.

Government of Sri Lanka, 4th of September 2021, Sri Lanka First and Updated NDC.

IPCC, 2014, Climate Change 2014: Impacts, Adaptation, and Vulnerability. Part A: Global and Sectoral Aspects. Contribution of Working Group II to the Fifth Assessment Report of the Intergovernmental Panel on Climate Change, Cambridge University Press, Cambridge, United Kingdom and New York, NY, USA, 1132 pages, https://www.ipcc.ch/site/assets/uploads/2018/02/WGIIAR5-PartA_FINAL.pdf

IPCC, 2014, Annex II: Glossary, in: Climate Change 2014: Impacts, Adaptation, and Vulnerability. Part A: Global and Sectoral Aspects. Contribution of Working Group II to the Fifth Assessment Report of the Intergovernmental Panel on Climate Change, Cambridge University Press, Cambridge, United Kingdom and New York, NY, USA, pp. 1757–1776, https://www.ipcc.ch/site/assets/uploads/2018/02/WGIIAR5-AnnexII_FINAL.pdf

IPCC Special Report on the Ocean and Cryosphere in a Changing Climate/ , pp. 447–587 https://doi.org/10.1017/9781009157964.007.

Potsdam Institute for Climate Impact Research and Climate Analytics, 2013, Turn Down the Heat: Climate Extremes, Regional Impacts, and the Case for Resilience, World Bank, Washington, https://openknowledge.worldbank.org/handle/10986/14000?locale-attribute=en

RMI, 2019, Climate Finance Access Network: Program Document, https://rmi.org/rmi-introduces-the-climate-finance-access-network-at-cop25/

Schmidt C., Krauth T., and Wagner S. 2017.,xport of Plastic Debris by Rivers into the Sea. *Environmental Science & Technology*, 51(21), pp. 12246–12253.

Sumaila, U.R., Walsh M., Hoareau K., Cox A., et al. 2020, Ocean Finance: Financing the Transition to a Sustainable Ocean Economy. Washington, DC: World Resources Institute. www.oceanpanel.org/bluepapers/ocean-finance-financing-transition-sustainable-ocean-economyOcean. Finance: Financing the Transition to a Sustainable Ocean Economy

Sumaila, U.R., Walsh, M., Hoareau, K. et al. Financing a sustainable ocean economy. Nature Communications 12, 3259 (2021). https://doi.org/10.1038/s41467-021-23168-y

United Nations Environment Programme, 2020, Adaptation Gap Report 2020, Nairobi, https://www.unep.org/resources/adaptation-gap-report-2020

United Nations Environment Programme, 2021, Adaptation Gap Report 2021: The gathering storm – adapting to climate change in a post-pandemic world, Nairobi, https://www.unep.org/resources/adaptation-gap-report-2021

UNFCCC, 17th of September 2021, Nationally determined contributions under the Paris Agreements: Synthesis report by the secretariat, https://unfccc.int/documents/306848

World Bank Group, 2021, Climate Change Action Report 2021–2025: Supporting Green, Resilient and Inclusive Development, https://openknowledge.worldbank.org/bitstream/handle/10986/35799/CCAP-2021-25.pdf?sequence=2&isAllowed=y

Websites

Asian Development Bank (ADB), www.adb.org//

OECD, Environment -OECD, www.oecd.org/environment/.

UNFCCC, National Adaptation Plans, https://unfccc.int/topics/adaptation-and-resilience/workstreams/national-adaptation-plans#:~:text=UNFCCC%20Nav&text=It%20enables%20Parties%20to%20formulate,programmes%20to%20address%20those%20needs

Conclusion

Chapter 12

Conclusion-Can Local and International Dialogue on Climate Adaptation Echo and Reinforce One Another? A Way Forward

Pascaline Gaborit PhD

Climate action is often tagged as 'think global, act local.' The current urgency of climate adaptation calls on the contrary for more 'local thinking' of adaptive approaches that consider the local geography, society, and culture, and more global action to counter climate change like international agreements, funding streams, political will, and accountability. As the previous chapters have demonstrated, some solutions do exist, such as grey solutions, early warning systems, nature-based solutions, and green infrastructure, and most of all the protection of global ecosystems such as forests and oceans. But we are not halfway on the path towards resilience to climate change. The way before climate action will get enough attention is very long. We also do not need only to 'talk the talk', but we also need to 'walk the walk' towards transformative change, and each walk has a strong price. More actions are needed to strengthen and enhance societies,' territories,' and ecosystems' resilience to climate change. Dialogue is an important element at the local level. It broadens the scope of decision-making and facilitates the interactions between the decision makers and civil society. Dialogue is also opening mindsets, visions, viewpoints, and horizons for actions. Dialogue is indeed a source of seeds for action, a place to exchange views and knowledge, and to capitalize on the tools of the represented stakeholders. Dialogue is, however, not sufficient to create action. Indeed, the presence of strong antagonistic interests, the lack of frame or funding streams, inconsistent policies, the lack of will or the absence of accountability are hampering climate adaptation from being fully implemented. What is needed are more bold global decisions, better shaped policies, a real 'care' in the calculation of climate impacts, and the emergence of strong trust community networks for the implementation.

I. Dialogue at the Local Level as Part of the Solution

Much has been written about how governance is key in climate adaptation and how political ecology is important in understanding future policy making (Basset, Fogelman 2013, Funtowicz 2020, Koch et al. 2021). Climate adaptation can also be the right entry point for climate action in specific territories such as cities, as the benefits can be estimated directly for the local communities, while actions towards climate mitigation have rather global impacts creating the need for simultaneous efforts of all countries. Territories with climate and resilience adaptation plans could be frontrunners in the future, creating examples for actions with different time frames. There is a consensus that the fight against climate adaptation necessitates smart cooperation among all of the different stakeholders (Rosenzweig et al. 2018, Wijaya et al. 2020, Gaborit 2021). In this framework dialogue among the local stakeholders can become a key element for action. Dialogue is understood here as the creation of spaces of institutional exchanges among stakeholders in the form of a problem-solving framework, leading to 'different way of thinking, talking, learning, and acting' (de Araujo et al. 2021). It is also a way of learning together and identifying solutions and hurdles in a problem-solving approach and a way to stimulate the exchange of knowledge. This dialogue can be organized on environmental issues, or more concretely about climate adaptation, acknowledging the needs, but also taking stock of possible disagreements on different values, ideas, or interests. It can also integrate the different narratives, resistances, antagonisms, and demands. There are illustrative examples on how local dialogue spilled over to national dialogue. The city of Paris had adopted a climate plan in 2007 like other cities such as Barcelona, Copenhagen, or Freiburg[1], far before the Paris Agreement had been signed at the international level. Among others, in Paris, the creation of a 'green transition fund' has benefited several hundreds of millions of Euros both from the public and the private sectors, to fund carbon-free initiatives. Green bonds in particular have issued more than 620 million Euros since 2015[2]. Many cities with illustrative examples in Europe or in California have created mechanisms to incentivize solar energy production on rooftops. We can list Barcelona

---

[1] In particular through the support of local authorities' networks such as ICLEI.
[2] https://www.fournisseur-energie.com/paris-fonds-vert/

Conclusion-Can Local and International Dialogue

(in 2017) and Brussels among the cities that have created a market of green certificates for individual solar energy producers.

The experience of the CRIC project and other sources of information from cities shows that there are consultation processes among stakeholders in cities and dialogue mechanisms in place at the local level. In Indonesia this dialogue is called *Musrenbang*. In most of the countries, national or regional representatives are also part of the dialogue. The CRIC project[3] shows, however, that the local decision-makers and stakeholders currently lack the capacity in terms of infrastructure, funding, and information in real time to tackle climate disasters and further adaptation. The first reason for this is the astounding number of simultaneous challenges that cities' governments must answer: to ensure development planning, to increase urban resilience, to maintain the continuity of urban ecosystems, to reduce the vulnerability of the coastal and exposed areas and the potential economic and human losses, and to develop contingency plans as well as disaster responses. The second reason hampering stakeholders' cooperation is that the different organizations are pursuing different interests in the absence of a strong national cohesive program for each of the concerned cities. The focus group discussions show that the dialogue among stakeholders, including with official representatives, is not only necessary but also insufficient when it comes to achieving climate response, adaptation, and resilience action. The research in several cities equally highlights the issue of a lack of trust among the cities' stakeholders and the local population towards the capacity of the different organizations to implement climate disasters' actions.

In addition to this, the question of climate or weather information transfer in real time and the anticipation of the risk level by the stakeholders has equally emerged[4] as a difficulty in the case of climate events e.g., a flood. This real time information transfer is needed between the meteorological agencies and the first aid responders, but also between the first aid responders and the local population – (See Fig. 3). A cooperation is also needed to agree on the levels of alerts and on the best alert system to be set up, such as the choices between broadband systems, SMS, sirens, or loudspeakers. Therefore, advocating for a 'multi-stakeholder' cooperation is not sufficient and should be accompanied by

---

[3] Through the focus groups, interviews, discussions and reports.
[4] Within the focus group discussions.

incentives, coordination with community representatives, information, accountability, and transparency to be efficient. The consultative process and the population's inclusion, including the vulnerable groups, in the decision-making process are also both recognized as a necessity, requiring more efficient disaster risk reduction and management (Ziervogel 2017, Djalante et al. 2020, Gaborit 2021). The work with community representatives, civil society organizations, and moral authorities seems equally important. Transparency and a local engagement process may also confront communities with the lack of available choices. The need for prioritization will continue to arise due to necessary trade-offs by the decision-makers between different priorities, levels of alerts, evacuations routes and contingency plans, possible relocations, and priority infrastructure or which neighborhoods should be protected. The increase in climate events is highlighting the lack of current global available solutions and the lack of land mechanisms that are easily activated by the local authorities (Wirawan et al. 2020, Afrizal et al. 2020, Gaborit 2021). Disinformation can unfortunately also be at stake.'

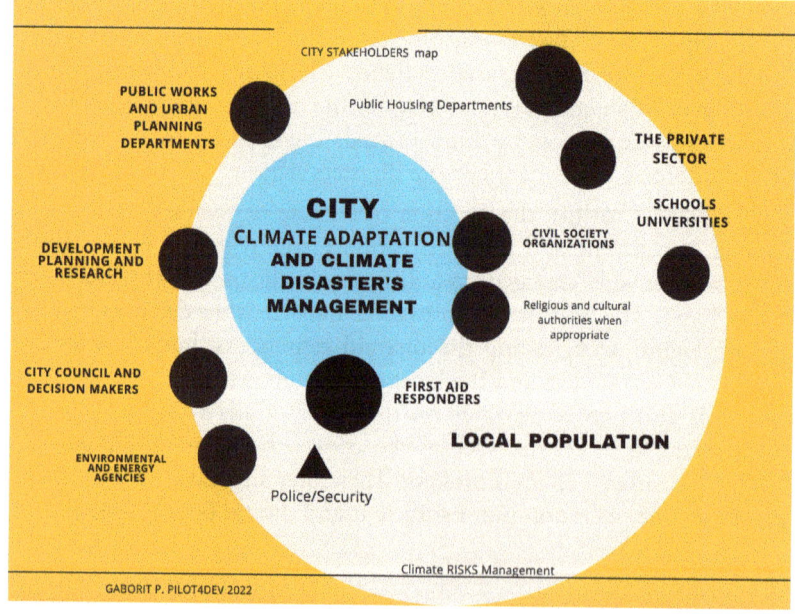

Fig. 1: The stakeholders' involvement in a dialogue at the local level

II. The Need of Strong Trust Community Networks to Implement Climate Adaptation Actions

Solving climate adaptation from a governance angle is going to be challenging. Governance is not only about creating new dialogue forums ad circles, but also about creating consistent policy making and frameworks for accountability and inclusiveness. We need to solve the challenge of 'we are in here together, we need to work together' by making it as concrete as possible.

Referring to one renowned author in sociology and social science such as Charles Tilly can elucidate a way out of the situation. 'Social life has always posed a fundamental problem for any group: in the face of strong incentives for individuals to serve themselves first and to avoid calls for cooperation in collective efforts, how nevertheless to produce collective benefits such as protection from predators, maintenance of the environment and guarantee of future food supplies?' These questions seem very accurate. According to Tilly, humans have created three different ways of creative collective benefits: authoritative organizations, collaborative institutions, and trust networks. It would be interesting to see how the three patterns would answer to the needs and urgency of climate adaptation alongside mitigation.

Authoritative organizations (or political regimes) guarantee top-down control and collective benefits over coercion, capital, and commitment in concentrated authorities. National States authorities long predominated among authoritative organizations. 'They exercised more extensive control over coercion, capital and commitment than any other of their rivals, and coordinated activities over far larger ranges of activity, population and resources' (Tilly 2005:40). As Tilly states 'As producers of collective benefits, authoritative organizations have obvious weaknesses: authorities often produce few collective benefits, impose large collective costs, use their control over resources chiefly to benefit themselves, and use those same means to perpetuate themselves to office, regardless of popular suffering or discontent. But when they work well, authoritative organizations can create collective benefits through top-down application of incentives: coercion, capital and commitment' (Tilly 2005:40). Considering current authoritarian regimes' work on climate adaptation, both at the national and at the city levels, seem, however far from convincing. Indeed, the increasing environmental and climate footprint of authoritative regimes, compared to the climate reduction targets by democratic inclusive regimes like Denmark show that authoritative regimes are not proposing any reliable answer to climate adaptation, although a vertical perspective on hierarchical powers could make it seem possible at first. The reason for this, is that the power 'head' of authoritative regimes, does not rely on enough 'bottom up' solutions, as the basis of elites is restricted by a strong submission to the regime, which is also hampering many to share and give solutions. This is contrary to how trust networks are formed.

Collaborative institutions: In the collaborative institutions, members rely on authority for coordination. They, however, rest on mutual consent with a right of exit. Use of common pooled resources is an example. Collaborative institutions could in our case of climate adaptation, be represented by international organizations. Indeed, the likelihood that people adopt rules and institutions to regulate the use of common resources is high in similar contexts. According to Ostrom (1998), sustained collective cooperation in the face of contrary individual self-interest depends on the mutual strengthening of three factors: interpersonal trust, investments made in reputation and trustworthiness, and adaption of reciprocity of norms. These three elements could also work in the creation of accountability mechanisms (trust, trustworthiness, and reciprocity of rules, or rule of law).

Trust networks: The third identified category by Tilly, 'trust networks,' would entirely reflect what is working at the local level, or within environmental or climate NGOs or transnational trust networks working against climate change. Trust networks are based on interpersonal relations, with a strong emphasis on ties. Simply, we will refer to the definition of both trust networks and *'relevant networks 'bounded, internally communicating sets of relations entailing mutual obligations'* (Tilly, 2005:44). At the local level, the adoption of local policies, but above all their implementation depends on the coexistence between collaborative institutions (including local governments, climate agencies, and civil society organizations), and trust networks for any further positive action. Diverse authors insist on the necessity of better participatory approaches and civic engagement in climate adaptation and in Early Warning Systems (Hegger et al 2017, Marchezini 2020). Hegger et al. show that local governments in the Netherlands have been involving residents in stormwater management, and in the prevention of heat climate-related stress. Other authors show that the climate and environmental action also results from cooperation from a wide range of stakeholders, like citizens' associations, forums, and not for profit organizations leading to changes through interactions and social transactions (Hamman 2011). The same article, however, shows that the level of commitment from citizens in case of floods seems very low and the author was wondering about the lack of solidarity networks in the phase of preparedness. The example of the COVID-19 Pandemic has also sadly shown that solidarity networks were not always developed and adaptive to the needs of people, especially of the aging population and of the youth, while on the contrary we can assume that solidarity and trust networks have heavily suffered from the top down measures. According to studies and analysis from recent floods, solidarity networks have emerged in the 'post disaster' phase, such as in the distribution of food, but they were considered scattered and not entirely well-planned. Some scams have also been noticed, like food vendors exaggerating prices and abusive contracts from insurance intermediaire[5]. This shows that resilience can only happen if there is enough preparedness both from the institutions in charge, the first aid responders and the communities in a complementary way. But the preparedness also necessitates a certain level of risk acceptance and the creation either of a collaborative system or a system of trust where cooperation takes precedence over conflict or inaction.

Similarly, the non-involvement of the private sector and finance in better clean energies and in low carbon solutions also reflects a lack of overall trust and trust networks in what the other stakeholders, including the governments and regulators, are going to decide.

In times of peace and stability, trust is a fundamental condition for a fair and cooperative society. According to the classical authors on trust (Luhman N. 1979, Braithwaite and Levi 1998, Hardin 2004), governments and institutions depend on the confidence of their citizens for the payment of taxes, the acceptance of regulations and justice, the implementation of social programs, and for the achievement of stability objectives, including the military. Other authors view trust as a social connection, where the emotional attachment to the government or to the institutions becomes more important than their rational choices, or than the analysis of their performance (Szescynski B., 1999). As a middle line between both theories, other authors have shown evidence that trust also

---

[5] Documentary Investigations Inondations: le temps des sinistrés – La nuit du déluge: December, RTBF La une 2021, https://www.rtbf.be/auvio/detail_investigation?id=2820945

plays an important part in contributing to social capital. Many authors, such as Robert Putnam (1993), or Charles Tilly (2005) have referred to trust as one component of social capital. 'Features of social organization, such as trust, norms, and networks can improve the efficiency of society by facilitating coordinated action' (Putnam, 1993:167). These authors have demonstrated that trust and social capital are often correlated with economic and civic cooperation, leading to development and progress, while the alternative is a vicious cycle of distrust, entailing negative consequences in terms of the economy, conflicts, and failed governance.

Local and more national collective efforts rely on cooperation, social networks and a more generalized trust to be able to implement the necessary steps for climate action. A system of multi-level governance can be adapted with more or less institutionalized forms of action and organizations to bring the efforts further both in terms of climate adaptation and mitigation.

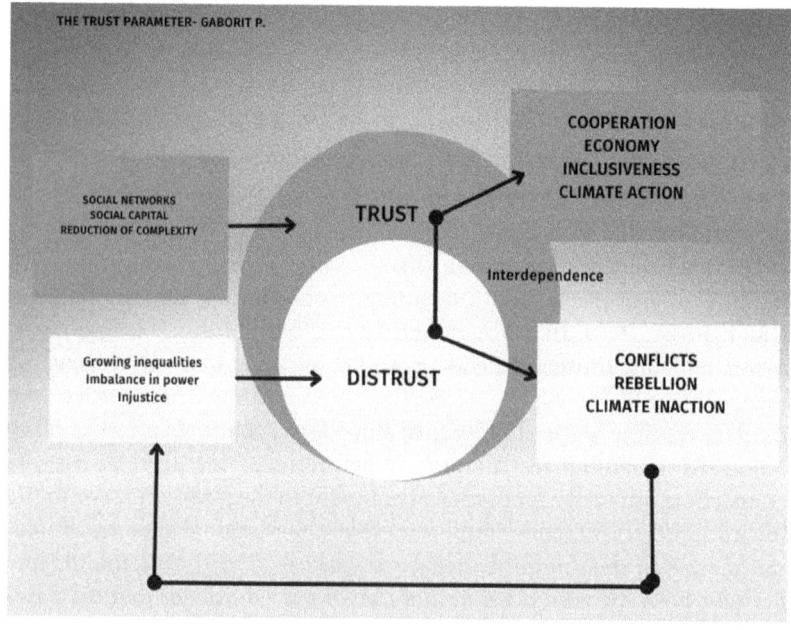

Fig. 2: The Trust Parameter (Source: Gaborit.P)

III. Towards More Consistent Global Policies, Funding, Accountability, and Political Will

The international dialogue on climate change gained momentum in Paris in 2015, when the national governments agreed and pledged to reduce climate change by 2°C and possibly 1.5°C compared to the preindustrial area. Since then, there is a common consensus that the different dialogue platforms have been disappointing in their scope and agreements. The 2021 report on the international panel on climate change is also raising the alarm that the trajectory of the GHG emissions could lead to an increase of 2°C sooner than expected. Global surface temperature will continue to increase until at least mid-century under all emissions scenarios considered. Global warming of 1.5°C and 2°C will be exceeded during the 21st century unless deep reductions in CO_2 and other greenhouse gas emissions occur in the coming decades (IPCC 2021). The advantage of the IPCC report is to base their assessments not only on 'promises' and National Determined Contributions (NDC) from the national governments, but also on the real (independently assessed) GHG emissions and scientifically observed changes, events and modifications led by climate change. The reasons for this lack of efficiency in the climate negotiations can be explained by several parameters linked to governance.

Climate Change Is Currently Given Less Attention

As noted by recent studies (Leil Filho et al. 2022) of many social and economic problems worldwide, the COVID-19 pandemic has also led to a sudden halt in face-to-face climate-related meetings in the framework of the UNFCCC. 'The study concluded that there is a negative impact of the COVID-19 pandemic on the UNFCCC process, more minor government priorities regarding climate action, loss of traction of the process, and a challenge to achieve the Paris Agreement, with less significant support from the respondents from less developed countries' (Leil Filho et al. 2022). This situation has been worsened by the current geopolitical situation including Russia's war in Ukraine, the increase in energy prices, the problems of food transportation, and the rising tensions between China and the USA. The findings and all the recent IPCC reports suggest that urgent action is needed to make up for the lost time and to place climate issues more prominently not only on the global agenda, but also on the national and on the local agendas.

A Decreased Trust in a Changing International Order

The failure of climate negotiations within the annual COPS, since the 2015 Paris Agreement, seems to reflect a lack of will of national governments to abide by the current international liberal order. According to Féron et al. (2020), there is indeed dissatisfaction with the current international order from the Southern countries through the angle of inequalities and climate justice, but also a dissatisfaction from key economic powers, most notably the BRICs, including the authoritative regimes of Russia and China. In this new untrusted but also competitive multilateral order, it will be increasingly difficult to create approved norms and codes of conduct leading to trusted mechanisms of accountability among national governments and blocks created by the Western World, or China, Russia, and emerging economies. A new multilateral complex and multipolar world is emerging, but, according to the authors, this is not the only current change. The international world order, whether understood as multiplex, or multipolar, integrates a fragmentation of significant actors such as cities, multinational companies, non-governmental organizations, and different communities enabled by new technological developments' (Féron et al. 2020: xxi). This situation is increased by 'how cross-cultural individual encounters are shaping the cultures of agencies, increasing the fragmentation of an internationalized or globalized environment.' This current context makes it increasingly complex to develop a holistic and global understanding of the needed change to identify solutions for the difficult times ahead, especially with so much uncertainty, fragmented governance, societies and looming climate change. The current Russia's aggression and war in Ukraine is on a top of this raising a real challenge in terms of the creation of a multilateral order which would be 'trusted' by all the parties. The crisis challenging the international communities also highlights the importance and reliance for European Countries on Russian gas imports, which are highly detrimental for climate change, because of methane leaks. The economic interrelations have highlighted a heavy reliance on fossil fuels interconnections and geopolitical interdependence among regions. Several authors had already identified that the COVID-19 pandemic has decreased the attention to climate change that is needed for action (Leil Filho et al. 2022). Faced with this situation, and the rapidly changing international order, there is an 'urgent need' to put the climate change agenda again at the center of attention and of action, before too many tipping points are crossed. But Trust is also a matter of relying on consistent policies and orientations. Climate change cannot be achieved at all costs, and especially not by sacrificing the rule of law or fundamental freedoms.

Inconsistent or Aantagonistic Policies

There are increasing climate initiatives based on smart technologies and innovations. In the area of the construction sector, for instance, the development of greener technologies like sustainable concrete, recycling, or earth materials bring interesting horizons for the development of low carbon solutions. For instance, the development of green cement (low carbon) and with a strong recycling potential in European countries is a promising evolution for the future of urban construction, which can use better, more resilient materials to counter rainfalls and heat. This optimism is, however, jeopardized by the development of often antagonistic policies, which happens when the negative effects of one policy on climate adaptation and mitigation (such as deforestation, fisheries, intensive agriculture and beef or chicken livestock farming, energy investment in fossil fuels, and coal mining), nullify the impacts of the climate related policies. This situation occurs both at the international level with the challenges of a transformative energy transition and at the local level where growing urbanization and agricultural needs are responsible for quick deforestation, land clearance, and land reconversion.

The complexity of an approach based on climate risks. The difficulty of communicating on complex climate-related problems is definitely a very challenging matter. As Moench et al. 2011, there is a myth among decision makers, that scientists will make research adapted to local situations and specificities. The responsibility to take action and to implement information-based climate adaptation strategies and actions lies primarily in the hands of national decision makers, but also city mayors, city teams, and urban practitioners. Communication and dialogue need to be made on adapted, and accurate information which is complex in the area of climate related risks. In cities for instance there are various interrelated problems linked to climate change and adaptation, such as water or waste management. At the national or territorial levels, the complexity may even be higher. Some of the problems, like droughts and strong winds, occur occasionally, while others such as flash floods can be seasonal and repetitive but can also occur in different areas leading to unpredictability. Coastal and low-lying territories in particular are faced with floods while the conversion of land for urbanization and agriculture is rising simultaneously. Consequently, many water catchment areas are sacrificed. On different territories, needed engineering plans will not be implemented due to budget constraints. The question of funding remains very central to the priorities for climate adaptation, in hope that national and international funding could complement the municipal and provincial or regional budgets on the matter. The different cities in our projects were confronted with diverse priorities and challenges: waste management, flood management, and early warning systems, or simply access to water and water management. Urbanization, transportation, waste, rivers clean-up, water, and also economic development and social inclusion were at the forefront of the local governments' concerns alongside disaster risk prevention.

The Lack of Easily Accessible Funding

In this context, the available cities' budgets including the funding of national government programs would quickly become insufficient to tackle the needs. Initiatives and funds have been set up: for instance for Indonesia the Indonesia Impact Fund (IFF), the Green bonds and Green Sukuk initiative, a clear budget tagging earmarked for climate for each ministry and region, and the Indonesia Disaster Resilience Initiatives Project (IDRIP).[6] The 100 billion dollars pledge for a fund on climate justice, which was pledged at the COP22, and never really materialized (despite the existence of a Global Green Climate Fund in the framework of UNFCCC since 2010). Despite this situation, it is difficult for cities to always propose bankable projects and follow all the constraints and requests from the donors and institutions. In addition to this, many other funds mainly focus on technical assistance rather than on investments, consultations, research, partnerships and actions. In different cities, the funding will be used for priority investments to increase the preparedness of selected local governments to manage natural hazards and to strengthen Early Warning Systems (EWS). These mechanisms

---

[6] World Bank Project: Indonesia Disaster Resilience Initiatives Project (IDRIP) – P170874.

aim, among others, to enhance and strengthen the urban resilience of cities which can be defined by their capacity to respond, adapt and recover from the pressures and crises related to climate change and other changes: subsidence, demographic growth, poor land management, insecurity or attacks, economic downturns, social unrest, unsustainable use of resources and declining ecosystems (Ziervogel 2017, Diab 2020, Gaborit et al. 2021). Furthermore, different studies show that mitigation actions, when compared to adaptation actions, receive the main portion of global climate finance flows by international multilateral development aid organizations and development banks (Grafakos et al. 2019). This portion was over 96 % between 2010 and 2011 (Buchner et al. 2012, Chan 2019) and still less than one fifth for the year 2015–2016 (Tall et al. 2021). This shows that medium term investments for the development of local climate adaptation strategies still rely on local political will and also on a limited local available funding.

This may lead to risky trade-offs, focusing on what the city can truly tackle. like domestic waste, to the detriment of more long-term climate adaptation policies that are aimed to protect the populations that are most exposed to hazards through contingency plans, evacuation routes, early warnings, and relocation programs. Indeed, a source of conflict generally arises at the urban level land planning and also contributes to more deforestation and climate impacts. This constitutes an illustrative point that reflects on the interrelation between factors and impacts (Fig. 2). It is explained below, yet it also falls under the approach of socioeconomic pathways: urbanization needs, deforestation, conflicts, and inequalities. K. Pathak reminded us in one of the previous chapters, based on a study from the Asian Development bank, that countries will need to invest around 9 % of their GDPs on infrastructure development during the period of 2016–2030 to deal with climate change. The question of funding is also linked to the question of bureaucratic burdens and technocratic rules to access the funds.

Bureaucracy

Many studies have shown that the accessible funding is often entirely 'blocked' by bureaucracies and complex reporting rules which are jeopardizing the actions and activities' sustainability. Some of the international and/or European programs are sadly often highlighted as a prototype for red tape to the detriment of project results. However similar problems are reported with funding from many agencies. As concrete example, the world bank report on the reconstruction of Aceh and Nias after the 2004 devastating tsunami reports the difficulties in transforming funds and donations into sustainable recovery and reconstruction programs (World bank 2005). At the opposite side of the large donors' bureaucracy, small donors are also exposed to corruption, misuse of funds, ethical problems and difficulties in the implementation. Accountability is indeed a difficult condition to achieve.

The Question of Accountability

In terms of climate adaptation and climate mitigation, the problem will soon become a question of accountability as detailed earlier. The pledge of 100 billion Dollars for climate justice decided by the international community at the COP 22 was not ever actualized (even though a global Green Climate Fund within UNFCCC exists since 2010)[7]. The legal trials where some European countries' governments have been recognized as guilty of inaction by their respective national courts have only partially uncovered the problem. National courts have indeed requested the governments to act for the climate in February 2019 in the Netherlands and in November 2020 in France. But the responsibilities of inaction are very much spread over the different stakeholders involved in climate change, in industries like trade, energy, public policies, industries, companies, and citizens' consumption patterns, amongst others. The revelations by the Kayrros company and the European Spatial agency of massive methane' leaks along the Russian natural gas pipelines has shed a light over the difficulties in trusting the reporting of national governments regarding GHG emissions[8]. Methane would indeed have 80 % more impact on climate than CO_2 over 20 years, and leaks would account for half of methane related emissions. In the area of climate adaptation, the question of accountability is likely to become even more critical. Indeed, as shown by the examples of the floods in the Liège region of Belgium, in July 2021, the responsibilities are difficult to identify in the aftermath of a disaster. The parliament inquiry which followed the floods has demonstrated that there was no main actor or stakeholder which could be held responsible

---

[7] Muhamed Zeeshan, Friends of Europe Debate, 25/02/2022.
[8] https://cleantechnica.com/2022/02/05/massive-methane-emissions-by-the-oil-gas-industry-detected-from-satellite-space/

for the floods and its impacts. Contrarily, a numerous chain of dysfunctions and a cascading system has been identified, especially for the first aid responders which were not properly equipped, but also for the cities' mayors which were waiting for further provincial instructions to start the evacuations, and finally for the authorities responsible for the dam which did not entirely optimize the floods' stream. This did not clarify first the main responsibilities in the face of an unprecedented flood, nor did it clarify the accountabilities in terms of recovery and reconstruction, as some victims are still waiting for compensation or financial support. This example shows that the future lies in the creation of transparent accountability systems, where all government tiers, as well as public and private actors, are all hold responsible and accountable for their actions, policies, and decisions.

Conclusion

'Urgent action is needed' to make climate action more prominent on the political agenda and to agree on solutions. Climate adaptation and mitigation at the national and at the local levels will be highly reliant on accountability systems, which entails the creation of collaborative institutions and trust networks where the codes, norms, rules of conduct, and accountability systems are clearly identified. This principle of accountability seems in strong contradiction with the practice of authoritarian regimes. The principles of authoritarian regimes indeed entail a lack of accountability of the ruling elites towards the population and citizens. These regimes also possibly utilize propaganda and lies in the data and figures. Nevertheless, democratic regimes also present shortcomings in terms of rule of law, bureaucracies, accountability, and policy frameworks. Overall, the interrelations between complex economies and governance systems are often hiding inconsistent or antagonistic policies towards climate in the areas of deforestation, fisheries, support to intensive agriculture, and heavy subsidies to fossil fuel companies. Climate Adaptation and resilience are also highly dependent on complex solutions, including preparedness, and adaptive planning relying on dialogue, anticipation, risks prevention, and climate disaster management inclusive mechanisms. Although environmental organizations and the press are authorized to investigate and to show the impacts of such policies, the complexities of climate change make it difficult for civil societies to follow up entirely the stakes, problems, solutions and responsibilities. In a global system with an increasing opposition between authoritarian and liberal regimes, and with a growing distrust among countries and blocks (authoritarian/liberal and north/south), the international dialogue on climate action and climate adaptation will be seriously disabled to the detriment of the most vulnerable areas exposed to climate change. The environmental forums created at the local level are very interesting examples of the way forward in the identification of solutions in a problem-solving approach. This concrete enabling problem-solving approach could echo at the international and the global level to bridge the divide and restore enough trust for constructive action and dialogue before too many tipping points are reached for both climate change and stability. Current disasters should be effectively acknowledged and lead to solidarity mechanisms, humanitarian aid, support to resilience, and help to build back better the affected areas. Transparency will be needed on climate change impacts, solutions, interrelations and problems, so that the driving forces including international organizations, leaders, practicionners, civil society movements, think tanks, green leaders, the media, students, the private sector, food producers, engineers, investors, banks, or creative industries can adapt their strategies in the light of the current situation.

References

Afrizal, Berenschot W. et al. 2020, 'Resolving Land Conflicts in Indonesia' Review essay, Bijdragen tot de taal, land en volkenkunde, Vol. 176 (2020), pp. 561–574

Basset T., Fogelman C. 2013, 'Déjà vu or something new? The adaptation concept in the climate change literature', *Geoforum*, Vol. 48, pp. 42–53

Braithwaite V., Levi M., 1998, Trust and Governance, Russell Sage Foundation series on Trust

De Arauso Arosa Monteiro R., Ferraz de Toledo R., Roberti Jacobi R., 2021, 'Dialogue method: a proposal to Foster intra and inter-community Dialogic Engagement', *Journal of Dialogue Studies*, Special issue, Dialogue with and among the Existing, Transforming and Emerging Communities, Vol. 9, pp. 165–183

Djalante R., Garschagen M., Thomalla F., Shaw R. (Eds). 2017, *Disaster Risk Reduction in Indonesia,* Springer International Publishing

Féron E., Juutinen M., Käkönen J., Maïche K. (Eds). 2020, Shedding light on a Changing International Order: Theoretical and Empirical Challenges', Tampere TAPRI Press

Funtowicz S. 2020, 'From risk calculations to narratives of danger', *Climate Risk Management*, Vol. 27, 100212 https://reader.elsevier.com/reader/sd/pii/S2212096320300024?token=B6CBEA02054999EE756B04F666D4701C3905CDF460510AD1F52DF26A57DFF692EC212672A58C4D7A49A536076F984FAB&originRegion=eu-west-1&originCreation=20210808171814 (2020), , last accessed 08.08.2021

Gaborit P., Aleksic A., Marengo P., Diab Y., Pathak K. 2020, Policy briefs ten Pilot Cities. https://www.resilient-cities.com/en/knowledge/175-policy-briefs-for-pilot-cities-2, last accessed 31.03.2021

Gaborit P. 2021, 'Vulnerabilities and Resilience to Climate Change in Tanzania', in Gaborit P. et Olomi D. (Eds.), *Learning from resilience strategies in Tanzania: An outlook of international development challenges*, Brussels, Peter Lang International, https://www.peterlang.com/document/1152350

Hamman P., Causer J.Y. 2011, *Villes, environnement et transactions démocratiques* (Dir.) Peter Lang, Ecopolis

Hardin R. *Distrust*, NYC, Russell Sage Foundation

Hegger D.L.T., Mees H.L.P., Driessen P.J., Runhaar A.C. 2017, 'The roles of residents in Climate Adaptation: A systematic Review in the case of the Netherlands', *Environmental Policy and Governance*. https://doi.org/10.1002/eet.1766

IPCC International Panel on Climate Change, 2021, 6th assessment report: Climate Change the Physical Assessment report, https://www.ipcc.ch/report/ar6/wg1/downloads/report/IPCC_AR6_WGI_SPM_final.pdf

Koch L., Gorris P., Pahl Wostl C. 2021, 'Narratives, Narration and Social Structure in environmental Governance', *Global Environmental Change*, Vol. 69, July 2021, 102317, https://doi.org/10.1016/j.gloenvcha.2021.102317, last accessed 23.07.2021

Leil Filho W., Hickmann W., Nagy J., Rimi Abubakar I., et al., 2022, 'The influence of the Corona Virus on Sustainable Development Goal 13 and the United Nations Framework Convention on Climate Change Processes' *Frontiers in Environment Science*, https://doi.org/10.3389/fenvs.2022.784466, last accessed 25/02/2022

Luhmann, N. 1979, *Trust and Power: Two Works by Niklas Luhmann*. Translation of German originals Vertrauen 1968 and Macht 1975. Chichester: John Wiley.

Marchezini V. 2020, 'What is a sociologist doing here?: an unconventional People centered approach to improve warning implementation in the

Sendai Framework for Disaster Risk Reduction', *International Journal of Disaster Risk Science* Vol. 11, pp. 218–229, https://doi.org/10.1007/s13753-020-00262-1, last accessed 12 February 2022

Misztal B.A. 1992, The Notion of Trust in Social Theory, Policy, Organisation and Society, Vol. 5, No.1, pp. 6–15, https://doi.org/10.1080/10349952.1992.11876774

Moench M., Tyler S., Lage J. 2011, 'Catalyzing urban governance: Applying resilience concepts to planning practice in the ACCCRN Program 2009–2011', ACCRN publication

Mulkhan U., Mayaguezz H., Tisnanta H.S., Kurniawan N. 2020, 'Urban Analysis Report Pangkal Pinang', https://www.resilient-cities.com/en/?preview=1&option=com_dropfiles&format=&task=frontfile.download&catid=41&id=40&Itemid=1000000000000, last accessed 31.03.2021

Ostrom E. 1990, *Governing the Commons: The Evolution of Institutions for Collective Action*, Cambridge University Press

Putnam R. 1993, *Making Democracy Work: Civic Traditions in Modern Italy*, Princeton University Press

Rosenzweig C., Solecki W., Romero-Lankao P., Mehrotra S., Dhakal S., Ali Ibrahim S. (Eds). *Climate Change and Cities: Second Assessment Report of the Urban Climate Change Research Network*, Cambridge University Press, New York

Szescynski B. 1999, 'Risk and Trust: The Performative Dimension', *Environmental Values*, pp. 239–252 The White Horse Press, Cambridge UK

Tall A., Lynagh S., Bianco Vecchi C., Bardouille P., Montoya Pino F., Shabahat E., Stenek V., Stewart F., Power S., Paladines C., Neves P., Kerr L. 'Enabling Finance in Climate Adaptation and Resilience: Current Statue, Barriers to Investment and Blueprint for Action' 2021 World Bank Group, and GFDRR Global Facility for Disaster Reduction and Recovery.

Tilly C. 2005, *Trust and Rule*, Cambridge University Press

Wijaya N., Nitivattanon V., Prasad Shrestha R., Minsun Kim S. 2020, 'Drivers and Benefits of Integrating Climate Adaptation Measures into Urban Development: Experience from Coastal Cities of Indonesia' Sustainability 2020, 12, 750, Mdpi.com

World Bank 2005, 'Rebuilding a better Aceh and Nias: stocktaking of the reconstruction effort', World Bank

Ziervogel G., Pelling M., Cartwright A.A., Chu E., Deshpande T., Harris L, ..., Zweig P. 2017, 'Inserting rights and justice into urban resilience: a focus on everyday risk', *Environment and Urbanization,* Vol. 29(1), pp. 123–138

Biographies and Acknowledgments

Biographies

Lead Author and Editor

Dr. Pascaline Gaborit is the director of Pilot4dev. The aim of PILOT4DEV is to connect stakeholders to boost sustainable development but also to create dialogue among institutions, stakeholders and civil society. Pascaline Gaborit graduated in social sciences with a PhD. on trust and conflicts including questions related to resilience. She is climate Chair for Belgium of the G100 women's network. She also was a member of WIIS Brussels: Women in international security.

She published books and articles on cities, climate international cooperation, culture, conflicts, security and gender. She worked for the Global relations forum (strategic Director), and for an international cities network www.pilotcities.eu .

She organizes, and moderates events, publishes reports and analysis, manages projects, coaches events and occasionally gives lectures at university. She is part of the scientific committee in several programs and European projects.

She is a Sea Ambassador, Professional Dive Guide, Diving assistant instructor, and instructor on marine ecosystems under the SSI federation (Scuba School International) in her free time. She is equally the happy mother of two children.

Pascaline currently works as expert for the project 'CRIC: Climate Resilient and Inclusive Cities' taking place in Southeast Asia and in India. The project focuses on Early Warnings, Climate adaptation, and Disasters' resilience strategies. She is involved in other projects in different countries and regions: in Europe, in Tanzania with the project Pilot 4 Research and Dialogue, www.pilot4dialogue.com, and in Guadeloupe/Caribbeans with the project LIFE Adapt Island. www.cayoli.fr/lifeadapt.

Her main books include: ' *Learning from Resilience Strategies in Tanzania, an outlook of international development challenges*' (2021) available in open Access: Learning from Resilience Strategies in Tanzania - Peter Lang Verlag, '*The strength of Culture for Development*' (2015); and '*European and Asian Sustainable Towns*' (2016), all published by Peter Lang International. She had published earlier a collective book on '*Gender Stereotypes*' (2008) and another on '*Trust in post conflict societies*' (2009) by publisher L'Harmattan.

She contributed to 2 international reports:

- Hidden Cities Report, World Health Organization-UN Habitat, 2010
- Second Urban Climate Change Research Network (UCCRN) 2018, Assessment Report on Climate Change and Cities (ARC3-2). Contribution to chapter 2: Civic Engagement

Chapters' Authors and Contributors

Diana Nur Afifah is a lecturer in the Department of Nutrition Science, Faculty of Medicine, Universitas Diponegoro. Her area of interest is nutrition and functional food. She is a researcher in CENURE (Center of Nutrition Research) Universitas Diponegoro. Her research mainly supported SDGs, especially Goal 2: No hunger and Goal 3: Good health and wellbeing. Her research was published in reputable international journals. She is also frequently invited as a speaker regarding this goal as community service activities on television and radio.

Dessy Ariyanti is a lecturer in the Department of Chemical Engineering, the Faculty of Engineering, Universitas Diponegoro. She holds doctoral degree in Chemical and Material Engineering from the University of Auckland, New Zealand. Her research focuses on the development of advanced materials for water and wastewater treatment, renewable energy, and energy storage system. She is a researcher in Center of Biomass and Renewable Energy as well as in Advanced Material Laboratory in Universitas Diponegoro. She also actively involved in the community developments program for plastic waste and organic waste conversion into valuable products.

Mr. Subbiah Arjunapermal is the Director of the Regional Integrated Multi-hazard Early Warning System (RIMES) and works with the RIMES Council in leading RIMES and delivering on its long-term vision of forearmed, forewarned and resilient communities. Subbiah has operational responsibility for the RIMES Program Unit and Regional Early Warning Center, including their staff and programs, expansion, and execution of RIMES' mission.

Subbiah has over 20 years of experience in establishing multi-institutional and multi-disciplinary mechanisms and building institutional capacities for generation and application of climate information at different timescales, as well as 39 years of experience in drought mitigation and management. He served as Reviewer of the IPCC Special Report on Managing Risks of Extreme Events and Disasters to Advance Climate Change Adaptation (2010–2012). Subbiah held senior positions with the Government of India, involved in policy formulation and implementation of development and disaster risk reduction programs until the late 1990s. Subsequently, Subbiah held the post of Director, Climate Risk Management and Team Leader for Early Warning Systems at Asian Disaster Preparedness Center, and evolved and implemented a multi-country Climate Risk Management program and regional multi-hazard project that eventually transformed into RIMES in 2009 as an inter-governmental and international institution.

He has guided capacity development programs for National Meteorological and Hydrological Services of over 20 countries in Asia, Africa and Pacific for better-resolution and more reliable weather and climate forecasts and in establishment of forecast provider-user forums to foster collaboration among National Meteorological and Hydrological Services and their stakeholders for increasing the use of climate information in planning and decision-making processes. He has facilitated the application and integration of science-based information on weather, climate, and hazards in sectoral planning and end-user decision-making, for resource management and reducing disaster risks in several countries.

He has expertise in assessment of climate variability impacts on communities and on biophysical and societal systems, and subsequent evaluation of application potential of long-lead forecasts for local resource management that led to the development, funding, and implementation of several climate forecast application programs. He has also developed and operationalized drought management strategies for 1987–1988 and 2000–2001 drought events in India.

Muhammad Attorik Falensky was born in Sungai Penuh, Jambi, Indonesia on July 11 1999. He achieved his Bachelor of Science degree from the Department of Geography, Universitas Indonesia in 2021. Recently, he focused on developing the education community 'Sungai Inspirasi' in Sungai Penuh Kerinci and now he is a hydrology research assistant at the University of Indonesia and has published several academic research in the utilization of GIS and Remote Sensing in hydrology, forestry and settlement. His latest publication is 'Evaluation of urban settlement sustainability in an inland municipality in Central Java, Indonesia', published in the IOP Conference Series: Earth Environmental Science.

Dr. Itesh Dashheads (Systems Research and Development Division at RIMES) has developed and supervised planning, design, development of several decision support systems, solutions, and processes that meet early warning and domain specific user needs effectively and efficiently. He also leads the implementation of the Impact based forecasting, data mining and analytics and prioritizes mainstreaming and application of weather and climate information in decision making for critical Sectors like Agriculture, Water Resources, Disaster Management. Itesh has more than 15 years of combined operational experience in the DSS development, and community level Capacity development of the end users in the RIMES member states in using these tools. He has operational working Experience in more than 15 countries including Africa, Asia and the Pacific.

He holds a Master's degree in Information Communication and Technologies and a Ph.D. degree in Disaster Management from Asian Institute of Technology, Thailand.

Gina Fitri was born in Sarilamak, Indonesia, on January 15, 1999. She obtained her bachelor's degree from the Department of Geography, Universitas Indonesia in 2021 and currently continued her career as a Risk Management Officer in PT Superintending Company of Indonesia (SUCOFINDO). She focuses on pursuing future opportunities related to risk management and tries to elaborate between geography and risk management.

Wiwandari Handayani is a professor at the Department of Urban and Regional Planning, Universitas Diponegoro. She graduated with a bachelor's degree in urban and regional planning (UNDIP), master's degrees in urban and regional planning (ITB) and population studies (ANU-Australia), and holds a doctoral degree from the University of Stuttgart – Germany in Regional Development Planning. Her research focuses on urban and regional resilience. She actively involves in works related to urban/region resilience since 2011. She was in the team of Asian Cities Climate Change Resilience Network (ACCCRN) in 2011–2014 and Deputy Chief Resilience Officer of Semarang City since 2015.

Muhamad Iko Kersapati was born in Bekasi, Indonesia, April 23, 1994. He obtained his bachelor's degree from the Department of Geography, Universitas Indonesia in 2017. In 2021, He continued his study as a master student at Data Science for Cultural Heritage, University College London. Latterly, he is developing the utilization and implementation of Geographical Information Systems (GIS) in arts, humanities, and social sciences as his specialty in the integrated method of research. In 2020, he was involved as the team leader for book writing of Spice Routes with the focused region of Banda Islands, held by the Ministry of Education and Culture as a part of the Spice Routes Program.

Dr. Satyagopal Korlapati, is an officer belonging to the Indian Administrative Service of 1987 batch, allotted to Tamil Nadu Cadre. Besides Master's degree, Doctor of Philosophy in Zoology, & P.G. Diploma in Environmental Studies, he also has Masters Degree in Public Administration (Maxwell School, Syracuse University, USA) & Graduate Certificate in Environmental Decision Making (State University of New York, College of Environmental Science & Forestry) USA. He is a recipient of International Honour of Phi Beta Delta Award for excellence in Academics.

He worked extensively on Environment related issues in the State of Tamil Nadu as well as in Government of India in various capacities. During his field assignments as Sub-Collector, Additional Collector (Development) and District Collector he was very active in promoting Soil & Moisture Conservation, Integrated Watershed Management, Rehabilitation & Restoration of Water Ways & Bodies, Drought Mitigation, Afforestation and Sustainable Rural Development besides promoting Community & NGO involvement in Interface forest management.

As Additional Chief Secretary/Commissioner Revenue Administration & State Relief Commissioner, he strengthened Preparedness, Rescue, Relief and Build Back Better Systems to promote Risk Reduction & to enhance Resilience to Disaster Risks. Several new practices were introduced during his tenure which, were identified as Best Practices by National Disaster Management Authority & recommended to other States for adoption. He prepared and secured approval for TN State Disaster Management Perspective Plan 2018–2030, which adopts Systems Approach for holistic treatment of river basins & protection of fragile ecosystems to reduce disaster risks and enhance resilience to fulfill the major goals identified under Sendai framework. As State Relief Commissioner, he managed the worst drought faced by Tamil Nadu during the years 2016 & 2017 and the severe water crisis during 2019 in Chennai and a few other Districts. He also initiated several measures for Flood Mitigation such as Cut & cover Canals, riverine reservoirs, diversion canals for Intra-basin transfer, substitution of concrete surplus weirs with shutters etc. which, provided permanent solution in many Areas of Very High Vulnerability in Chennai & its neighborhood. He successfully handled severe Cyclones and minimized loss of lives during Vardha, Ockhi & Gaja Cyclones. He took personal interest in implementing, on a large scale, Rain water harvesting works, conversion of defunct recharge bore-wells & open-wells as recharge structures and renovation of Temple Tanks.

Ilya Moeliono has more than 20 years of experience in various development program including participatory action research, conflict resolution, participatory planning and networking in Indonesia. He graduated from Cornell University with a thesis that focused on conflict resolution management. He had written several publications in natural resource management, conflict management, and participatory appraisal. He currently serves as Senior Advisor in CRU.

Dr. Unang Mulkhan is the Director of the SDGs Center at the University of Lampung in Indonesia. Currently, he is also one of teachers for a course called Co-Design for Inclusive Public Spaces and Services (CIPSS) under the project of Lund University, Sweden for three countries: Indonesia, Lebanon and Turkey. He is also a research consultant for Raoul Wallenberg Institute (RWI) of Asia Pacific office in designing handbook and modules for localizing human rights especially gender and environmental aspects in the context of SDGs in Asia Pacific. In 2020, he conducted an Urban Analysis Report of Pangkalpinang city in the island of Bangka Belitung in Indonesia, under the Climate Resilient and Inclusive Cities (CRIC) project funded and supported by European Union (EU) and ACR+ Association of Cities and Regions for sustainable Resource management. Mr Mulkhan received his doctoral degree in Management, Work and Organisation from the University of Stirling in the UK and postgraduate diploma from the Institute for Housing and Urban Development Studies (IHS) at Erasmus University, the Netherlands.

Mr. Ramraj Narasimhan is the Chief, Special Programs Management at the Regional Integrated Multi-hazard Early Warning System (RIMES). He has over 20 years of experience in Disaster Risk Reduction with expertise straddling development & humanitarian domains.

With United Nations Development Program (in India, Sri Lanka and Nepal) and intergovernmental organizations such as ADPC and RIMES, he has worked in 15 Asian and African nations on designing and implementing varied national/regional disaster risk reduction programs. In several countries, he has provided advise to, and worked with national governments, Planning Commission, Federal Parliament on development issues with focus on disaster and climate risk management and in formulation of national roadmaps/strategies for disaster risk reduction (DRR) aligning with Agenda 2030, Sendai Framework for DRR.

He has played significant roles in responses to 3 major crises- 2015- Nepal Earthquake, 2004- Indian Ocean Tsunami and the 2003- Sri Lanka Floods, formulating and managing programs on post-disaster recovery and reconstruction, early warning systems and climate risk management leading to long-term resilience and development. He is an Architect-Planner by qualification, with specialization in housing.

Kamlesh Kumar Pathak has over 17 years of experience in climate change policy and in managing projects and programs in low-emission developments and climate resilient investments with focus on manufacturing, energy infrastructure, urban and related cross-sectoral issues. He provided expertise and technical assistance to National Government, Sub-National Government, private and non-profit sectors in designing and implementation project and program governed by international conventions and agreements. His engagement is prioritized around topics of climate change, SDG goals and providing expertise on projects with DFIs (Banking and Insurance Sector), International Development Organisations and multilateral development banks.

He is actively engaged in providing expertise on projects funded by European Union & member institutions and UN within the procurement framework of Grant and RFPs to accelerate action on multilateral environmental agreements and mechanism (NDC, GCF etc)

Bulan Prabawani is a lecturer as well as a researcher at the Department of Business Administration, Faculty of Social and Political Sciences. She graduated with a bachelor's degree in Business Administration (UNDIP), a master's degree in Management (UNDIP), and holds a doctoral degree from Edith Cowan University, Australia in Business. Bulan is the Head of the Business Administration Department since 2016 and is actively involved in research at Education for Sustainable Development (ESD) and Business Sustainability in which she has obtained national competitive grants for research and publication since 2014 for ESD, redenomination, business sustainability.

Heri Purwanto was born in Karanganyar, Indonesia, October 8, 1995. He graduated with a Bachelor of Archaeology, Udayana University, Bali in 2017 with the specification of classic archaeology. Recently, He continued his master's degree in the Faculty of Brahma Widya, Hindu State University of I Gusti Bagus Sugriwa, Bali. He has published several academic research and his recent work are 'Mandala Kadewaguruan: Religious Educational Site in the Western Slope of Lawu Mountain in XIV-XV Century', published by the Indonesian Journal of Education and Culture 2020.

Dr. Jothiganesh Shanmugasundaram is a Climate Application Expert. He focused on risk assessments, evaluated and facilitated development of appropriate climate application products and services based on user need assessments, and facilitated capacity building activities of RIMES Member States in the application of these products and services for improved planning and decision-making to manage resources and disaster risks in climate sensitive sectors. He has more than 10 years of experience in implementing climate risk management, disaster risk assessment and management projects for more than 20 countries in Asia-Pacific region, with several government, non-government, inter-government, international, and UN agencies.

He holds Ph.D. in Geography with specialization in Climatology from West Virginia University, USA; M.E in Remote Sensing and Geographic Information System from Asian Institute of Technology, Thailand; and B.E in Civil Engineering from Annamalai University, India.

Zoé Thouvenot studied political sciences at the Free University of Brussels and the Higher School of Economics of Moscow where she obtained her bachelor's degree. Then, she studied at Trinity College Dublin in International Politics (MSc) with a particular focus on environmental politics, economic inequalities in their relationship to democracy, the functioning of the European Union and the concept of political illiberalism. She currently works at Pilot4DEV as project officer. Her email is the following: zoe.thouvenot@gmail.com

Arief Wicaksono has more than 30 years of experience in climate change, land and natural resource management, Strategic Environmental Studies (SEA) and strengthen local communities' capacities to be able manage their environment with several national and international non-governmental organizations (NGOs), and several international development projects in Indonesia. He currently serves as Director of Conflict Resolution Unit (CRU).

Pilot4dev Key Experts

Prof Youssef Diab is Director of Research at the University Gustave Eiffel. He is a recognized expert on cities' resilience, governance and climate adaptation. He obtained his Ph.D. in Civil and Environmental Engineering in 1992 and his Habilitation Diploma for Research in 2000 in the field of Urban Engineering and Environment. He is Professor of Urban Engineering at UPEM and also scientific director of the EIVP where he is in charge of the R/D strategy and management. During the last 10 years he managed a research program of an amount of 1 Million Euro/year in the field of sustainable and resilient urban planning and design. He already supervised 38 Ph.D. theses and has more than 100 papers in international conferences and referred journals. His research work is related to the field of sustainable and resilient cities and sustainable urban policy by developing making decision tools.

Prof. Isabelle Milbert graduated in Law and Political Sciences at the University of Paris, where she specialized on Indian law and urban management. She arrived in Geneva in 1988 to teach at the Graduate

Institute of Development Studies (IUED) and carry out research and consultations for various assignments (Swiss Development Corporation, French Ministries, and International Organisations). From 2001 to 2004, she headed project team on 'Governance, Human Development and Environment', under the NCCR North South. She is the head of the Geneva Federation for Cooperation's Commission for knowledge sharing and is the President of Pilot4dev since 2018.

Contributions and Acknowledgments

We thank all the authors, contributors and reviewers of this book.

We would like to thank all the people who have made this book possible and especially **Vishnu Rao, Dr. Alain Simon, Jeffrey Raven and Bertrand Ginet** for their peer reviews. I also thank the publisher Peter Lang and in particular Thierry Waser, Editorial Director, and Suresh Selvamani from the Production Team.

For the content, we would like to thank the team at UCLG ASPAC, in particular **Aniessa Sari, Hamidah Nur, Hizbullah Arief, Putra Dwitama, Asih Budiati, Arie Setiawan, Maria Serenade, Dr. Bernadia Irawati Tjandradewi, the field officers** and their colleagues. We would like to **thank all the experts of the urban analysis reports, the project partners, and the cities' representatives**. We would also like to thank the whole team of the **LIFE Adapt Island project in particular Mélissa Dalle and Léna Jardin**.

Finally, we are grateful to **Addison Adkins** for the proofreading, editing and improvements of the language. **Addison Adkins** has been educated at the University of Southern California, the London School of Economics and Political Science, and obtained her MSc in International Politics from Trinity College Dublin. Her email is adkinsaddison@gmail.com.

www.ingramcontent.com/pod-product-compliance
Ingram Content Group UK Ltd.
Pitfield, Milton Keynes, MK11 3LW, UK
UKHW021828140426
5217IPUK00017B/1255